# Particles, Sources, and Fields

*Volume III*

Advanced Book Classics
David Pines, Series Editor

Anderson, P.W., *Basic Notions of Condensed Matter Physics*
Bethe H. and Jackiw, R., *Intermediate Quantum Mechanics, Third Edition*
Feynman, R., *Photon-Hadron Interactions*
Feynman, R., *Quantum Electrodynamics*
Feynman, R., *Statistical Mechanics*
Feynman, R., *The Theory of Fundamental Processes*
Negele, J. W. and Orland, H., *Quantum Many-Particle Systems*
Nozières, P., *Theory of Interacting Fermi Systems*
Parisi, G., *Statistical Field Theory*
Pines, D., *The Many-Body Problem*
Quigg, C., *Gauge Theories of the Strong, Weak, and Electromagnetic Interactions*
Schwinger, J., *Particles, Sources, and Fields, Volume I*
Schwinger, J., *Particles, Sources, and Fields, Volume II*
Schwinger, J., *Particles, Sources, and Fields, Volume III*

# PARTICLES, SOURCES, AND FIELDS

## *Volume III*

JULIAN SCHWINGER
late, University of California at Los Angeles

CRC Press
Taylor & Francis Group
Boca Raton London New York

CRC Press is an imprint of the
Taylor & Francis Group, an **informa** business

ADVANCED BOOK PROGRAM

First published 1973 by Perseus Books Publishing

Published 2018 by CRC Press
Taylor & Francis Group
6000 Broken Sound Parkway NW, Suite 300
Boca Raton, FL 33487-2742

*CRC Press is an imprint of the Taylor & Francis Group, an informa business*



Visit the Taylor & Francis Web site at
http://www.taylorandfrancis.com

and the CRC Press Web site at
http://www.crcpress.com

Library of Congress Catalog Card Number: 98-87896

ISBN 13: 978-0-7382-0055-2 (pbk)

Cover design by Suzanne Heiser

# Editor's Foreword

Perseus Books's *Frontiers in Physics* series has, since 1961, made it possible for leading physicists to communicate in coherent fashion their views of recent developments in the most exciting and active fields of physics—without having to devote the time and energy required to prepare a formal review or monograph. Indeed, throughout its nearly forty-year existence, the series has emphasized informality in both style and content, as well as pedagogical clarity. Over time, it was expected that these informal accounts would be replaced by more formal counterparts—textbooks or monographs—as the cutting-edge topics they treated gradually became integrated into the body of physics knowledge and reader interest dwindled. However, this has not proven to be the case for a number of the volumes in the series: Many works have remained in print on an on-demand basis, while others have such intrinsic value that the physics community has urged us to extend their life span.

The *Advanced Book Classics* series has been designed to meet this demand. It will keep in print those volumes in *Frontiers in Physics* or its sister series, *Lecture Notes and Supplements in Physics*, that continue to provide a unique account of a topic of lasting interest. And through a sizable printing, these classics will be made available at a comparatively modest cost to the reader.

These lecture notes by Julian Schwinger, one of the most distinguished theoretical physicists of this century, provide both beginning graduate students and experienced researchers with an invaluable introduction to the author's perspective on quantum electrodynamics and high-energy particle physics. Based on lectures delivered during the period 1966 to 1973, in which Schwinger developed a point of view (the physical source concept) and a technique that emphasized the unity of particle physics, electrodynamics, gravitational theory, and many-body theory, the notes serve as both a textbook on source theory and an informal historical record of the author's approach to many of the central problems in physics. I am most pleased that *Advanced Book Classics* will make these volumes readily accessible to a new generation of readers.

David Pines
Aspen, Colorado
July 1998

# Vita

**Julian Schwinger**

University Professor, University of California, and Professor of Physics at the University of California, Los Angeles since 1972, was born in New York City on February 12, 1918. Professor Schwinger obtained his Ph.D. in physics from Columbia University in 1939. He has also received honorary doctorates in science from four institutions: Purdue University (1961), Harvard University (1962), Brandeis University (1973), and Gustavus Adolphus College (1975). In addition to teaching at the University of California, Professor Schwinger has taught at Purdue University (1941–43), and at Harvard University (1945–72). Dr. Schwinger was a Research Associate at the University of California, Berkeley, and a Staff Member of the Massachusetts Institute of Technology Radiation Laboratory. In 1965 Professor Schwinger became a co-recipient (with Richard Feynman and Sin Itiro Tomonaga) of the Nobel Prize in Physics for work in quantum electrodynamics. A National Research Foundation Fellow (1939–40) and a Guggenheim Fellow (1970), Professor Schwinger was also the recipient of the C. L. Mayer Nature of Light Award (1949); the First Einstein Prize Award (1951); a J. W. Gibbs Honorary Lecturer of the American Mathematical Society (1960); the National Medal of Science Award for Physics (1964); a Humboldt Award (1981); the Premio Citta di Castiglione de Sicilia (1986); the Monie A. Ferst Sigma Xi Award (1986); and the American Academy of Achievement Award (1987).

# Vita

# Special Preface

Isaac Newton used his newly invented method of fluxious (the calculus) to compare the implications of the inverse square law of gravitation with Kepler's empirical laws of planetary motion. Yet, when the time came to write the *Principia*, he resorted entirely to geometrical demonstrations. Should we conclude that calculus is superfluous?

*Source theory*—to which the concept of renormalization is foreign—and *renormalized operator field theory* have both been found to yield the same answers to electrodynamic problems (which disappoints some people who would prefer that source theory produce new—and wrong—answers). Should we conclude that source theory is thus superfluous?

Both questions merit the same response: the simpler, more intuitive formation, is preferable.

This edition of *Particles, Sources, and Fields* is more extensive than the original two volumes of 1970 and 1973. It now contains four additional sections that finish the chapter entitled, "Electrodynamics II." These sections were written in 1973, but remained in partially typed form for fifteen years. I am again indebted to Mr. Ronald Bohm, who managed to decipher my fading scribbles and completed the typescript. Particular attention should be directed to Section 5–9, where, in a context somewhat larger than electrodynamics, a disagreement between source theory and operator field theory finally does appear.

Readers making their first acquaintance with source theory should consult the Appendix in Volume I. This Appendix contains suggestions for threading one's way through the sometimes cluttered pages.

*Los Angeles, California*
*April 1988*

J. S.

# Contents

*If you can't join 'em,*
*beat 'em.*

# Particles, Sources, and Fields

# 5

# ELECTRODYNAMICS II

For some time now we have been occupied with the implications of two-particle exchange. This leaves several important areas unexplored, however. There is the obvious question of extending the procedures to more elaborate multiparticle exchange mechanisms. And the practical applications of the results have been essentially limited to the idealization of a particle moving in a prescribed field, avoiding the relativistic two-body problem. This chapter is concerned with both types of investigations. But, in order to prevent too heavy a concentration of the often ponderous calculations involved in the higher order multiparticle exchange processes, such discussions will be interspersed among the two-body considerations, somewhat as dictated by the relevance to comparison with experiment.

## 5-1 TWO-PARTICLE INTERACTIONS. NON-RELATIVISTIC DISCUSSION

It is helpful to set the stage for two-particle relativistic theory by first assuming the simpler nonrelativistic context. Let us consider two kinds of particles, labeled 1 and 2 (no confusion with causal labels should occur here). The vacuum amplitude that describes them under conditions of non-interaction is

$$\langle 0_+|0_-\rangle^\eta = \exp[iW(\eta)],$$

$$W(\eta)_{\text{nonint.}} = -\int (d\mathbf{r})\, dt\, (d\mathbf{r}')\, dt'\, \eta^*(\mathbf{r}t)G(\mathbf{r}-\mathbf{r}', t-t')\eta(\mathbf{r}'t')|_1$$

$$-\int (d\mathbf{r})\, dt\, (d\mathbf{r}')\, dt'\, \eta^*(\mathbf{r}t)G(\mathbf{r}-\mathbf{r}', t-t')\eta(\mathbf{r}'t')|_2. \qquad (5\text{–}1.1)$$

To avoid writing out all these space-time coordinates, we shall often convey such an expression by the notation indicated in

$$W(\eta)_{\text{nonint.}} = -\int d1\, d1'\, \eta^*(1)G(1, 1')\eta(1') - \int d2\, d2'\, \eta^*(2)G(2, 2')\eta(2'). \qquad (5\text{–}1.2)$$

The particular term in the expansion of $\exp[iW]$ that represents two particles, one of each type, is

$$-\int d1 \cdots d2'\, \eta^*(2)\eta^*(1)G(1, 1')G(2, 2')\eta(1')\eta(2'), \qquad (5\text{–}1.3)$$

which displays the propagation function of the noninteracting two-particle system as the product of the individual propagation functions:

$$G(12, 1'2')_{\text{nonint.}} = G(1, 1')G(2, 2'). \tag{5–1.4}$$

Utilizing the individual differential equations [cf. (4–11.4)], which we shall write as

$$\begin{aligned}(E - T)_1 G(1, 1') &= \delta(1, 1'),\\ (E - T)_2 G(2, 2') &= \delta(2, 2'),\end{aligned} \qquad E = i\frac{\partial}{\partial t}, \tag{5–1.5}$$

one derives the differential equation for the two-particle propagation function,

$$(E - T)_1(E - T)_2 G(12, 1'2')_{\text{nonint.}} = \delta(1, 1')\, \delta(2, 2'). \tag{5–1.6}$$

The explicit expression (5–1.4) can be recovered from the differential equation by adjoining the retarded boundary conditions that are exhibited in Eq. (4–11.3).

A related version of the differential equation emerges on introducing the individual particle fields,

$$\psi(1) = \int d1'\, G(1, 1')\eta(1'), \qquad \psi(2) = \int d2'\, G(2, 2')\eta(2'), \tag{5–1.7}$$

which obey

$$(E - T)_1\psi(1) = \eta(1), \qquad (E - T)_2\psi(2) = \eta(2). \tag{5–1.8}$$

Then, the two-particle field, defined under noninteraction circumstances by

$$\psi(12)_{\text{nonint.}} = \psi(1)\psi(2) = \int d1'\, d2'\, G(1, 1')G(2, 2')\eta(1')\eta(2'), \tag{5–1.9}$$

is governed by

$$(E - T)_1(E - T)_2\psi(12)_{\text{nonint.}} = \eta(1)\eta(2). \tag{5–1.10}$$

The basic characteristic of a nonrelativistic theory is the meaningfulness of absolute simultaneity. Accordingly, it is natural to consider the specialization of these multi-time fields and propagation functions to the equal time situation. The explicit propagation function construction of Eq. (4–11.3) can be presented as

$$G(\mathbf{r} - \mathbf{r}', t - t') = -i\eta(t - t')\exp[-iT(t - t')]\, \delta(\mathbf{r} - \mathbf{r}'), \tag{5–1.11}$$

and thus

$$\begin{aligned}&G_1(\mathbf{r}_1 - \mathbf{r}'_1, t - t')G_2(\mathbf{r}_2 - \mathbf{r}'_2, t - t')\\ &\quad = -\eta(t - t')\exp[-i(T_1 + T_2)(t - t')]\, \delta(\mathbf{r}_1 - \mathbf{r}'_1)\, \delta(\mathbf{r}_2 - \mathbf{r}'_2).\end{aligned} \tag{5–1.12}$$

The function defined by

$$G(\mathbf{r}_1\mathbf{r}_2 t, \mathbf{r}'_1\mathbf{r}'_2 t') = iG(\mathbf{r}_1 t\mathbf{r}_2 t, \mathbf{r}'_1 t'\mathbf{r}'_2 t'), \tag{5–1.13}$$

then obeys

$$(E - T_1 - T_2)G(\mathbf{r}_1\mathbf{r}_2t, \mathbf{r}'_1\mathbf{r}'_2t')_{\text{nonint.}} = \delta(\mathbf{r}_1 - \mathbf{r}'_1)\,\delta(\mathbf{r}_2 - \mathbf{r}'_2)\,\delta(t - t'), \quad (5\text{–}1.14)$$

which is a more familiar two-particle generalization of the one-particle Green's function equation of Eq. (5–1.5).

To examine further the relation between the two types of propagation functions, it is convenient to adopt a matrix notation with regard to spatial variables, while time variables are made explicit. Thus we present the Eqs. (5–1.4) and (5–1.12, 13) as

$$\begin{aligned} &G_{12}(t_1t_2, t'_1t'_2) \\ &= G_1(t_1, t'_1)G_2(t_2, t'_2) \\ &= -\eta(t_1 - t'_1)\eta(t_2 - t'_2)\exp[-iT_1(t_1 - t'_1)]\exp[-iT_2(t_2 - t'_2)], \end{aligned} \quad (5\text{–}1.15)$$

and

$$G_{1+2}(t, t') = -i\eta(t - t')\exp[-i(T_1 + T_2)(t - t')], \quad (5\text{–}1.16)$$

where the latter notation, $G_{1+2}$, emphasizes that the equal time version regards particles 1 and 2 as parts of a single system. Suppose, for example, that $t_1 > t_2$ and $t'_1 > t'_2$. Then

$$\begin{aligned} &G_{12}(t_1t_2, t'_1t'_2) \\ &= -\exp[-iT_1(t_1 - t_2)]\,\eta(t_2 - t'_1)\exp[-i(T_1 + T_2)(t_2 - t'_1)] \\ &\quad \times \exp[-iT_2(t'_1 - t'_2)] \\ &= iG_1(t_1, t_2)G_{1+2}(t_2, t'_1)G_2(t'_1, t'_2), \end{aligned} \quad (5\text{–}1.17)$$

which is an example of the general relation (assuming $t_< > t'_>$)

$$G_{12}(t_1t_2, t'_1t'_2) = iG(t_>, t_<)G_{1+2}(t_<, t'_>)G(t'_>, t'_<), \quad (5\text{–}1.18)$$

where, on the right side, the first single particle Green's function refers to the particle with the larger ($t_>$) of the time variables $t_1, t_2$, while the other single particle function is associated with the particle having the lesser ($t'_<$) of the time values $t'_1, t'_2$. This is made explicit in the constructions (remember that these are retarded functions)

$$\begin{aligned} G(t_>, t_<) &= G_1(t_1, t_2) + G_2(t_2, t_1), \\ G(t'_>, t'_<) &= G_2(t'_1, t'_2) + G_1(t'_2, t'_1). \end{aligned} \quad (5\text{–}1.19)$$

The physical picture of the multi-time propagation function that (5–1.18) supplies is quite simple. At time $t'_<$, one of the particles is created. This single-particle situation lasts until time $t'_>$, when the other particle is emitted. The two-particle configuration endures until time $t_<$, when one of the particles is detected. And eventually, at time $t_>$, the final particle is also detected. Let us also note that Eq. (5–1.18) includes the original definition of (5–1.13), or

$$G_{1+2}(t, t') = iG_{12}(tt, t't'), \tag{5–1.20}$$

since

$$iG(t, t')\big|_{t-t'\to +0} = 1. \tag{5–1.21}$$

The equal time two-particle field is correspondingly defined by

$$\psi_{1+2}(t) = \psi_{12}(tt). \tag{5–1.22}$$

On using the following special example of (5–1.18),

$$G_{12}(tt, t'_1t'_2) = G_{1+2}(t, t'_>)G(t'_>, t'_<), \tag{5–1.23}$$

this becomes

$$\psi_{1+2}(t) = \int dt'_1\, dt'_2\, G_{1+2}(t, t'_>)G(t'_>, t'_<)\eta_1(t'_1)\eta_2(t'_2), \tag{5–1.24}$$

which, with the definition

$$\eta_{1+2}(t') = \int dt'_<\, G(t'_>, t'_<)\eta_1(t'_1)\eta_2(t'_2), \qquad t' = t'_>, \tag{5–1.25}$$

reads

$$\psi_{1+2}(t) = \int dt'\, G_{1+2}(t, t')\eta_{1+2}(t'). \tag{5–1.26}$$

The equivalent field differential equation is

$$(E - T_1 - T_2)\psi_{1+2}(t)_{\text{nonint.}} = \eta_{1+2}(t), \tag{5–1.27}$$

and this identifies $\eta_{1+2}(t)$ as an equal time two-particle source. The coordinate indices are made explicit in writing the latter equation as

$$(E - T_1 - T_2)\psi(\mathbf{r}_1\mathbf{r}_2t)_{\text{nonint.}} = \eta(\mathbf{r}_1\mathbf{r}_2t), \tag{5–1.28}$$

which also illustrates the practice of omitting subscripts when the necessary information is amply evident in the arguments of the functions. With the analogous source definition

$$\eta^*_{1+2}(t) = \int dt_>\, \eta^*{}_2(t_2)\eta^*{}_1(t_1)G(t_>, t_<), \qquad t = t_<, \tag{5–1.29}$$

one can present the two-particle vacuum amplitude of (5–1.3) in the form applicable to a single system:

$$-i\int dt\,dt'\,\eta^*_{1+2}(t)G_{1+2}(t,t')\eta_{1+2}(t'). \tag{5–1.30}$$

One should note, however, that $\eta_{1+2}(t)$ and $\eta^*_{1+2}(t)$ are a complex conjugate pair of functions only when the earlier acting emission source of $\eta_{1+2}$ and the later acting detection source of $\eta^*_{1+2}$ are extended sources, dealing with virtual, rather than real particles. This is reasonable, for these are the conditions for the degree of time locality that permits an effective description by just one time variable. To see it in detail, let us write out the sources, using Eq. (5–1.19):

$$\eta_{1+2}(t') = \int dt'_1\,G_1(t',t'_1)\eta_1(t'_1)\eta_2(t') + \int dt'_2\,G_2(t',t'_2)\eta_1(t')\eta_2(t'_2),$$

$$\eta^*_{1+2}(t) = \int dt_1\,\eta^*{}_2(t)\eta^*{}_1(t_1)G_1(t_1,t) + \int dt_2\,\eta^*(t_2)\eta^*{}_1(t)G_2(t_2,t), \tag{5–1.31}$$

where each propagation function can be represented as

$$G(t,t') = \int_{-\infty}^{\infty}\frac{dE}{2\pi}\frac{\exp[-iE(t-t')]}{E+i\varepsilon-T}. \tag{5–1.32}$$

The complex conjugate of the first structure in (5–1.31) does reproduce the form of the second one, except that, instead of the typical function

$$G(t',t) = \int_{-\infty}^{\infty}\frac{dE}{2\pi}\frac{\exp[iE(t-t')]}{E+i\varepsilon-T}, \tag{5–1.33}$$

one finds the transposed, complex conjugate, or adjoint function

$$G(t,t')^\dagger = \int_{-\infty}^{\infty}\frac{dE}{2\pi}\frac{\exp[iE(t-t')]}{E-i\varepsilon-T}. \tag{5–1.34}$$

But the two are equivalent if the sign of $i\varepsilon$ is irrelevant, that is, if the relation $E-T=0$, the condition for real particle propagation, is effectively not satisfied in virtue of the nature of the sources.

Now let two particles, one of each type, approach each other and scatter, in a primitive interaction act. The nonrelativistic concept of a primitive interaction is an instantaneous process, which is not localized spatially, in general. The scattered particles can be described by an effective two-particle source, which is measured by the strength of the excitation—the product of the two individual fields—and by a function $V$, intrinsic to the mechanism. Thus, we write

$$(1/i)\eta(\mathbf{r}_1t_1)\eta(\mathbf{r}_2t_2)|_{\text{eff.}} = \delta(t_1-t_2)V(\mathbf{r}_1-\mathbf{r}_2)\psi(\mathbf{r}_1t_1)\psi(\mathbf{r}_2t_2), \tag{5–1.35}$$

where, besides the explicitly stated translationally invariant dependence upon spatial coordinates, the function $V$ may involve momenta, spins, and other particle attributes. The field produced by the combination of the emission sources and the effective source of (5–1.35) is, according to Eq. (5–1.9),

$$\psi(12) = \psi(1)\psi(2) + i\int d1'\, d2'\, G(1, 1')G(2, 2')V(1'2')\psi(1')\psi(2'), \quad (5\text{–}1.36)$$

where

$$V(12) = \delta(t_1 - t_2)V(\mathbf{r}_1 - \mathbf{r}_2). \quad (5\text{–}1.37)$$

An important property of the interaction function $V$ can be inferred from the structure of the addition to $W$ that describes the exchange of a pair of particles between the effective emission source (5–1.35) and detection sources. On referring to Eq. (5–1.3), one sees that the addition, $\delta W$, can be obtained as

$$i\,\delta W = -\int d1\, d2\, \eta^*(2)\eta^*(1)\, \delta\psi(12), \quad (5\text{–}1.38)$$

where $\delta\psi(12)$ represents the interaction-induced field of (5–1.36). Thus,

$$\delta W = -\int d1\, d2\, \psi^*(2)\psi^*(1)V(12)\psi(1)\psi(2), \quad (5\text{–}1.39)$$

according to the constructions

$$\psi^*(1') = \int d1\, \eta^*(1)G(1, 1'), \qquad \psi^*(2') = \int d2\, \eta^*(2)G(2, 2'). \quad (5\text{–}1.40)$$

The more explicit form of (5–1.39) is

$$\delta W = -\int (d\mathbf{r}_1)(d\mathbf{r}_2)\, dt\, \psi^*(\mathbf{r}_2 t)\psi^*(\mathbf{r}_1 t)V(\mathbf{r}_1 - \mathbf{r}_2)\psi(\mathbf{r}_1 t)\psi(\mathbf{r}_2 t). \quad (5\text{–}1.41)$$

Suppose we consider circumstances in which the sources are incapable of emitting real particles ($E - T \neq 0$). Then [cf. Eq. (4–11.11)]

$$G(\mathbf{r} - \mathbf{r}', t - t') = \int \frac{(d\mathbf{p})\, dE}{(2\pi)^4} \exp\{i[\mathbf{p}\cdot(\mathbf{r} - \mathbf{r}') - E(t - t')]\} \frac{1}{E - T(\mathbf{p})} \quad (5\text{–}1.42)$$

obeys

$$G(\mathbf{r} - \mathbf{r}', t - t')^* = G(\mathbf{r}' - \mathbf{r}, t' - t), \quad (5\text{–}1.43)$$

which restates more concretely the relation between (5–1.33) and (5–1.34). In consequence, each $\psi^*(\mathbf{r}t)$ is the complex conjugate of the corresponding $\psi(\mathbf{r}t)$. Under these circumstances, also, the vacuum persistence probability must remain unity, or the quantity $W$ real. That is true of the individual particle contributions, Eq. (5–1.1), and it will be true of $\delta W$ as well if $V(\mathbf{r}_1 - \mathbf{r}_2)$ is a real, or more generally, Hermitian function of its variables.

The repetition of the primitive interaction will add further terms to the field (5–1.36). But these effects are easily summarized. The complete field $\psi(12)$ is the superposition of that representing noninteracting particles, $\psi(1)\psi(2)$, with the field representing particles coming from their last collision, as excited by the field generated by all sources, namely $\psi(12)$. Thus, the replacement, under the integration sign, of $\psi(1')\psi(2')$ by $\psi(1'2')$ produces an integral equation that describes unlimited repetitions of the primitive interaction:

$$\psi(12) = \psi(1)\psi(2) + i\int d1'\, d2'\, G(1, 1')G(2, 2')V(1'2')\psi(1'2'). \qquad (5\text{–}1.44)$$

The equivalent differential equation is

$$(E - T)_1(E - T)_2\psi(12) = \eta(1)\eta(2) + iV(12)\psi(12), \qquad (5\text{–}1.45)$$

which is also obtained directly from (5–1.35) by replacing the field of noninteracting particles with the total field. The determining differential equation for $\psi(12)$ is, therefore,

$$[(E - T)_1(E - T)_2 - iV(12)]\psi(12) = \eta(1)\eta(2). \qquad (5\text{–}1.46)$$

We write its Green's function solution as

$$\psi(12) = \int d1'\, d2'\, G(12, 1'2')\eta(1')\eta(2'), \qquad (5\text{–}1.47)$$

where

$$[(E - T)_1(E - T)_2 - iV(12)]G(12, 1'2') = \delta(1, 1')\,\delta(2, 2') \qquad (5\text{–}1.48)$$

is a generalization of Eq. (5–1.6) to interacting particles.

Since the effective source of (5–1.35) is only operative at equal times, the single-time, two-particle source defined in Eq. (5–1.25) is particularly simple. The delta function extracts the equal time limit of $G(t'_>, t'_<)$ which is taken from the side of positive time difference [Eq. (5–1.21)]. Thus,

$$\eta(\mathbf{r}_1\mathbf{r}_2t)\big|_{\text{eff.}} = V(\mathbf{r}_1 - \mathbf{r}_2)\psi(\mathbf{r}_1t)\psi(\mathbf{r}_2t), \qquad (5\text{–}1.49)$$

and the same line of argument, now applied to the differential equation (5–1.28), gives

$$(E - T_1 - T_2)\psi(\mathbf{r}_1\mathbf{r}_2t) = \eta(\mathbf{r}_1\mathbf{r}_2t) + V(\mathbf{r}_1 - \mathbf{r}_2)\psi(\mathbf{r}_1\mathbf{r}_2t), \qquad (5\text{–}1.50)$$

or

$$[E - T_1 - T_2 - V(\mathbf{r}_1 - \mathbf{r}_2)]\psi(\mathbf{r}_1\mathbf{r}_2t) = \eta(\mathbf{r}_1\mathbf{r}_2t). \tag{5–1.51}$$

The Green's function solution is

$$\psi(\mathbf{r}_1\mathbf{r}_2t) = \int (d\mathbf{r}'_1)(d\mathbf{r}'_2)\, dt'\, G(\mathbf{r}_1\mathbf{r}_2t, \mathbf{r}'_1\mathbf{r}'_2t')\eta(\mathbf{r}'_1\mathbf{r}'_2t'), \tag{5–1.52}$$

where

$$[E - T_1 - T_2 - V(\mathbf{r}_1 - \mathbf{r}_2)]G(\mathbf{r}_1\mathbf{r}_2t, \mathbf{r}'_1\mathbf{r}'_2t') = \delta(\mathbf{r}_1 - \mathbf{r}'_1)\, \delta(\mathbf{r}_2 - \mathbf{r}'_2)\, \delta(t - t') \tag{5–1.53}$$

extends (5–1.14) to the situation of interacting particles. We recognize that the primitive interaction function $V$ plays the role customarily assigned to potential energy.

It is intuitively evident that the relation (5–1.18) between the two types of propagation functions should persist in the presence of the interaction $V$, since its action is relevant only when both particles exist. Nevertheless, let us prove this directly. The Green's function integral equation equivalent to (5–1.44) is

$$G(12, 1'2') = G(1, 1')G(2, 2') + i\int d\bar{1}\, d\bar{2}\, G(1, \bar{1})G(2, \bar{2})V(\bar{1}\bar{2})G(\bar{1}\bar{2}, 1'2'). \tag{5–1.54}$$

Consider the situation with $t_1 > t_2$. Then, switching to the matrix notation, we can write (provided $t_2 > t'_1$)

$$G_1(t_1, t'_1) = \exp[-iT_1(t_1 - t_2)]\, G_1(t_2, t'_1) = iG_1(t_1, t_2)G_1(t_2, t'_1), \tag{5–1.55}$$

and (5–1.54) will read

$$G(t_1t_2, t'_1t'_2)$$
$$= iG_1(t_1, t_2)\left[G_1(t_2, t'_1)G_2(t_2, t'_2) + i\int d\bar{t}\, G_1(t_2, \bar{t})G_2(t_2, \bar{t})VG(\bar{t}\bar{t}, t'_1t'_2)\right], \tag{5–1.56}$$

where $V$ indicates $V(\mathbf{r}_1 - \mathbf{r}_2)$. Again, if $t'_2 < t'_1$, we have

$$G_1(t_2, t'_1)G_2(t_2, t'_2) = G_1(t_2, t'_1)G_2(t_2, t'_1)iG_2(t'_1, t'_2). \tag{5–1.57}$$

Accordingly, if we now define $G_{1+2}$ by the (more generally stated) relation

$$G(t_1t_2, t'_1t'_2) = iG(t_>, t_<)G_{1+2}(t_<, t'_>)G(t'_>, t'_<), \tag{5–1.58}$$

that function obeys the integral equation

$$G_{1+2}(t, t') = iG_1(t, t')G_2(t, t') + \int d\bar{t}\, iG_1(t, \bar{t})G_2(t, \bar{t})VG_{1+2}(\bar{t}, t'). \tag{5–1.59}$$

We recognize that

$$iG_1(t, t')G_2(t, t') = (1/i)\eta(t - t') \exp[- i(T_1 + T_2)(t - t')] \qquad (5\text{–}1.60)$$

is the equal-time Green's function for the noninteracting system, which obeys the differential equation [Eq. (5–1.14)]

$$(E - T_1 - T_2)iG_1(t, t')G_2(t, t') = \delta(t - t'). \qquad (5\text{–}1.61)$$

Therefore,

$$(E - T_1 - T_2)G_{1+2}(t, t') = \delta(t - t') + VG_{1+2}(t, t'), \qquad (5\text{–}1.62)$$

which is the differential equation (5–1.53), in matrix notation:

$$[E - T_1 - T_2 - V]G_{1+2}(t, t') = \delta(t - t'). \qquad (5\text{–}1.63)$$

It should be emphasized here that this discussion assumes that there is a time interval during which the particles coexist ($t_< > t'_>$). If that is not the situation ($t_< < t'_>$), we have a noninteracting arrangement where the Green's function is simply the product of the single particle functions. The Green's functions for the two domains join continuously.

We now proceed to set up action principles that will characterize the fully interacting system, at least in its two-particle interaction aspects. Some ingredients are already available—the action expression for noninteracting particles [Eq. (4–11.12)],

$$W_{\text{nonint.}} = -\int d1\,[\eta^*(1)\psi(1) + \psi^*(1)\eta(1) - \psi^*(1)(E - T)_1\psi(1)]$$
$$- \int d2\,[\eta^*(2)\psi(2) + \psi^*(2)\eta(2) - \psi^*(2)(E - T)_2\psi(2)], \qquad (5\text{–}1.64)$$

and the primitive interaction of Eq. (5–1.39),

$$W_{\text{prim. int.}} = -\int d1\,d2\,\psi^*(2)\psi^*(1)V(12)\psi(1)\psi(2). \qquad (5\text{–}1.65)$$

The action principle should also involve the two-particle field $\psi(12)$, but this should occur in such a way that nothing of the kind is required when interactions are absent, since the fields $\psi(1)$ and $\psi(2)$ would then provide a complete description. That suggests the introduction of the field

$$\chi(12) = \psi(12) - \psi(1)\psi(2). \qquad (5\text{–}1.66)$$

On combining the differential equations (5–1.8) and (5–1.46) we infer the following equation for this difference field,

$$[(E-T)_1(E-T)_2 - iV(12)]\chi(12) = iV(12)\psi(1)\psi(2). \qquad (5\text{–}1.67)$$

It identifies the source of the field $\chi$ as the first interaction of previously non-interacting particles.

A suitable action principle will now be stated where, for simplicity, no additional source for the $\chi$ field has been exhibited:

$$W = W_{\text{nonint.}} + W_{\text{prim. int.}} + W_\chi, \qquad (5\text{–}1.68)$$

in which

$$W_\chi = -i\int d1\, d2\, \chi^*(12)[(E-T)_1(E-T)_2 - iV(12)]\chi(12)$$
$$-\int d1\, d2\, [\chi^*(12)V(12)\psi(1)\psi(2) + \psi^*(2)\psi^*(1)V(12)\chi(12)]. \qquad (5\text{–}1.69)$$

This structure will be justified by its consequences. The field equation obtained by varying $\chi^*(12)$ is just Eq. (5–1.67), and the variation of $\chi(12)$ provides the analogous equation

$$\chi^*(12)[(E-T)_1(E-T)_2 - iV(12)] = \psi^*(2)\psi^*(1)iV(12). \qquad (5\text{–}1.70)$$

Their solutions can be stated with the aid of the Green's function $G(12, 1'2')$,

$$\chi(12) = \int d1'\, d2'\, G(12, 1'2')iV(1'2')\psi(1')\psi(2'),$$
$$\chi^*(1'2') = \int d1\, d2\, \psi^*(2)\psi^*(1)iV(12)G(12, 1'2'). \qquad (5\text{–}1.71)$$

The latter one uses the alternative presentation of (5–1.48) as

$$G(12, 1'2')[(E-T)'_1(E-T)'_2 - iV(1'2')] = \delta(1, 1')\,\delta(2, 2'), \qquad (5\text{–}1.72)$$

and the consistency of this system is confirmed by the fact that $W_\chi$ can be evaluated in two alternative ways to give

$$W_\chi = -\int d1 \cdots d2'\, \psi^*(2)\psi^*(1)V(12)iG(12, 1'2')V(1'2')\psi(1')\psi(2'). \qquad (5\text{–}1.73)$$

The sum of $W_{\text{prim. int.}}$ and $W_\chi$ involves the following combination standing between products of single-particle fields:

$$iV(12)G(12, 1'2')iV(1'2') + iV(12)\,\delta(1, 1')\,\delta(2, 2')$$
$$= (E-T)_1(E-T)_2G(12, 1'2')iV(1'2')$$

$$= (E - T)_1(E - T)_2G(12, 1'2')(E - T)'_1(E - T)'_2$$
$$- (E - T)_1(E - T)_2\, \delta(1, 1')\, \delta(2, 2'), \qquad (5\text{–}1.74)$$

which makes successive use of the Eqs. (5–1.48) and (5–1.72) for the Green's function. One can also write the latter form as

$$(E - T)_1(E - T)_2[G(12, 1'2') - G(1, 1')G(2, 2')](E - T)'_1(E - T)'_2. \quad (5\text{–}1.75)$$

The explicit expressions associated with the first right-hand version of (5–1.74) and with (5–1.75) are, respectively,

$$W_{\text{prim. int.}} + W_\chi = -\int d1 \cdots d2'\, \psi^*(2)\psi^*(1)(E - T)_1(E - T)_2G(12, 1'2')$$
$$\times V(1'2')\psi(1')\psi(2') \qquad (5\text{–}1.76)$$

and

$$W_{\text{prim. int.}} + W_\chi = i\int d1 \cdots d2'\, \psi^*(2)\psi^*(1)(E - T)_1(E - T)_2[G(12, 1'2')$$
$$- G(1, 1')G(2, 2')](E - T_1)'_1(E - T_2)'_2\psi(1')\psi(2'). \quad (5\text{–}1.77)$$

After the elimination of $\chi$ and $\chi^*$, the action principle still applies to variations of $\psi$ and $\psi^*$. Thus, the field equation for $\psi(2)$ derived by using the form (5–1.76) would be

$$(E - T)_2\psi(2) - \eta(2) = \int d1 \cdots d2'\, \psi^*(1)(E - T)_1(E - T)_2G(12, 1'2')$$
$$\times V(1'2')\psi(1')\psi(2'). \qquad (5\text{–}1.78)$$

It is an example of a set of nonlinear equations that could be solved by successive iteration. Evidently the right-hand side of (5–1.78) is at least cubic in the sources (counting both emission and absorption sources). If it were omitted, the error in evaluating $W$ would be, not quartic, but sextic in the sources owing to the stationary nature of the action. Accordingly, if we confine attention to the quadratic and quartic source terms in the expansion of the vacuum amplitude, descriptive of a single particle or a pair of particles, it suffices to use the non-interacting solutions for the fields $\psi(1)$ and $\psi(2)$. Then, we have

$$W \to -\int d1\, d1'\, \eta^*(1)G(1, 1')\eta(1') - \int d2\, d2'\, \eta^*(2)G(2, 2')\eta(2')$$
$$+ i\int d1 \cdots d2'\, \eta^*(2)\eta^*(1)[G(12, 1'2') - G(1, 1')G(2, 2')]\eta(1')\eta(2'), \quad (5\text{–}1.79)$$

and

$$\langle 0_+|0_-\rangle^\eta = 1 - i\int d1\, d1'\, \eta^*(1)G(1,1')\eta(1') - i\int d2\, d2'\, \eta^*(2)G(2,2')\eta(2')$$

$$- \int d1 \cdots d2'\, \eta^*(2)\eta^*(1)G(12,1'2')\eta(1')\eta(2') + \cdots. \qquad (5\text{–}1.80)$$

The effect of the quartic term in (5–1.79), which we are now verifying, has been to substitute, for the Green's function of two noninteracting particles, the function $G(12, 1'2')$ that contains the full account of the interaction.

In the situation under discussion, where interactions are instantaneous, the action principle can also be formulated using the equal time field $\psi(\mathbf{r}_1\mathbf{r}_2t)$, or rather,

$$\chi(\mathbf{r}_1\mathbf{r}_2t) = \psi(\mathbf{r}_1\mathbf{r}_2t) - \psi(\mathbf{r}_1t)\psi(\mathbf{r}_2t). \qquad (5\text{–}1.81)$$

The structure (5–1.68) is maintained, where we might now write

$$W_{\text{prim. int.}} = -\int (d\mathbf{r}_1)(d\mathbf{r}_2)\, dt\, \psi^*(\mathbf{r}_2t)\psi^*(\mathbf{r}_1t)V(\mathbf{r}_1-\mathbf{r}_2)\psi(\mathbf{r}_1t)\psi(\mathbf{r}_2t), \qquad (5\text{–}1.82)$$

but

$$W_\chi = \int (d\mathbf{r}_1)(d\mathbf{r}_2)\, dt\, [\chi^*(\mathbf{r}_1\mathbf{r}_2t)(E - T_1 - T_2 - V(\mathbf{r}_1-\mathbf{r}_2))\chi(\mathbf{r}_1\mathbf{r}_2t)$$

$$- \chi^*(\mathbf{r}_1\mathbf{r}_2t)V(\mathbf{r}_1-\mathbf{r}_2)\psi(\mathbf{r}_1t)\psi(\mathbf{r}_2t) - \psi^*(\mathbf{r}_2t)\psi^*(\mathbf{r}_1t)V(\mathbf{r}_1-\mathbf{r}_2)\chi(\mathbf{r}_1\mathbf{r}_2t)]. \qquad (5\text{–}1.83)$$

The $\chi$ field equations are $[V = V(\mathbf{r}_1 - \mathbf{r}_2)]$

$$(E - T_1 - T_2 - V)\chi(\mathbf{r}_1\mathbf{r}_2t) = V\psi(\mathbf{r}_1t)\psi(\mathbf{r}_2t) \qquad (5\text{–}1.84)$$

and

$$\chi^*(\mathbf{r}_1\mathbf{r}_2t)(E - T_1 - T_2 - V) = \psi^*(\mathbf{r}_2t)\psi^*(\mathbf{r}_1t)V, \qquad (5\text{–}1.85)$$

which are solved by

$$\chi(\mathbf{r}_1\mathbf{r}_2t) = \int (d\mathbf{r}'_1)(d\mathbf{r}'_2)\, dt'\, G(\mathbf{r}_1\mathbf{r}_2t, \mathbf{r}'_1\mathbf{r}'_2t')V(\mathbf{r}'_1-\mathbf{r}'_2)\psi(\mathbf{r}'_1t')\psi(\mathbf{r}'_2t'),$$

$$\chi^*(\mathbf{r}'_1\mathbf{r}'_2t') = \int (d\mathbf{r}_1)(d\mathbf{r}_2)\, dt\, \psi^*(\mathbf{r}_2t)\psi^*(\mathbf{r}_1t)V(\mathbf{r}_1-\mathbf{r}_2)G(\mathbf{r}_1\mathbf{r}_2t, \mathbf{r}'_1\mathbf{r}'_2t'). \qquad (5\text{–}1.86)$$

The use of either solution presents $W_\chi$ as

$$W_\chi = -\int (d\mathbf{r}_1) \cdots (d\mathbf{r}'_2)\, dt\, dt'\, \psi^*(\mathbf{r}_2 t)\psi^*(\mathbf{r}_1 t) V(\mathbf{r}_1 - \mathbf{r}_2) G(\mathbf{r}_1\mathbf{r}_2 t, \mathbf{r}'_1\mathbf{r}'_2 t')$$
$$\times V(\mathbf{r}'_1 - \mathbf{r}'_2)\psi(\mathbf{r}'_1 t')\psi(\mathbf{r}'_2 t'). \qquad (5\text{–}1.87)$$

The addition of the primitive interaction introduces the combination (matrix notation is used)

$$VG_{1+2}(t, t')V + \delta(t - t')V = (E - T_1 - T_2)G_{1+2}(t, t')V, \qquad (5\text{–}1.88)$$

where the latter form involves the differential equation of (5–1.63). The alternative version of this equation,

$$G_{1+2}(t, t')[E' - T_1 - T_2 - V] = \delta(t - t'), \qquad (5\text{–}1.89)$$

completes the elimination of $V$ to give

$$(E - T_1 - T_2)G_{1+2}(t, t')(E' - T_1 - T_2) - (E - T_1 - T_2)\, \delta(t - t')$$
$$= (E - T_1 - T_2)[G_{1+2}(t, t') - iG_1(t, t')G_2(t, t')](E' - T_1 - T_2). \qquad (5\text{–}1.90)$$

The last form is written out as

$$W_{\text{prim. int.}} + W_\chi = -\int (d\mathbf{r}_1) \cdots dt'\, \psi^*(\mathbf{r}_2 t)\psi^*(\mathbf{r}_1 t)(E - T_1 - T_2)[G(\mathbf{r}_1\mathbf{r}_2 t, \mathbf{r}'_1\mathbf{r}'_2 t')$$
$$- iG(\mathbf{r}_1 t, \mathbf{r}'_1 t')G(\mathbf{r}_2 t, \mathbf{r}'_2 t')](E' - T'_1 - T'_2)\psi(\mathbf{r}'_1 t')\psi(\mathbf{r}'_2 t').$$
$$(5\text{–}1.91)$$

As discussed before, we now use fields that obey field equations without interaction. As a result,

$$W \to W_{\text{nonint.}} - \int (d\mathbf{r}_1) \cdots dt'\, \eta^*(\mathbf{r}_1\mathbf{r}_2 t)[G(\mathbf{r}_1\mathbf{r}_2 t, \mathbf{r}'_1\mathbf{r}'_2 t')$$
$$- iG(\mathbf{r}_1 t, \mathbf{r}'_1 t')G(\mathbf{r}_2 t, \mathbf{r}'_2 t')]\eta(\mathbf{r}'_1\mathbf{r}'_2 t'), \qquad (5\text{–}1.92)$$

where the added term again serves to introduce the Green's function of the interacting system in the relevant term of the vacuum amplitude [cf. Eq. (5–1.30)].

As is usual with nonrelativistic systems, advantageous use can be made of center of mass and relative coordinates,

$$\mathbf{R} = \frac{m_1\mathbf{r}_1 + m_2\mathbf{r}_2}{m_1 + m_2}, \qquad \mathbf{r} = \mathbf{r}_1 - \mathbf{r}_2. \qquad (5\text{–}1.93)$$

The associated momenta are

$$\mathbf{P} = \mathbf{p}_1 + \mathbf{p}_2, \qquad \mathbf{p} = \frac{m_2\mathbf{p}_1 - m_1\mathbf{p}_2}{m_1 + m_2}. \qquad (5\text{–}1.94)$$

From the inverse relations ($M = m_1 + m_2$)

$$\mathbf{r}_1 = \mathbf{R} + \frac{m_2}{M}\mathbf{r}, \qquad \mathbf{r}_2 = \mathbf{R} - \frac{m_1}{M}\mathbf{r},$$

$$\mathbf{p}_1 = \frac{m_1}{M}\mathbf{P} + \mathbf{p}, \qquad \mathbf{p}_2 = \frac{m_2}{M}\mathbf{P} - \mathbf{p}, \tag{5–1.95}$$

we infer that

$$T_1 = \frac{m_1}{M}\frac{\mathbf{P}^2}{2M} + \frac{1}{M}\mathbf{P}\cdot\mathbf{p} + \frac{\mathbf{p}^2}{2m_1}, \qquad T_2 = \frac{m_2}{M}\frac{\mathbf{P}^2}{2M} - \frac{1}{M}\mathbf{P}\cdot\mathbf{p} + \frac{\mathbf{p}^2}{2m_2}, \tag{5–1.96}$$

which implies the familiar decomposition of the total kinetic energy,

$$T_1 + T_2 = \frac{\mathbf{P}^2}{2M} + \frac{\mathbf{p}^2}{2\mu} \equiv T_\mathbf{P} + T, \tag{5–1.97}$$

where

$$1/\mu = 1/m_1 + 1/m_2 \tag{5–1.98}$$

defines the reduced mass $\mu$. The independence of the center of mass and relative motions is conveyed, in Green's function language, by the factorization

$$G(\mathbf{r}_1\mathbf{r}_2t, \mathbf{r}'_1\mathbf{r}'_2t') = iG(\mathbf{R}t, \mathbf{R}'t')G(\mathbf{r}t, \mathbf{r}'t'). \tag{5–1.99}$$

We shall verify this, beginning with the Green's function equation of (5–1.53) which is now written as

$$\left[i\frac{\partial}{\partial t} - T_\mathbf{P} - T - V(\mathbf{r})\right]G(\mathbf{R}\mathbf{r}t, \mathbf{R}'\mathbf{r}'t') = \delta(t-t')\delta(\mathbf{R}-\mathbf{R}')\,\delta(\mathbf{r}-\mathbf{r}'). \tag{5–1.100}$$

The introduction of the Fourier representation

$$G(\mathbf{R}\mathbf{r}t, \mathbf{R}'\mathbf{r}'t') = \int \frac{(d\mathbf{P})}{(2\pi)^3}\exp\{i[\mathbf{P}\cdot(\mathbf{R}-\mathbf{R}') - T_\mathbf{P}(t-t')]\}\,G(\mathbf{r}t, \mathbf{r}'t') \tag{5–1.101}$$

yields the following **P**-independent equation,

$$\left[i\frac{\partial}{\partial t} - T - V(\mathbf{r})\right]G(\mathbf{r}t, \mathbf{r}'t') = \delta(t-t')\,\delta(\mathbf{r}-\mathbf{r}'). \tag{5–1.102}$$

That is the content of Eq. (5–1.99), where $G(\mathbf{R}t, \mathbf{R}'t')$ is identified as the Green's function of a free particle with mass $M$.

Eigenfunctions, labeled by energy $E$, and other quantum numbers collectively called $a$,

$$\psi_{Ea}(\mathbf{r}t) = \psi_{Ea}(\mathbf{r})\exp(-iEt), \tag{5–1.103}$$

are solutions of the homogeneous Green's function equation,

$$(E - T - V)\psi_{Ea}(\mathbf{r}) = 0, \qquad (5\text{–}1.104)$$

that have the orthonormality property (for discretely labeled states)

$$\int (d\mathbf{r})\psi^*_{Ea}(\mathbf{r})\psi_{E'a'}(\mathbf{r}) = \delta_{EE'}\,\delta_{aa'}. \qquad (5\text{–}1.105)$$

With a knowledge of the Green's function, all the eigenfunctions can be exhibited and, conversely, the Green's function can be constructed in terms of them. As for the latter, let the Green's function equation of (5–1.102) be multiplied by $\psi^*_{Ea}(\mathbf{r})$ and integrated. In view of the adjoint form of (5–1.104),

$$\psi^*_{Ea}(\mathbf{r})(E - T - V) = 0, \qquad (5\text{–}1.106)$$

that gives

$$\left(i\frac{\partial}{\partial t} - E\right)\int (d\mathbf{r})\psi^*_{Ea}(\mathbf{r})G(\mathbf{r}t, \mathbf{r}'t') = \delta(t - t')\psi^*_{Ea}(\mathbf{r}'). \qquad (5\text{–}1.107)$$

The solution of this Green's function equation is

$$\int (d\mathbf{r})\psi^*_{Ea}(\mathbf{r})G(\mathbf{r}t, \mathbf{r}'t') = (1/i)\eta(t - t')\exp[-\,iE(t - t')]\,\psi^*_{Ea}(\mathbf{r}'). \qquad (5\text{–}1.108)$$

Use of the completeness property, as expressed by

$$\sum_{Ea} \psi_{Ea}(\mathbf{r})\psi^*_{Ea}(\mathbf{r}') = \delta(\mathbf{r} - \mathbf{r}'), \qquad (5\text{–}1.109)$$

then supplies the eigenfunction construction

$$G(\mathbf{r}t, \mathbf{r}'t') = \sum_{Ea} \psi_{Ea}(\mathbf{r})(1/i)\eta(t - t')\exp[-\,iE(t - t')]\,\psi^*_{Ea}(\mathbf{r}') \qquad (5\text{–}1.110)$$

and, conversely, the completeness property is recovered by comparison of (5–1.110) with the limiting value deduced from the differential equation and the retarded boundary condition:

$$iG(\mathbf{r}t, \mathbf{r}'t')\big|_{t-t'\to+0} = \delta(\mathbf{r} - \mathbf{r}'). \qquad (5\text{–}1.111)$$

The center of mass motion can be reinstated, in accordance with Eq. (5–1.99), to produce the Green's function expansion

$$G(\mathbf{r}_1\mathbf{r}_2t, \mathbf{r}'_1\mathbf{r}'_2t')$$
$$= \sum_{\mathbf{P}Ea} \psi_{\mathbf{P}Ea}(\mathbf{r}_1\mathbf{r}_2)(1/i)\eta(t - t')\exp\{-\,i[(\mathbf{P}^2/2M) + E](t - t')\}\,\psi^*_{\mathbf{P}Ea}(\mathbf{r}'_1\mathbf{r}'_2),$$
$$(5\text{–}1.112)$$

where

$$\psi_{\mathbf{P}Ea}(\mathbf{r}_1\mathbf{r}_2) = \left(\frac{(d\mathbf{P})}{(2\pi)^3}\right)^{1/2} \exp(i\mathbf{P}\cdot\mathbf{R})\,\psi_{Ea}(\mathbf{r}). \tag{5–1.113}$$

These eigenfunctions have the orthonormality property

$$\int (d\mathbf{r}_1)(d\mathbf{r}_2)\psi^*_{\mathbf{P}Ea}(\mathbf{r}_1\mathbf{r}_2)\psi_{\mathbf{P}'E'a'}(\mathbf{r}_1\mathbf{r}_2) = \delta_{\mathbf{P}\mathbf{P}'}\,\delta_{EE'}\,\delta_{aa'}. \tag{5–1.114}$$

We can also write

$$G(\mathbf{r}_1\mathbf{r}_2t, \mathbf{r}'_1\mathbf{r}'_2t') = (1/i)\eta(t-t')\sum_{\mathbf{P}Ea}\psi_{\mathbf{P}Ea}(\mathbf{r}_1\mathbf{r}_2t)\psi^*_{\mathbf{P}Ea}(\mathbf{r}'_1\mathbf{r}'_2t'), \tag{5–1.115}$$

with

$$\psi_{\mathbf{P}Ea}(\mathbf{r}_1\mathbf{r}_2t) = \psi_{\mathbf{P}Ea}(\mathbf{r}_1\mathbf{r}_2)\exp\{-i[(\mathbf{P}^2/2M)+E]t\}. \tag{5–1.116}$$

Eigenfunction expansions for the multi-time Green's functions will now be considered. We recall that $(t_< > t'_>)$

$$G_{12}(t_1t_2, t'_1t'_2) = iG(t_>, t_<)G_{1+2}(t_<, t'_>)G(t'_>, t'_<), \tag{5–1.117}$$

where

$$G(t_>, t_<) = G_1(t_1, t_2) + G_2(t_2, t_1),$$
$$G(t'_>, t'_<) = G_2(t'_1, t'_2) + G_1(t'_2, t'_1), \tag{5–1.118}$$

and the matrix notation implies integrations over all spatial coordinates. The introduction of the expansion (5–1.115) for $G_{1+2}(t_<, t'_>)$ gives

$$t_< > t'_>:\quad G(\mathbf{r}_1t_1\mathbf{r}_2t_2, \mathbf{r}'_1t'_1\mathbf{r}'_2t'_2) = -\sum_{\mathbf{P}Ea}\psi_{\mathbf{P}Ea}(\mathbf{r}_1t_1\mathbf{r}_2t_2)\psi^*_{\mathbf{P}Ea}(\mathbf{r}'_1t'_1\mathbf{r}'_2t'_2), \tag{5–1.119}$$

with

$$\psi_{\mathbf{P}Ea}(\mathbf{r}_1t_1\mathbf{r}_2t_2) = i\int(d\bar{\mathbf{r}}_1)G_1(\mathbf{r}_1t_1, \bar{\mathbf{r}}_1t_2)\psi_{\mathbf{P}Ea}(\bar{\mathbf{r}}_1\mathbf{r}_2t_2)$$
$$+ i\int(d\bar{\mathbf{r}}_2)G_2(\mathbf{r}_2t_2, \bar{\mathbf{r}}_2t_1)\psi_{\mathbf{P}Ea}(\mathbf{r}_1\bar{\mathbf{r}}_2t_1) \tag{5–1.120}$$

and

$$\psi^*_{\mathbf{P}Ea}(\mathbf{r}_1t_1\mathbf{r}_2t_2) = i\int(d\bar{\mathbf{r}}_2)\psi^*_{\mathbf{P}Ea}(\mathbf{r}_1\bar{\mathbf{r}}_2t_1)G_2(\bar{\mathbf{r}}_2t_1, \mathbf{r}_2t_2)$$
$$+ i\int(d\bar{\mathbf{r}}_1)\psi^*_{\mathbf{P}Ea}(\bar{\mathbf{r}}_1\mathbf{r}_2t_2)G_1(\bar{\mathbf{r}}_1t_2, \mathbf{r}_1t_1). \tag{5–1.121}$$

The factors of $i$ have been introduced so that we shall have

$$\psi_{\mathbf{P}Ea}(\mathbf{r}_1 t\, \mathbf{r}_2 t) = \psi_{\mathbf{P}Ea}(\mathbf{r}_1\mathbf{r}_2 t), \qquad \psi^*_{\mathbf{P}Ea}(\mathbf{r}_1 t \mathbf{r}_2 t) = \psi^*_{\mathbf{P}Ea}(\mathbf{r}_1\mathbf{r}_2 t). \tag{5–1.122}$$

As the construction (5–1.119) suggests, the multi-time functions are eigenfunctions of the homogeneous version of the Green's function equation in (5–1.48). To verify that, consider, for example (the quantum number labels are omitted),

$$\begin{aligned}(E-T)_1(E-T)_2\psi(\mathbf{r}_1t_1\mathbf{r}_2t_2) &= \left(i\frac{\partial}{\partial t_2} - T_2\right) i\,\delta(t_1-t_2)\psi(\mathbf{r}_1\mathbf{r}_2t_2)\\ &\quad + \left(i\frac{\partial}{\partial t_1} - T_1\right) i\,\delta(t_2-t_1)\psi(\mathbf{r}_1\mathbf{r}_2t_1)\\ &= i\,\delta(t_1-t_2)\left(2i\frac{\partial}{\partial t} - T_1 - T_2\right)\psi(\mathbf{r}_1\mathbf{r}_2t)\big|_{t=t_1=t_2}\\ &\quad - \delta'(t_1-t_2)(\psi(\mathbf{r}_1\mathbf{r}_2t_1) - \psi(\mathbf{r}_1\mathbf{r}_2t_2)).\end{aligned} \tag{5–1.123}$$

Now,

$$\delta'(t_1-t_2)(f(t_1)-f(t_2)) = -\,\delta(t_1-t_2)f'(t)\big|_{t=t_1=t_2}, \tag{5–1.124}$$

and thus the right-hand side of (5–1.123) is

$$i\,\delta(t_1-t_2)\left(i\frac{\partial}{\partial t} - T_1 - T_2\right)\psi(\mathbf{r}_1\mathbf{r}_2t)\big|_{t=t_1=t_2} = i\,\delta(t_1-t_2)V(\mathbf{r}_1-\mathbf{r}_2)\psi(\mathbf{r}_1\mathbf{r}_2t)\big|_{t=t_1=t_2}. \tag{5–1.125}$$

In view of the first relation in Eq. (5–1.122), this is the anticipated equation:

$$[(E-T)_1(E-T)_2 - i\,\delta(t_1-t_2)\,V(\mathbf{r}_1-\mathbf{r}_2)]\psi(\mathbf{r}_1t_1\mathbf{r}_2t_2) = 0, \tag{5–1.126}$$

and a similar procedure shows that

$$\psi^*(\mathbf{r}_1t_1\mathbf{r}_2t_2)[(E-T)_1(E-T)_2 - i\,\delta(t_1-t_2)\,V(\mathbf{r}_1-\mathbf{r}_2)] = 0. \tag{5–1.127}$$

Analogous operations can be exploited to give alternative form to the eigenfunctions. Thus

$$G_1(t_1,t)G_2(t_2,t)(E-T_1-T_2) = \delta(t_1-t)\,G_2(t_2,t_1) + \delta(t_2-t)\,G_1(t_1,t_2), \tag{5–1.128}$$

from which we infer that (matrix notation, with labels omitted)

$$\begin{aligned}\psi(t_1t_2) &= i\int dt\,G_1(t_1,t)G_2(t_2,t)(E-T_1-T_2)\psi(t)\\ &= i\int dt\,G_1(t_1,t)G_2(t_2,t)V\psi(t).\end{aligned} \tag{5–1.129}$$

The latter version supplies a physical interpretation of the multi-time eigenfunctions in terms of measurements performed, after the last interaction, on free particles. Similarly, we have

$$\psi^*(t_1t_2) = i\int dt\, \psi^*(t)VG_1(t, t_1)G_2(t, t_2). \qquad (5\text{–}1.130)$$

Next, we are going to discuss how the multi-time eigenfunctions are used to express orthonormality. As a first step in an empirical investigation of this property, consider the product of [matrix notation]

$$\begin{aligned}\psi^*_{\mathbf{P}Ea}(t_1t_2) &= \psi^*_{\mathbf{P}Ea}(t_1)\eta(t_1 - t_2)\exp[-iT_2(t_1 - t_2)]\\ &\quad + \psi^*_{\mathbf{P}Ea}(t_2)\eta(t_2 - t_1)\exp[-iT_1(t_2 - t_1)] \qquad (5\text{–}1.131)\end{aligned}$$

and

$$\begin{aligned}\psi_{\mathbf{P}'E'a'}(t_1t_2) &= \eta(t_1 - t_2)\exp[-iT_1(t_1 - t_2)]\,\psi_{\mathbf{P}'E'a'}(t_2)\\ &\quad + \eta(t_2 - t_1)\exp[-iT_2(t_2 - t_1)]\,\psi_{\mathbf{P}'E'a'}(t_1), \qquad (5\text{–}1.132)\end{aligned}$$

namely,

$$\begin{aligned}&\psi^*_{\mathbf{P}Ea}(t_1t_2)\psi_{\mathbf{P}'E'a'}(t_1t_2)\\ &= \eta(t_1 - t_2)\psi^*_{\mathbf{P}Ea}(t_1)\exp[-i(T_1 + T_2)(t_1 - t_2)]\,\psi_{\mathbf{P}'E'a'}(t_2)\\ &\quad + \eta(t_2 - t_1)\psi^*_{\mathbf{P}Ea}(t_2)\exp[-i(T_1 + T_2)(t_2 - t_1)]\,\psi_{\mathbf{P}'E'a'}(t_1). \qquad (5\text{–}1.133)\end{aligned}$$

Ordinary orthonormality statements, whether written as (5–1.114) or in the equivalent form

$$\int (d\mathbf{r}_1)(d\mathbf{r}_2)\psi^*_{\mathbf{P}Ea}(\mathbf{r}_1\mathbf{r}_2t)\psi_{\mathbf{P}'E'a'}(\mathbf{r}_1\mathbf{r}_2t) = \delta_{\mathbf{P}\mathbf{P}'}\,\delta_{EE'}\,\delta_{aa'}, \qquad (5\text{–}1.134)$$

do not involve time integrations. Here, however, there are two time variables, $t_1$ and $t_2$, or, alternatively,

$$t = \tfrac{1}{2}(t_1 + t_2), \qquad \tau = t_1 - t_2, \qquad (5\text{–}1.135)$$

which suggests that an integration over the relative time variable $\tau$ is required. We note that

$$\begin{aligned}&\psi^*_{\mathbf{P}Ea}(t + \tfrac{1}{2}\tau)\exp[-i(T_1 + T_2)\tau]\,\psi_{\mathbf{P}'E'a'}(t - \tfrac{1}{2}\tau)\\ &= \psi^*_{\mathbf{P}Ea}(t)\exp\left[i\left(\frac{E + E'}{2} - T_1 - T_2\right)\tau\right]\psi_{\mathbf{P}'E'a'}(t), \qquad (5\text{–}1.136)\end{aligned}$$

and

$$\psi^*_{\mathbf{P}Ea}(t - \tfrac{1}{2}\tau) \exp[i(T_1 + T_2)\tau]\, \psi_{\mathbf{P}'E'a'}(t + \tfrac{1}{2}\tau)$$
$$= \psi^*_{\mathbf{P}Ea}(t) \exp\left[-i\left(\frac{E + E'}{2} - T_1 - T_2\right)\tau\right] \psi_{\mathbf{P}'E'a'}(t), \qquad (5\text{–}1.137)$$

where $\tau$ is required to be positive and negative, respectively, in the two forms. Now we see that, after compensating the effect of the $\tau$ integration from $-\infty$ to $\infty$ by a factor proportional to $\frac{1}{2}(E + E') - T_1 - T_2$, the ordinary orthonormality statement is recovered:

$$\left(\frac{1}{2i}\right) \int d\tau (d\mathbf{r}_1)(d\mathbf{r}_2) \psi^*_{\mathbf{P}Ea}(\mathbf{r}_1 t_1 \mathbf{r}_2 t_2) \left(\frac{E + E'}{2} - T_1 - T_2\right) \psi_{\mathbf{P}'E'a'}(\mathbf{r}_1 t_1 \mathbf{r}_2 t_2) = \delta_{\mathbf{P}\mathbf{P}'}\, \delta_{EE'}\, \delta_{aa'}.$$
$$(5\text{–}1.138)$$

This structure is suggested more directly by considering the eigenfunction equation of (5–1.126), where we now write

$$E_1 = \tfrac{1}{2}E + E_\tau, \qquad E_2 = \tfrac{1}{2}E - E_\tau, \qquad (5\text{–}1.139)$$

with

$$E = i(\partial/\partial t), \qquad E_\tau = i(\partial/\partial \tau). \qquad (5\text{–}1.140)$$

That gives

$$\left[\tfrac{1}{4}(E' - T_1 - T_2)^2 - \left(E_\tau - \frac{T_1 - T_2}{2}\right)^2 - i\,\delta(\tau) V\right] \psi_{\mathbf{P}'E'a'} = 0 \quad (5\text{–}1.141)$$

and, similarly,

$$\psi^*_{\mathbf{P}Ea} \left[\tfrac{1}{4}(E - T_1 - T_2)^2 - \left(E_\tau - \frac{T_1 - T_2}{2}\right)^2 - i\,\delta(\tau) V\right] = 0, \quad (5\text{–}1.142)$$

where the differential operator $E$ has been replaced by energy eigenvalues, since the latter determine the response to a rigid translation of both the variables, which leaves $\tau$ fixed. The remaining variables in Eqs. (5–1.141, 142) are $\mathbf{r}_1$, $\mathbf{r}_2$, and $\tau$. We now proceed conventionally by cross-multiplying the two equations, which are then subtracted and integrated over all variables to produce

$$\int d\tau (d\mathbf{r}_1)(d\mathbf{r}_2) \psi^*_{\mathbf{P}Ea}(\mathbf{r}_1 t_1 \mathbf{r}_2 t_2) [(E - T_1 - T_2)^2$$
$$- (E' - T_1 - T_2)^2] \psi_{\mathbf{P}'E'a'}(\mathbf{r}_1 t_1 \mathbf{r}_2 t_2) = 0, \qquad (5\text{–}1.143)$$

or

$$(E - E') \int d\tau (d\mathbf{r}_1)(d\mathbf{r}_2) \psi^*_{\mathbf{P}Ea}(\mathbf{r}_1 t_1 \mathbf{r}_2 t_2) \left(\frac{E + E'}{2} - T_1 - T_2\right) \psi_{\mathbf{P}'E'a'}(\mathbf{r}_1 t_1 \mathbf{r}_2 t_2) = 0.$$
$$(5\text{–}1.144)$$

The form of (5–1.138) is hereby recognized. But, to fix the absolute factor of the normalization statement, one falls back on the preceding development.

The multi-time Green's function is now known explicitly for the two disjoint time regions $t_< > t'_>$ and $t_< < t'_>$. In the first region the two particles coexist for a finite time interval; the Green's function can be represented by the eigenfunction expansion (5–1.119). The other time region is such that the particles do not coexist, and therefore do not interact:

$$t_< < t'_>: \quad G(12, 1'2') = G(1, 1')G(2, 2'). \tag{5–1.145}$$

It would be desirable to obtain these two forms in a unified way by proceeding from a single expression. We shall do this by applying the integral equations that are equivalent to the differential equations of (5–1.48) and (5–1.72), namely

$$G_{12} = G_1G_2 + G_1G_2iV(12)G_{12} \tag{5–1.146}$$

and

$$G_{12} = G_1G_2 + G_{12}iV(12)G_1G_2, \tag{5–1.147}$$

which are written in a four-dimensional matrix notation [cf. Eq. (3–12.21)]. The combination of the two gives

$$G_{12} = G_1G_2 + G_1G_2iV(12)G_1G_2 + G_1G_2iV(12)G_{12}iV(12)G_1G_2, \tag{5–1.148}$$

which is also effectively contained in Eq. (5–1.74). We then switch to three-dimensional matrix notation and write out this equation as

$$\begin{aligned} &G_{12}(t_1t_2, t'_1t'_2) \\ &= G_1(t_1, t'_1)G_2(t_2, t'_2) + \int dt\, G_1(t_1, t)G_2(t_2, t)iVG_1(t, t'_1)G_2(t, t'_2) \\ &\quad + \int dt\, dt'\, G_1(t_1, t)G_2(t_2, t)iV(1/i)G_{1+2}(t, t')iVG_1(t', t'_1)G_2(t', t'_2), \end{aligned} \tag{5–1.149}$$

in which it has been recognized that the last term involves the equal-time Green's function

$$G_{1+2}(t, t') = (1/i)\eta(t - t') \sum \psi(t)\psi^*(t'). \tag{5–1.150}$$

The $\psi(t)$ appearing here are the eigenfunctions (5–1.116) with the labels omitted, for simplicity. We now observe that

$$\begin{aligned} V\eta(t - t') \sum \psi(t)\psi^*(t') &= \eta(t - t')(E - T_1 - T_2) \sum \psi(t)\psi^*(t') \\ &= (E - T_1 - T_2)\eta(t - t') \sum \psi(t)\psi^*(t') - i\,\delta(t - t'), \end{aligned} \tag{5–1.151}$$

which uses the homogeneous equation obeyed by the eigenfunctions,

$$(E - T_1 - T_2 - V)\psi(t) = 0, \qquad E = i(\partial/\partial t), \tag{5–1.152}$$

and the expression of completeness,

$$\sum \psi(t)\psi^*(t) = 1. \tag{5–1.153}$$

The additional delta function term obtained in this way cancels the linear $V$ term of (5–1.149). Furthermore, we recall that

$$G_1(t_1, t)G_2(t_2, t)(E - T_1 - T_2) = \delta(t_1 - t)G_2(t_2, t_1) + \delta(t_2 - t)G_1(t_1, t_2), \tag{5–1.154}$$

which gives the reduced form

$$\begin{aligned} &G_{12}(t_1t_2, t'_1t'_2) \\ &= G_1(t_1, t'_1)G_2(t_2, t'_2) + \int dt\, dt'\, [\delta(t_1 - t)G_2(t_2, t_1) + \delta(t_2 - t)G_1(t_1, t_2)] \\ &\quad \times \eta(t - t') \sum \psi(t)\psi^*(t')VG_1(t', t'_1)G_2(t', t'_2). \end{aligned} \tag{5–1.155}$$

We proceed analogously to complete the elimination of $V$, in its explicit manifestation,

$$\begin{aligned} \sum \psi(t)\psi^*(t')\eta(t - t')V &= \sum \psi(t)\psi^*(t')(E' - T_1 - T_2)\eta(t - t') \\ &= \sum \psi(t)\psi^*(t')\eta(t - t')(E' - T_1 - T_2) - i\,\delta(t - t'). \end{aligned} \tag{5–1.156}$$

This combines with the relation

$$(E' - T_1 - T_2)G_1(t', t'_1)G_2(t', t'_2) = \delta(t' - t'_1)G_2(t'_1, t_2) + \delta(t' - t'_2)G_1(t'_2, t'_1), \tag{5–1.157}$$

to give

$$\begin{aligned} &G_{12}(t_1t_2, t'_1t'_2) \\ &= G_1(t_1, t'_1)G_2(t_2, t'_2) - iG_1(t_1, t'_1)G_2(t_2, t_1)G_2(t_1, t'_2) \\ &\quad - iG_1(t_1, t_2)G_1(t_2, t'_1)G_2(t_2, t'_2) + \int dt\, dt'\, [\delta(t_1 - t)G_2(t_2, t_1) + \delta(t_2 - t)G_1(t_1, t_2)] \\ &\quad \times \eta(t - t') \sum \psi(t)\psi^*(t')[\delta(t' - t'_1)G_2(t'_1, t'_2) + \delta(t' - t'_2)G_1(t'_2, t'_1)], \end{aligned} \tag{5–1.158}$$

or

$$G_{12}(t_1t_2, t'_1t'_2)$$
$$= [G_1(t_1, t'_1) - iG_1(t_1, t_2)G_1(t_2, t'_1)][G_2(t_2, t'_2) - iG_2(t_2, t_1)G_2(t_1, t'_2)]$$
$$- \eta(t_< - t'_>) \sum \psi(t_1t_2)\psi^*(t'_1t'_2). \qquad (5\text{–}1.159)$$

In writing the latter form we have used the vanishing of the product $G_1(t_1, t_2)G_2(t_2, t_1)$, noted that the double time integration assigns to $t$ and $t'$ the values $t_<$ and $t'_>$, respectively, and recognized the constructions (5–1.120, 121) of the multi-time eigenfunctions. Concerning the combination of free particle Green's functions that appears here, we recall that

$$t > t' > t'': \quad G(t, t'') = iG(t, t')G(t', t''), \qquad (5\text{–}1.160)$$

whereas the product on the right-hand side vanishes if the time variables are not in the indicated sequence. Accordingly, the free particle term of (5–1.159) disappears if $t_1 > t_2 > t'_1$, or if $t_2 > t_1 > t'_2$, which are summarized by $t_< > t'_>$, while, in the opposite situation, $t_< < t'_>$, the products of two Green's functions referring to the same particle are zero. The result is the anticipated one,

$$G_{12}(t_1t_2, t'_1t'_2) = \eta(t'_> - t_<)G_1(t_1, t'_1)G_2(t_2, t'_2)$$
$$- \eta(t_< - t'_>) \sum \psi(t_1t_2)\psi^*(t'_1t'_2). \qquad (5\text{–}1.161)$$

The states of the two-particle system fall into two distinct categories: those with $E > 0$, which constitute scattering situations, and those with $E < 0$, the bound states. Each example of the latter constitutes a composite particle which, in the present simplified description, appears as a stable particle. We must check a consistency aspect of our theory—the composite nature of a particle should be irrelevant to its phenomenological description. Let us return to Eq. (5–1.79) and pick out, in the quartic source term of $W$, the contribution of a particular bound state to $G(12, 1'2')$, using the construction of (5–1.148, 149) for this purpose. In a somewhat mixed notation, that gives

$$W_{\text{comp. part.}} = i \int d1 \cdots d2' \, \eta^*(2)\eta^*(1)G_1G_2iV(12)(1/i)G_{1+2}|_{\text{bd. state}} \, iV(12)$$
$$\times G_1G_2\eta(1')\eta(2'), \qquad (5\text{–}1.162)$$

where, according to (5–1.99) and (5–1.110),

$$G_{1+2}|_{\text{bd. state}} = G(\mathbf{R}t, \mathbf{R}'t')\psi(\mathbf{r}t)\psi^*(\mathbf{r}'t'), \qquad (5\text{–}1.163)$$

and

$$\psi(\mathbf{r}t) = \psi(\mathbf{r}) \exp(-iEt) \qquad (5\text{–}1.164)$$

is the eigenfunction of the specific bound state under consideration. Isolating the motion of the composite particle as a whole then gives

$$W_{\text{comp. part.}} = -\int (d\mathbf{R})\, dt\, (d\mathbf{R}')\, dt'\, \eta^*(\mathbf{R}t) G(\mathbf{R}t, \mathbf{R}'t') \eta(\mathbf{R}'t'), \quad (5\text{–}1.165)$$

where

$$\eta(\mathbf{R}t) = \int (d\mathbf{r})\, d1'\, d2'\, \psi^*(\mathbf{r}t) V(\mathbf{r}) G_1(\mathbf{r}_1 t, 1') G_2(\mathbf{r}_2 t, 2') \eta(1')\eta(2'),$$

$$\eta^*(\mathbf{R}t) = \int (d\mathbf{r})\, d1'\, d2'\, \eta^*(2')\eta^*(1') G_1(1', \mathbf{r}_1 t) G_2(2', \mathbf{r}_2 t) V(\mathbf{r}) \psi(\mathbf{r}t), \quad (5\text{–}1.166)$$

and [Eq. (5–1.95)]

$$\mathbf{r}_1 = \mathbf{R} + \frac{m_2}{M}\mathbf{r}, \qquad \mathbf{r}_2 = \mathbf{R} - \frac{m_1}{M}\mathbf{r}. \quad (5\text{–}1.167)$$

The form of (5–1.165) is correct [cf. Eq. (5–1.1)], but the complete phenomenological structure is attained only if $\eta(\mathbf{R}t)$ and $\eta^*(\mathbf{R}t)$ are indeed complex conjugate quantities. That will be true if each of the single-particle Green's functions effectively obeys

$$G(\mathbf{r}t, \mathbf{r}'t')^* = G(\mathbf{r}'t', \mathbf{r}t). \quad (5\text{–}1.168)$$

As in the discussion following Eq. (5–1.30), and also (5–1.41), the condition for this is that no real single-particle propagation shall occur under the circumstances that characterize the functioning of the composite particle sources, which is surely satisfied if neither single-particle source is capable of emitting, or absorbing, real particles. Let us also display the sources for a particular composite particle,

$$\eta_{\mathbf{P}} = \left[\frac{(d\mathbf{P})}{(2\pi)^3}\right]^{1/2} \int (d\mathbf{R})\, dt\, \exp[-i(\mathbf{P}\cdot\mathbf{R} - T_{\mathbf{P}}t)]\, \eta(\mathbf{R}t),$$

$$\eta^*_{\mathbf{P}} = \left[\frac{(d\mathbf{P})}{(2\pi)^3}\right]^{1/2} \int (d\mathbf{R})\, dt\, \eta^*(\mathbf{R}t) \exp[i(\mathbf{P}\cdot\mathbf{R} - T_{\mathbf{P}}t)], \quad (5\text{–}1.169)$$

which are given by

$$\eta_{\mathbf{P}} = \int (d\mathbf{r}_1)(d\mathbf{r}_2)\, dt\, d1'\, d2'\, \psi^*_{\mathbf{P}}(\mathbf{r}_1\mathbf{r}_2 t) V(\mathbf{r}) G_1(\mathbf{r}_1 t, 1') G_2(\mathbf{r}_2 t, 2') \eta(1')\eta(2') \quad (5\text{–}1.170)$$

and

$$\eta^*_{\mathbf{P}} = \int (d\mathbf{r}_1)(d\mathbf{r}_2)\, dt\, d1'\, d2'\, \eta^*(2')\eta^*(1') G_1(1', \mathbf{r}_1 t) G_2(2', \mathbf{r}_2 t) V(\mathbf{r}) \psi_{\mathbf{P}}(\mathbf{r}_1\mathbf{r}_2 t), \quad (5\text{–}1.171)$$

where $\psi_{\mathbf{P}}(\mathbf{r}_1\mathbf{r}_2t)$ is the eigenfunction of (5–1.116). We recognize here the structure of the multi-time eigenfunctions in Eqs. (5–1.129, 130) [remember that the latter use matrix notation], and thus

$$i\eta_{\mathbf{P}} = \int d1\, d2\, \psi^*{}_{\mathbf{P}}(12)\eta(1)\eta(2),$$

$$i\eta^*{}_{\mathbf{P}} = \int d1\, d2\, \eta^*(2)\eta^*(1)\psi_{\mathbf{P}}(12). \tag{5–1.172}$$

These multi-time constructions are also in evidence when the eigenfunction expansion of Eq. (5–1.119) is inserted in (5–1.79).

No specific reference to the B.E. or F.D. nature of the sources has occurred in this section. But we should comment on the statistics of a composite particle in relation to those of its constituents [cf. Section 3–9, p. 252]. The products of two commuting numbers or of two anticommuting numbers are completely commuting objects: if the two constituents have the same statistics, the composite is a B.E. particle. The product of a commuting number with an anticommuting one is an anticommuting object: constituents of opposite statistics produce a composite F.D. particle.

There is an interesting way of exhibiting symbolically the solution of the multi-time Green's function equation of (5–1.48). It is suggested by the first two terms of the construction (5–1.148), which are also the initial terms of an iterative solution,

$$G(12, 1'2') = G(1, 1')G(2, 2') + \int d\bar{1}\, d\bar{2}\, G(1, \bar{1})G(2, \bar{2})iV(\bar{1}\bar{2})G(\bar{1}, 1')G(\bar{2}, 2') + \cdots. \tag{5–1.173}$$

Let us return to the single-particle Green's function equation, and introduce an arbitrary potential energy term, a function of space and time:

$$(E - T - V(1))G^V(1, 1') = \delta(1, 1'). \tag{5–1.174}$$

The effect of an infinitesimal variation in $V(1)$ is given by

$$(E - T - V(1))\, \delta G^V(1, 1') = \delta V(1)\, G^V(1, 1'), \tag{5–1.175}$$

and the solution of the differential equation is

$$\delta G^V(1, 1') = \int d\bar{1}\, G^V(1, \bar{1})\, \delta V(\bar{1})\, G^V(\bar{1}, 1'). \tag{5–1.176}$$

A functional derivative notation will be used to convey this differential expression,

$$\frac{\delta G^V(1, 1')}{\delta V(\bar{1})} = G^V(1, \bar{1})G^V(\bar{1}, 1'). \tag{5–1.177}$$

If the auxiliary function $V(1)$ is set equal to zero after the differentiation of (5–1.177), we encounter just the product of two free-particle Green's functions that occurs (for each particle) in Eq. (5–1.173). Accordingly, we can write the latter as

$$G(12, 1'2') = G(1, 1')G(2, 2') + i\int d\bar{1}\, d\bar{2}\, \frac{\delta}{\delta V(\bar{1})}\, V(\bar{1}\bar{2})\, \frac{\delta}{\delta V(\bar{2})}\, G^V(1, 1')$$
$$\times\, G^V(2, 2')|_{V=0} + \cdots. \tag{5–1.178}$$

It is a natural presumption that the effect of the indefinite repetition of the interaction is expressed by the exponential operator that has its first terms exhibited in (5–1.178):

$$G(12, 1'2') = \exp\left[i\int d\bar{1}\, d\bar{2}\, \frac{\delta}{\delta V(\bar{1})}\, V(\bar{1}\bar{2})\, \frac{\delta}{\delta V(\bar{2})}\right] G^V(1, 1')G^V(2, 2')|_{V=0}. \tag{5–1.179}$$

We shall verify this. [For a related quantum mechanical discussion using the action principle, see *Quantum Kinematics and Dynamics*, Section 7.9.] According to the equations illustrated in (5–1.174), we have

$$\begin{aligned} &(E - T)_1(E - T)_2G(12, 1'2') \\ &= \exp[\ ]\, (\delta(1, 1') + V(1)G^V(1, 1'))(\delta(2, 2') + V(2)G^V(2, 2'))|_{V=0} \\ &= \delta(1, 1')\, \delta(2, 2') + \exp[\ ]\, V(1)V(2)G^V(1, 1')G^V(2, 2')|_{V=0}, \end{aligned} \tag{5–1.180}$$

where the bracket indicates the functional differential operator of (5–1.179), and simplifications associated with terms that do not contain both $V(1)$ and $V(2)$ have been inserted. Now observe that

$$\exp[\ ]\, V(1) = V(1)\exp[\ ] + \exp[\ ]\, i\int d\bar{2}\, V(1\bar{2})\, \frac{\delta}{\delta V(\bar{2})}, \tag{5–1.181}$$

according to the functional derivative relation

$$\delta V(1)/\delta V(\bar{1}) = \delta(1, \bar{1}) \tag{5–1.182}$$

that expresses the identity

$$\delta V(1) = \int d\bar{1}\, \delta V(\bar{1})\, \delta(1, \bar{1}). \tag{5–1.183}$$

After $V(1)$ has been moved to the left of all functional derivatives, it is set equal to zero. The first stage in performing the same service for $V(2)$ involves

$$i\int d\bar{2}\, V(1\bar{2})\, \frac{\delta}{\delta V(\bar{2})}\, V(2) = V(2)i\int d\bar{2}\, V(1\bar{2})\, \frac{\delta}{\delta V(\bar{2})} + iV(12). \tag{5–1.184}$$

Now the use of the relation analogous to (5–1.181) gives

$$[(E-T)_1(E-T)_2 - iV(12)]G(12, 1'2') - \delta(1, 1')\,\delta(2, 2')$$

$$= -\exp[\ ]\int d\bar{1}\, d\bar{2}\; V(1\bar{2})V(2\bar{1}) \frac{\delta}{\delta V(\bar{1})}\frac{\delta}{\delta V(\bar{2})} G^V(1, 1')G^V(2, 2')|_{V=0}. \qquad (5\text{–}1.185)$$

The double functional derivative appearing here is evaluated as [Eq. (5–1.177)]

$$\frac{\delta}{\delta V(\bar{1})}\frac{\delta}{\delta V(\bar{2})} G^V(1, 1')G^V(2, 2') = G^V(1, \bar{1})G^V(\bar{1}, 1')G^V(2, \bar{2})G^V(\bar{2}, 2'). \qquad (5\text{–}1.186)$$

At this point, the instantaneous nature of the interaction and the retarded character of the Green's functions become decisive. The time delta functions in $V$ demand that

$$\bar{t}_2 = t_1, \qquad \bar{t}_1 = t_2 \qquad (5\text{–}1.187)$$

Hence, the Green's function product in (5–1.186) contains the factor

$$G_1{}^V(t_1, t_2)G_2{}^V(t_2, t_1) = 0, \qquad (5\text{–}1.188)$$

since

$$\eta(t_1 - t_2)\eta(t_2 - t_1) = 0, \qquad (5\text{–}1.189)$$

apart from an isolated point at $t_1 - t_2 = 0$, which does not yield a nonzero time integral. That completes the verification of Eq. (5–1.179).

The instantaneous character of the interaction is made explicit on writing (5–1.179) as

$$G(12, 1'2') = \exp\left[i\int d\bar{t}(d\bar{\mathbf{r}}_1)(d\bar{\mathbf{r}}_2)\frac{\delta}{\delta V_1(\bar{\mathbf{r}}_1\bar{t})}V(\bar{\mathbf{r}}_1 - \bar{\mathbf{r}}_2)\frac{\delta}{\delta V_2(\bar{\mathbf{r}}_2\bar{t})}\right]$$

$$\times\, G^{V_1}(1, 1')G^{V_2}(2, 2')|_{V_{1,2}=0}. \qquad (5\text{–}1.190)$$

One can also specialize to the equal-time Green's function:

$$G(\mathbf{r}_1\mathbf{r}_2t, \mathbf{r}'_1\mathbf{r}'_2t') = \exp[\ ]\, G^{V_{1,2}}(\mathbf{r}_1\mathbf{r}_2t, \mathbf{r}'_1\mathbf{r}'_2t')|_{V_{1,2}=0}, \qquad (5\text{–}1.191)$$

where

$$(E - T_1 - T_2 - V_1 - V_2)G^{V_{1,2}}(\mathbf{r}_1\mathbf{r}_2t, \mathbf{r}'_1\mathbf{r}'_2t') = \delta(t-t')\,\delta(\mathbf{r}_1 - \mathbf{r}'_1)\,\delta(\mathbf{r}_2 - \mathbf{r}'_2). \qquad (5\text{–}1.192)$$

For a direct derivation of the differential equation in (5–1.53), we apply the preceding equation for $G^{V_{1,2}}$,

$$(E - T_1 - T_2)G(\mathbf{r}_1\mathbf{r}_2t, \mathbf{r}'_1\mathbf{r}'_2t') - \delta(t-t')\,\delta(\mathbf{r}_1 - \mathbf{r}'_1)\,\delta(\mathbf{r}_2 - \mathbf{r}'_2)$$

$$= \exp[\ ]\,(V_1(\mathbf{r}_1t) + V_2(\mathbf{r}_2t))G^{V_{1,2}}(\mathbf{r}_1\mathbf{r}_2t, \mathbf{r}'_1\mathbf{r}'_2t')|_{V_{1,2}=0}, \qquad (5\text{–}1.193)$$

and then use (5–1.181) to present the right-hand side as

$$\exp[\ ]\left\{\int (d\bar{\mathbf{r}}_2)V(\mathbf{r}_1-\bar{\mathbf{r}}_2)i\frac{\delta}{\delta V_2(\bar{\mathbf{r}}_2 t)}+\int (d\bar{\mathbf{r}}_1)V(\bar{\mathbf{r}}_1-\mathbf{r}_2)i\frac{\delta}{\delta V_1(\bar{\mathbf{r}}_1 t)}\right\}$$
$$\times\, G^{V_{1,2}}(\mathbf{r}_1\mathbf{r}_2t,\mathbf{r}'_1\mathbf{r}'_2t')|_{V_{1,2}=0}. \qquad (5\text{–}1.194)$$

Next, we consider the two-particle analogue of Eq. (5–1.176),

$$\delta G^{V_{1,2}}(\mathbf{r}_1\mathbf{r}_2t,\mathbf{r}'_1\mathbf{r}'_2t')=\int d\bar{t}\,(d\bar{\mathbf{r}}_1)(d\bar{\mathbf{r}}_2)G^{V_{1,2}}(\mathbf{r}_1\mathbf{r}_2t,\bar{\mathbf{r}}_1\bar{\mathbf{r}}_2\bar{t})[\delta V_1(\bar{\mathbf{r}}_1\bar{t})$$
$$+\,\delta V_2(\bar{\mathbf{r}}_2\bar{t})]G^{V_{1,2}}(\bar{\mathbf{r}}_1\bar{\mathbf{r}}_2\bar{t},\mathbf{r}'_1\mathbf{r}'_2t'), \qquad (5\text{–}1.195)$$

which is conveyed by the functional derivatives

$$\frac{\delta G^{V_{1,2}}(\mathbf{r}_1\mathbf{r}_2t,\mathbf{r}'_1\mathbf{r}'_2t')}{\delta V_1(\bar{\mathbf{r}}_1\bar{t})}=\int(d\bar{\mathbf{r}}_2)G^{V_{1,2}}(\mathbf{r}_1\mathbf{r}_2t,\bar{\mathbf{r}}_1\bar{\mathbf{r}}_2\bar{t})G^{V_{1,2}}(\bar{\mathbf{r}}_1\bar{\mathbf{r}}_2\bar{t},\mathbf{r}'_1\mathbf{r}'_2t') \qquad (5\text{–}1.196)$$

and

$$\frac{\delta G^{V_{1,2}}(\mathbf{r}_1\mathbf{r}_2t,\mathbf{r}'_1\mathbf{r}'_2t')}{\delta V_2(\bar{\mathbf{r}}_2\bar{t})}=\int(d\bar{\mathbf{r}}_1)G^{V_{1,2}}(\mathbf{r}_1\mathbf{r}_2t,\bar{\mathbf{r}}_1\bar{\mathbf{r}}_2\bar{t})G^{V_{1,2}}(\bar{\mathbf{r}}_1\bar{\mathbf{r}}_2\bar{t},\mathbf{r}'_1\mathbf{r}'_2t'). \qquad (5\text{–}1.197)$$

What is required in (5–1.194) is the evaluation of these functional derivatives at $\bar{t}=t$, which means the equally weighted average of the two limits $\bar{t}\to t\pm 0$. Recalling that

$$G(\mathbf{r}_1\mathbf{r}_2t,\bar{\mathbf{r}}_1\bar{\mathbf{r}}_2t+0)=0,$$
$$G(\mathbf{r}_1\mathbf{r}_2t,\bar{\mathbf{r}}_1\bar{\mathbf{r}}_2t-0)=(1/i)\,\delta(\mathbf{r}_1-\bar{\mathbf{r}}_1)\,\delta(\mathbf{r}_2-\bar{\mathbf{r}}_2), \qquad (5\text{–}1.198)$$

we have

$$\lim_{\bar{t}\to t} iG(\mathbf{r}_1\mathbf{r}_2t,\bar{\mathbf{r}}_1\bar{\mathbf{r}}_2\bar{t})=\tfrac{1}{2}\delta(\mathbf{r}_1-\bar{\mathbf{r}}_1)\,\delta(\mathbf{r}_2-\bar{\mathbf{r}}_2), \qquad (5\text{–}1.199)$$

and thus

$$i\frac{\delta G^{V_{1,2}}(\mathbf{r}_1\mathbf{r}_2t,\mathbf{r}'_1\mathbf{r}'_2t')}{\delta V_1(\bar{\mathbf{r}}_1t)}=\tfrac{1}{2}\delta(\mathbf{r}_1-\bar{\mathbf{r}}_1)G^{V_{1,2}}(\mathbf{r}_1\mathbf{r}_2t,\mathbf{r}'_1\mathbf{r}'_2t'),$$
$$i\frac{\delta G^{V_{1,2}}(\mathbf{r}_1\mathbf{r}_2t,\mathbf{r}'_1\mathbf{r}'_2t')}{\delta V_2(\bar{\mathbf{r}}_2t)}=\tfrac{1}{2}\delta(\mathbf{r}_2-\bar{\mathbf{r}}_2)G^{V_{1,2}}(\mathbf{r}_1\mathbf{r}_2t,\mathbf{r}'_1\mathbf{r}'_2t'). \qquad (5\text{–}1.200)$$

The outcome for the right-hand side of (5–1.193) is

$$V(\mathbf{r}_1-\mathbf{r}_2)G(\mathbf{r}_1\mathbf{r}_2t,\mathbf{r}'_1\mathbf{r}'_2t'), \qquad (5\text{–}1.201)$$

as anticipated.

The discussion thus far has been concerned exclusively with the interaction of two different particles. Some words are in order concerning the modifications needed when the two particles are identical. Beginning with

$$W_{\text{nonint.}} = -\int d1\, d1'\, \eta^*(1)G(1, 1')\eta(1'), \qquad (5\text{–}1.202)$$

the quadratic term in the expansion of $\exp[iW_{\text{nonint.}}]$ is

$$-\tfrac{1}{2}\int d1 \cdots d2'\, \eta^*(2)\eta^*(1)G(1, 1')G(2, 2')\eta(1')\eta(2'), \qquad (5\text{–}1.203)$$

where the integers no longer have the context of different particle types. This is rewritten as

$$-\int \frac{d1\, d2}{2}\frac{d1'\, d2'}{2}\, \eta^*(2)\eta^*(1)G(12, 1'2')_{\text{nonint.}}\eta(1')\eta(2'), \qquad (5\text{–}1.204)$$

in which

$$G(12, 1'2')_{\text{nonint.}} = G(1, 1')G(2, 2') \pm G(1, 2')G(2, 1') \qquad (5\text{–}1.205)$$

makes explicit reference to the statistics of the particles under consideration (+, B.E.; −, F.D.). Correspondingly the two-particle Green's function has definite symmetry properties (in general, not only under noninteracting circumstances)

$$G(12, 1'2') = \pm\, G(21, 1'2') = \pm\, G(12, 2'1'). \qquad (5\text{–}1.206)$$

The factors of $\frac{1}{2}$ in the differential volume elements are thereby understood as avoiding repetitious counting of the identical particles. The Green's function (5–1.205) obeys the differential equation

$$(E - T)_1(E - T)_2G(12, 1'2')_{\text{nonint.}} = \delta(1, 1')\,\delta(2, 2') \pm \delta(1, 2')\,\delta(2, 1'). \qquad (5\text{–}1.207)$$

It should now be sufficiently clear that, as a general rule, all previous results are translated into the identical particle situation be replacing delta functions with the appropriately symmetrized combinations illustrated above, and by avoiding duplicate counting in all integrations.

## 5-2 TWO-PARTICLE INTERACTIONS. RELATIVISTIC THEORY I

Before embarking on the first stages of a relativistic theory of electromagnetically interacting particles, let us review some aspects of skeletal interaction theory, as discussed in Section 3–12. We are going to be interested in the multi-photon annihilation of a spin $\frac{1}{2}$ particle-antiparticle pair, and in the inverse arrangement.

These processes are described by the skeletal interaction terms of Eq. (3-12.17), as detailed in Eq. (3-12.24). The first examples of the latter can be written as the vacuum amplitude

$$iW_{21} = i\int (dx)A^\mu(x)J_\mu(x)|_{\text{eff.}},$$

$$iW_{22} = \tfrac{1}{2}i^2\int (dx)(dx')A^\mu(x)A^\nu(x')J_\mu(x)J_\nu(x')|_{\text{eff}}, \tag{5-2.1}$$

where the effective photon sources are

$$J_\mu(x)|_{\text{eff.}} = \tfrac{1}{2}\psi(x)\gamma^0 eq\gamma_\mu\psi(x) \tag{5-2.2}$$

and

$$iJ_\mu(x)J_\nu(x')|_{\text{eff.}} = \tfrac{1}{2}[\psi(x)\gamma^0 eq\gamma_\mu G_+(x-x')eq\gamma_\nu\psi(x') + \psi(x')\gamma^0 eq\gamma_\nu G_+(x'-x)eq\gamma_\mu\psi(x)]. \tag{5-2.3}$$

Since these structures originate in the expansion of the interaction expression

$$W_{2\ldots} = \tfrac{1}{2}\int (dx)(dx')\eta(x)\gamma^0 G_+^A(x,x')\eta(x'), \tag{5-2.4}$$

they are given a more compact and unified presentation in the notation of functional derivatives:

$$J_\mu(x)|_{\text{eff.}} = \frac{1}{i}\frac{\delta}{\delta A^\mu(x)}iW_{2\ldots}|_{A=0},$$

$$J_\mu(x)J_\nu(x')|_{\text{eff.}} = \frac{1}{i}\frac{\delta}{\delta A^\mu(x)}\frac{1}{i}\frac{\delta}{\delta A^\nu(x')}iW_{2\ldots}|_{A=0}. \tag{5-2.5}$$

We see here the sense in which $(1/i)\,\delta/\delta A^\mu(x)$ plays a symbolic role as the source of the multi-photon emission, or absorption. All expressions of this type are comprised in the functional form of a Taylor series expansion,

$$iW_{2\ldots} = \exp\left[i\int (d\xi)A^\mu(\xi)\frac{1}{i}\frac{\delta}{\delta A^\mu(\xi)'}\right]\frac{i}{2}\int (dx)(dx')\eta(x)\gamma^0 G_+^{A'}(x,x')\eta(x')|_{A'=0}. \tag{5-2.6}$$

It has also been noted in Section 3-12 that the question of photon radiation from the charged particle sources can be avoided by using the photon propagation functions of a certain class of gauges. That is expressed here by writing

$$A^\mu(x) = \int (dx')D_+(x-x')^{\mu\nu}J_\nu(x'), \tag{5-2.7}$$

where $D_+^{\mu\nu}$ is of the form described in Eqs. (3–12.8, 9). This differs from the simpler version, $g^{\mu\nu}D_+(x-x')$, by gauge terms that are associated with one, or both, of the vector indices $\mu$ and $\nu$. Specific details will be recalled later.

The system of interest in this section contains two spin $\frac{1}{2}$ charged particles of different types, which are labeled 1 and 2. The vacuum amplitude that describes them in the absence of interaction is

$$\left[i\tfrac{1}{2}\int(dx)(dx')\eta(x)\gamma^0G_+(x-x')\eta(x')\right]_1\left[i\tfrac{1}{2}\int(dx)(dx')\eta(x)\gamma^0G_+(x-x')\eta(x')\right]_2$$

$$= -\tfrac{1}{4}\int(dx_1)\cdots(dx'_2)\eta(x_2)\eta(x_1)\gamma_1{}^0\gamma_2{}^0G_+(x_1-x'_1)G_+(x_2-x'_2)\eta(x'_1)\eta(x'_2), \quad (5\text{–}2.8)$$

which we express by means of a two-particle Green's function,

$$G_+(x_1x_2, x'_1x'_2)_{\text{nonint.}} = G_+(x_1-x'_1)G_+(x_2-x'_2). \qquad (5\text{–}2.9)$$

The effect of interactions can be variously introduced by considering different causal arrangements. We have prepared the way for the following one. Particle and antiparticle of type 2 annihilate into an arbitrary number of photons, which subsequently recombine to form the particle and antiparticle of type 1, or vice versa. The skeletal description of these processes is given in (5–2.6); it is the omission of form factors for the various acts that constitutes the skeletal nature of the description. And, in the characterization of the exchanged photons by the simple propagation function $D_+$ (gauge questions aside) we also employ a skeletal description. The unlimited exchange of noninteracting photons is expressed symbolically by the vacuum amplitude factor

$$\exp\left[i\int(d\xi)(d\xi')\,\frac{1}{i}\frac{\delta}{\delta A_1{}^\mu(\xi)}\,D_+(\xi-\xi')^{\mu\nu}\,\frac{1}{i}\frac{\delta}{\delta A_2{}^\nu(\xi')}\right], \qquad (5\text{–}2.10)$$

acting upon the particle part of the vacuum amplitude. The latter is (5–2.8), with the propagation functions replaced by those that contain the effects of the electromagnetic fields $A^\mu_{1,2}$, as described by

$$\left[\gamma\left(\frac{1}{i}\partial-eqA\right)+m\right]_1 G_+^{A_1}(x_1,x'_1) = \delta(x_1-x'_1),$$

$$\left[\gamma\left(\frac{1}{i}\partial-eqA\right)+m\right]_2 G_+^{A_2}(x_2,x'_2) = \delta(x_2-x'_2). \qquad (5\text{–}2.11)$$

What emerges is a symbolic expression for the two-particle propagation function that incorporates the skeleton interactions being considered:

$$G_+(x_1x_2, x'_1x'_2)$$
$$= \exp\left[-i\int (d\xi)(d\xi')\frac{\delta}{\delta A_1{}^\mu(\xi)}D_+(\xi-\xi')^{\mu\nu}\frac{\delta}{\delta A_2{}^\nu(\xi')}\right]G_+^{A_1}(x_1, x'_1)G_+^{A_2}(x_2, x'_2)|_{A_{1,2}=0},$$
(5-2.12)

where the space-time presentation has removed the reference to the initial causal arrangement.

There is an evident resemblance here to the nonrelativistic construction in Eq. (5-1.179). We can proceed analogously in the first stages of deriving a differential equation

$$\left(\gamma\frac{1}{i}\partial+m\right)_1\left(\gamma\frac{1}{i}\partial+m\right)_2 G_+(x_1x_2, x'_1x'_2)$$
$$= \exp[\ ]\,[\delta(x-x') + eq\gamma A(x)G_+{}^A(x, x')]_1[\delta(x-x') + eq\gamma A(x)G_+{}^A(x, x')]_2|_{A_{1,2}=0}$$
$$= \delta(x_1-x'_1)\,\delta(x_2-x'_2) + \exp[\ ]\,[eq\gamma A(x)G_+{}^A(x, x')]_1[eq\gamma A(x)G_+{}^A(x, x')]_2|_{A_{1,2}=0}.$$
(5-2.13)

The next step involves the rearrangement

$$\exp[\ ]\,A_1{}^\mu(\xi) = A_1{}^\mu(\xi)\exp[\ ] + \exp[\ ]\int(d\xi')D_+(\xi-\xi')^{\mu\nu}\frac{1}{i}\frac{\delta}{\delta A_2{}^\nu(\xi')}, \quad (5\text{-}2.14)$$

which, with a similar statement referring to $A_2{}^\nu$, leads to

$$\left[\left(\gamma\frac{1}{i}\partial+m\right)_1\left(\gamma\frac{1}{i}\partial+m\right)_2 - I^{(1)}(x_1x_2)\right]G_+(x_1x_2, x'_1x'_2) - \delta(x_1-x'_1)\,\delta(x_2-x'_2)$$
$$= -\exp[\ ]\int(d\xi)(d\xi')(eq\gamma^\mu)_1(eq\gamma^\nu)_2D_+(x_1-\xi')_{\mu\lambda}D_+(x_2-\xi)_{\nu\kappa}$$
$$\times\frac{\delta}{\delta A_{1\kappa}(\xi)}\frac{\delta}{\delta A_{2\lambda}(\xi')}G_+^{A_1}(x_1, x'_1)G_+^{A_2}(x_2, x'_2)|_{A_{1,2}=0}, \quad (5\text{-}2.15)$$

where

$$I^{(1)}(x_1x_2) = -i(eq\gamma^\mu)_1\,D_+(x_1-x_2)_{\mu\nu}\,(eq\gamma^\nu)_2. \quad (5\text{-}2.16)$$

The differential relation inferred from Eq. (5-2.11),

$$(\gamma\Pi+m)\delta G_+{}^A(x, x') = eq\gamma\delta A(x)G_+{}^A(x, x'), \quad (5\text{-}2.17)$$

is solved by

$$\delta G_+{}^A(x, x') = \int(d\xi)G_+{}^A(x, \xi)eq\gamma\delta A(\xi)G_+{}^A(\xi, x'), \quad (5\text{-}2.18)$$

and this result is conveyed by the functional derivative

$$\frac{\delta G_+{}^A(x, x')}{\delta A_\mu(\xi)} = G_+{}^A(x, \xi) eq\gamma^\mu G_+{}^A(\xi, x'). \tag{5–2.19}$$

[An example of this relation is the equivalence of the first statement in Eq. (5–2.5) with (5–2.2).] Accordingly, the right-hand side of (5–2.15) becomes

$$\begin{aligned} -\exp[\;] \int (d\xi)(d\xi')(eq\gamma^\mu)_1(eq\gamma^\nu)_2 \, D_+(x_1 - \xi')_{\mu\lambda} \, D_+(x_2 - \xi)_{\nu\kappa} \\ \times G_+^{A_1}(x_1, \xi)(eq\gamma^\kappa)_1 G_+^{A_1}(\xi, x'_1) G_+^{A_2}(x_2, \xi')(eq\gamma^\lambda)_2 G_+^{A_2}(\xi', x'_2)|_{A_{1,2}=0}, \end{aligned} \tag{5–2.20}$$

in analogy with (5–1.185, 186).

But here the resemblance to the nonrelativistic discussion ceases. The photon propagation function does not transmit an instantaneous interaction, and the particle propagation functions do not obey retarded boundary conditions. The appearance of four particle propagation functions in (5–2.20) means that new classes of propagation functions are being introduced in the process of finding $G_+(x_1x_2, x'_1x'_2)$. Thus, the two-particle equation of nonrelativistic theory has no strict counterpart in the relativistic domain, except in the inexact sense of approximation schemes that relate (5–2.20) directly to the two-particle Green's function. An illustration of this emerges on distinguishing the propagation function factor $G_+^{A_1}(\xi, x'_1)G_+^{A_2}(\xi', x'_2)$ from $G_+^{A_1}(x_1, \xi)G_+^{A_2}(x_2, \xi')$. The former describes the propagation of the particles from their initial creation region up to the domain of the two-photon exchange process considered in (5–2.20), while the latter represents the particles during the interaction process. It is plausible that circumstances should exist where the additional interactions between the particles (symbolized by exp[ ]) during the two-photon exchange process would be relatively negligible, whereas they certainly cannot be ignored throughout the previous history of the particles. Accepting this argument gives the following approximate evaluation for (5–2.20),

$$\int (d\bar{x}_1)(d\bar{x}_2) I^{(2)}(x_1x_2, \bar{x}_1\bar{x}_2) G_+(\bar{x}_1\bar{x}_2, x'_1x'_2), \tag{5–2.21}$$

in which

$$\begin{aligned} I^{(2)}(x_1x_2, \bar{x}_1\bar{x}_2) = -(eq\gamma^\mu)_1(eq\gamma^\nu)_2 \, D_+(x_1 - \bar{x}_2)_{\mu\lambda} \, D_+(x_2 - \bar{x}_1)_{\nu\kappa} \\ \times G_+(x_1 - \bar{x}_1) G_+(x_2 - \bar{x}_2)(eq\gamma^\kappa)_1(eq\gamma^\lambda)_2, \end{aligned} \tag{5–2.22}$$

and leads to the symbolically presented two-particle equation

$$[(\gamma p + m)_1(\gamma p + m)_2 - I_{12}]G_{12} = 1, \tag{5–2.23}$$

where

$$I_{12} = I_{12}^{(1)} + I_{12}^{(2)} + \cdots . \qquad (5\text{–}2.24)$$

This discussion has maintained the generality of

$$D_+(x - x')_{\mu\nu} = g_{\mu\nu}D_+(x - x') + \text{gauge terms}. \qquad (5\text{–}2.25)$$

But clearly our principal concern is with the physical implications of the formalism, which must be independent of the specific choice of gauge terms. To examine what the latter influence, consider the result of changing $D_+(\xi - \xi')^{\mu\nu}$ by a $\mu$-dependent gauge transformation. This alters (5–2.10) by a factor of the form

$$\exp\left[\int (d\xi)\,\partial^\mu\lambda(\xi)\,\frac{\delta}{\delta A_1{}^\mu(\xi)}\right], \qquad (5\text{–}2.26)$$

where $\lambda(\xi)$ is also a linear functional of $\delta/\delta A_2$. The effect of the operator (5–2.26) is to produce a translation of $A^\mu$:

$$A_1{}^\mu(\xi) \to A_1{}^\mu(\xi) + \partial^\mu\lambda(\xi). \qquad (5\text{–}2.27)$$

This is a gauge transformation, and its consequence for $G_+^{A_1}(x_1, x'_1)$ is given by

$$G_+^{A_1}(x_1, x'_1) \to \exp[ieq\lambda(x_1)]\,G_+^{A_1}(x_1, x'_1)\exp[-ieq\lambda(x'_1)]. \qquad (5\text{–}2.28)$$

The thing to appreciate is that the alteration involves the terminal points of the Green's function, which is not surprising when one recalls that the gauge terms appear as an alternative way of representing the electromagnetic model of the source, and the source radiation that it characterizes. Such aspects of the Green's function are generally not of physical interest, and we must learn to separate them from the information that is desired. This situation is not new, of course. It is encountered in any scattering arrangement, but in such circumstances there are intuitively evident theoretical counterparts for the experimental shielding that absorbs direct electromagnetic radiation from the particle sources.

Our concern in this section is with energy spectra. To illustrate the problem in a very simple context, we consider a limit in which the particles are very massive and remain relatively at rest. In these circumstances, the particles should be describable by the photon source formalism. This is evident in the reduced form the Green's function equation acquires when all reference to spatial momentum is deleted:

$$[-\gamma^0(i\partial_0 - eqA^0) + m]G_+{}^A(x, x') = \delta(x^0 - x^{0\prime})\,\delta(\mathbf{x} - \mathbf{x}'), \qquad (5\text{–}2.29)$$

for $(x^0 > x^{0\prime})$

$$G_+{}^A(x, x') = \delta(\mathbf{x} - \mathbf{x}')\exp\left[-ieq\int_{x^{0\prime}}^{x^0} dt\,A^0(\mathbf{x}, t)\right]\frac{i}{2}(1 + \gamma^0)\exp[-im(x^0 - x^{0\prime})] \qquad (5\text{–}2.30)$$

exhibits the properties of the charge $eq$, which is located at the point $\mathbf{x}$ during the time interval from $x^{0\prime}$ to $x^0$. Whether we use Eq. (5–2.12) or apply the source description directly, the interaction between the particles is expressed by the factor

$$\exp\left[i(eq)_1(eq)_2\int_{x_1^{0\prime}}^{x_1^0} dt_1 \int_{x_2^{0\prime}}^{x_2^0} dt_2\, D_+(\mathbf{x}_1 - \mathbf{x}_2, t_1 - t_2)^{00}\right]. \tag{5–2.31}$$

In the radiation gauge, where [Eq. (3–15.51)]

$$A^0(\mathbf{x}t) = \int (d\mathbf{x}')\mathscr{D}(\mathbf{x} - \mathbf{x}')J^0(\mathbf{x}'t),$$

$$\mathscr{D}(\mathbf{x}) = \frac{1}{4\pi|\mathbf{x}|}, \tag{5–2.32}$$

we have

$$D_+(\mathbf{x}_1 - \mathbf{x}_2, t_1 - t_2)^{00} = -\,\delta(t_1 - t_2)\,\mathscr{D}(\mathbf{x}_1 - \mathbf{x}_2). \tag{5–2.33}$$

Then (5–2.31) reduces to

$$\exp[-\,iET], \tag{5–2.34}$$

in which $T$ is the duration of the interval that the particles coexist, and

$$E = (eq)_1(eq)_2\,\mathscr{D}(\mathbf{x}_1 - \mathbf{x}_2) \tag{5–2.35}$$

is the anticipated Coulomb interaction energy of the charges.

Now let us compare this elementary result with that obtained by omitting all gauge terms and working directly with $g^{\mu\nu}D_+$:

$$D_+(x - x')^{00} = -\,D_+(x - x') = -\int \frac{(d\mathbf{k})}{(2\pi)^3}\frac{i}{2|\mathbf{k}|}\exp[i\mathbf{k}\cdot(\mathbf{x} - \mathbf{x}') - i|\mathbf{k}|\,|x^0 - x^{0\prime}|\,. \tag{5–2.36}$$

In carrying out the time integrations of (5–2.31), it is helpful to use the differential equation

$$(\partial_0^2 + |\mathbf{k}|^2)\frac{i}{2|\mathbf{k}|}\exp(-\,i|\mathbf{k}|\,|x^0 - x^{0\prime}|) = \delta(x^0 - x^{0\prime}). \tag{5–2.37}$$

Thus,

$$|\mathbf{k}|^2\int_{x_1^{0\prime}}^{x_1^0} dt_1 \int_{x_2^{0\prime}}^{x_2^0} dt_2 \frac{i}{2|\mathbf{k}|}\exp(-\,i|\mathbf{k}|\,|t_1 - t_2|)$$

$$= T + \frac{i}{2|\mathbf{k}|}\exp(-\,i|\mathbf{k}|\,|t_1 - t_2|)\Big|_{x_1^{0\prime}}^{x_1^0}\Big|_{x_2^{0\prime}}^{x_2^0}.$$

$$= T + \frac{i}{2|\mathbf{k}|}[\exp(-i|\mathbf{k}|\,|x_1{}^0 - x_2{}^0|) - \exp(-i|\mathbf{k}|\,|x_1{}^0 - x_2^{0\prime}|) - \exp(-i|\mathbf{k}|\,|x_2{}^0 - x_1^{0\prime}|)$$

$$+ \exp(-i|\mathbf{k}|\,|x_1^{0\prime} - x_2^{0\prime}|)], \qquad (5\text{–}2.38)$$

where $T$ is again the coexistence interval for the two particles. For the equal-time situation represented by

$$x_1{}^0 = x_2{}^0, \qquad x_1^{0\prime} = x_2^{0\prime}, \qquad x_1{}^0 - x_1^{0\prime} = T, \qquad (5\text{–}2.39)$$

this reduces to

$$T + \frac{i}{|\mathbf{k}|}[1 - \exp(-i|\mathbf{k}|T)]. \qquad (5\text{–}2.40)$$

In this circumstance, we have

$$-\int dt_1\, dt_2\, D_+^{00} = \int \frac{(d\mathbf{k})}{(2\pi)^3} \frac{\exp[i\mathbf{k}\cdot(\mathbf{x}_1 - \mathbf{x}_2)]}{|\mathbf{k}|^2}\left[T + \frac{i}{|\mathbf{k}|}[1 - \exp(-i|\mathbf{k}|T)]\right]$$

$$= T\mathscr{D}(\mathbf{x}_1 - \mathbf{x}_2) + i\int \frac{(d\mathbf{k})}{(2\pi)^3} \frac{\exp[i\mathbf{k}\cdot(\mathbf{x}_1 - \mathbf{x}_2)]}{|\mathbf{k}|^3}[1 - \exp(-i|\mathbf{k}|T)], \qquad (5\text{–}2.41)$$

and the factor (5–2.31) becomes

$$\exp[-iET]\exp\left[(eq)_1(eq)_2\int \frac{(d\mathbf{k})}{(2\pi)^3} \frac{\exp[i\mathbf{k}\cdot(\mathbf{x}_1 - \mathbf{x}_2)]}{|\mathbf{k}|^3}[1 - \exp(-i|\mathbf{k}|T)]\right], \qquad (5\text{–}2.42)$$

where $E$ retains the meaning given in (5–2.35).

This is a generating function for an energy spectrum. The notation

$$a_{\mathbf{k}} = -(eq)_1(eq)_2 \frac{(d\mathbf{k})}{(2\pi)^3} \frac{\exp[i\mathbf{k}\cdot(\mathbf{x}_1 - \mathbf{x}_2)]}{|\mathbf{k}|^3}, \qquad (5\text{–}2.43)$$

and

$$A = \sum_{\mathbf{k}} a_{\mathbf{k}}, \qquad (5\text{–}2.44)$$

displays it as

$$\exp(-iET)\exp(-A)\exp\left[\sum_{\mathbf{k}} a_{\mathbf{k}} \exp(-i|\mathbf{k}|T)\right]$$

$$= \exp(-A)\exp(-iET) + \sum_{\mathbf{k}} \exp(-A)\, a_{\mathbf{k}} \exp[-i(E + |\mathbf{k}|)T]$$

$$+ \tfrac{1}{2}\sum_{\mathbf{k}\mathbf{k}'} \exp(-A)\, a_{\mathbf{k}} a_{\mathbf{k}'} \exp[-i(E + |\mathbf{k}| + |\mathbf{k}'|)T] + \cdots. \qquad (5\text{–}2.45)$$

We recognize that the system has the ground state energy $E$, and excited states in which an arbitrary number of photons are present. The latter are an artifact of the particular way that the two-particle system has been created (a nonphysical one). The only physical information contained in the generating function (5–2.42) is the energy of the system without photons, the Coulomb energy $E$. One will ask how this single bit of meaningful information might have been identified, without otherwise knowing it. The answer is found on minimizing the irrelevant terminal effects by considering a very long time interval. The oscillatory character of $\exp(-i|\mathbf{k}|T)$ ensures that only values of $|\mathbf{k}| \lesssim 1/T$ contribute to that portion of the momentum integral in (5–2.42), which serves as an effective infrared cut-off to the $T$-independent part of the integral. Thus, the asymptotic form of (5–2.42) is

$$\exp(-A_T)\exp(-iET), \qquad (5\text{–}2.46)$$

where, roughly,

$$A_T = \frac{(eq)_1(eq)_2}{4\pi}\frac{2}{\pi}\int_{\sim 1/T}^{\infty}\frac{dk^0}{k^0}\frac{\sin k^0 R}{k^0 R}, \qquad R = |\mathbf{x}_1 - \mathbf{x}_2|. \qquad (5\text{–}2.47)$$

The factor $\exp(-A_T)$ has the appearance of the (infrared sensitive) probability that no photon will be emitted during the creation process, but even this is not physical information since $\exp(-A_T)$ exceeds unity for opposite signs of the charges. What remains is the energy $E$.

The radiation gauge has qualified for further consideration, at least in predominantly nonrelativistic situations, by showing two advantages. It simplifies the problem of extracting physically significant information, and it improves the convergence of the series (5–2.24). Both properties stem from the presumed dominance of the instantaneous component of the propagation function tensor, displayed in Eq. (5–2.33). The remaining components can be extracted from the complete construction given in Eqs. (3–12.8) and (3–15.48), as combined in

$$\begin{aligned} D_+(k)_{\mu\nu} &= \left[g_{\mu\nu} - \frac{k_\mu k_\nu + (nk)(k_\mu n_\nu + n_\mu k_\nu)}{k^2 + (nk)^2}\right] D_+(k) \\ &= -\frac{n_\mu n_\nu}{k^2 + (nk)^2} + \left[g_{\mu\nu} + n_\mu n_\nu - \frac{(k_\mu + n_\mu nk)(k_\nu + n_\nu nk)}{k^2 + (nk)^2}\right] D_+(k), \qquad (5\text{–}2.48) \end{aligned}$$

or, less covariantly, from the transverse field equations of (3–15.52, 53). The construction of the divergenceless part of $\mathbf{J}(x)$ that is given in the latter equation can be presented symbolically as

$$\mathbf{J}_T(x) = \left(1 - \frac{\nabla\nabla}{\nabla^2}\right)\cdot \mathbf{J}(x) \qquad (5\text{–}2.49)$$

and then

$$\mathbf{A}(x) = \left(1 - \frac{\boldsymbol{\nabla}\boldsymbol{\nabla}}{\nabla^2}\right) \cdot \int (dx') D_+(x - x') \mathbf{J}(x'). \tag{5–2.50}$$

This exhibits the spatial components of the propagation function tensor,

$$D_+(x - x')_{kl} = \left(\delta_{kl} - \frac{\partial_k \partial_l}{\nabla^2}\right) D_+(x - x'), \tag{5–2.51}$$

which is also the content of the last term in Eq. (5–2.48), considered in the coordinate system where $n_\mu$ coincides with the time axis.

It is useful to extract an instantaneous part from (5–2.51) as well. This is accomplished by the rearrangement indicated in

$$\frac{1}{k^2} = \frac{1}{k^2 + (nk)^2} + \frac{(nk)^2}{k^2 + (nk)^2} \frac{1}{k^2}, \tag{5–2.52}$$

or

$$D_+(x - x') = \delta(x^0 - x^{0\prime}) \mathscr{D}(\mathbf{x} - \mathbf{x}') + \partial_0{}^2 \frac{1}{\nabla^2} D_+(x - x'). \tag{5–2.53}$$

Thus, the instantaneous part is given by

$$D_+(x - x')_{kl}^{\text{inst.}} = \delta(x^0 - x^{0\prime}) \int \frac{(d\mathbf{k})}{(2\pi)^3} \left(\delta_{kl} - \frac{k_k k_l}{\mathbf{k}^2}\right) \frac{\exp[i\mathbf{k} \cdot (\mathbf{x} - \mathbf{x}')]}{\mathbf{k}^2}, \tag{5–2.54}$$

with a noninstantaneous remainder of

$$D_+(x - x')_{kl}^{\text{noninst.}} = \int \frac{(dk)}{(2\pi)^4} \left(\delta_{kl} - \frac{k_k k_l}{\mathbf{k}^2}\right) \frac{(k^0)^2}{\mathbf{k}^2} \frac{\exp[ik(x - x')]}{k^2 - i\varepsilon}. \tag{5–2.55}$$

To give an explicit spatial form to the instantaneous function, we note that

$$\frac{1}{\nabla^2} \mathscr{D}(\mathbf{x}) = \frac{1}{\nabla^2} \frac{1}{4\pi |\mathbf{x}|} = \frac{1}{4\pi} \tfrac{1}{2} |\mathbf{x}|, \tag{5–2.56}$$

according to (4–15.48), where the possibility of an added constant is without interest since we are only concerned with

$$\partial_k \partial_l \frac{1}{\nabla^2} \mathscr{D}(\mathbf{x}) = \frac{1}{4\pi} \frac{1}{2} \left(\frac{\delta_{kl}}{|\mathbf{x}|} - \frac{x_k x_l}{|\mathbf{x}|^3}\right). \tag{5–2.57}$$

This yields

$$D_+(x - x')_{kl}^{\text{inst.}} = \delta(x^0 - x^{0\prime}) \frac{1}{4\pi} \frac{1}{2} \left[\frac{\delta_{kl}}{|\mathbf{x} - \mathbf{x}'|} + \frac{(\mathbf{x} - \mathbf{x}')_k (\mathbf{x} - \mathbf{x}')_l}{|\mathbf{x} - \mathbf{x}'|^3}\right]. \tag{5–2.58}$$

Let us begin with a discussion of the two-particle equation of (5–2.23), in which only the instantaneous part of $I_{12}^{(1)}$ is retained. We write

$$I_{12}^{(1)}(x_1x_2)^{\text{inst.}} = i\,\delta(x_1{}^0 - x_2{}^0)\gamma_1{}^0\gamma_2{}^0 V(\mathbf{x}_1 - \mathbf{x}_2), \tag{5–2.59}$$

with

$$V(\mathbf{x}_1 - \mathbf{x}_2) = \frac{(eq)_1(eq)_2}{4\pi}\left[\frac{1}{|\mathbf{x}_1 - \mathbf{x}_2|} - \frac{1}{2}\frac{\boldsymbol{\alpha}_1\cdot\boldsymbol{\alpha}_2}{|\mathbf{x}_1 - \mathbf{x}_2|} - \frac{1}{2}\frac{\boldsymbol{\alpha}_1\cdot(\mathbf{x}_1 - \mathbf{x}_2)\boldsymbol{\alpha}_2\cdot(\mathbf{x}_1 - \mathbf{x}_2)}{|\mathbf{x}_1 - \mathbf{x}_2|^3}\right], \tag{5–2.60}$$

where we have returned to the use of the Hermitian matrices

$$\boldsymbol{\alpha} = \gamma^0\boldsymbol{\gamma}. \tag{5–2.61}$$

The equivalent integral equation is

$$G_{12}^{\text{inst.}} = G_1G_2 + G_1G_2I_{12}^{\text{inst.}}G_{12}^{\text{inst.}}. \tag{5–2.62}$$

We shall work with the equal-time functions

$$G_{1+2}(\mathbf{x}_1\mathbf{x}_2t, \mathbf{x}'_1\mathbf{x}'_2t') = iG_{12}(\mathbf{x}_1t\mathbf{x}_2t, \mathbf{x}'_1t'\mathbf{x}'_2t')^{\text{inst.}}\gamma_1{}^0\gamma_2{}^0, \tag{5–2.63}$$

and

$$G_{1+2}(\mathbf{x}_1\mathbf{x}_2t, \mathbf{x}'_1\mathbf{x}'_2t')_{\text{nonint.}} = iG_+(\mathbf{x}_1t, \mathbf{x}'_1t')\gamma_1{}^0G_+(\mathbf{x}_2t, \mathbf{x}'_2t')\gamma_2{}^0. \tag{5–2.64}$$

The corresponding specialization of the integral equation is

$$\begin{aligned} &G_{1+2}(\mathbf{x}_1\mathbf{x}_2t, \mathbf{x}'_1\mathbf{x}'_2t') \\ &= G_{1+2}(\mathbf{x}_1\mathbf{x}_2t, \mathbf{x}'_1\mathbf{x}'_2t')_{\text{nonint.}} \\ &\quad + \int (d\bar{\mathbf{x}}_1)(d\bar{\mathbf{x}}_2)\,d\bar{t}\,G_{1+2}(\mathbf{x}_1\mathbf{x}_2t, \bar{\mathbf{x}}_1\bar{\mathbf{x}}_2\bar{t})_{\text{nonint.}}V(\bar{\mathbf{x}}_1 - \bar{\mathbf{x}}_2)G_{1+2}(\bar{\mathbf{x}}_1\bar{\mathbf{x}}_2\bar{t}, \mathbf{x}'_1\mathbf{x}'_2t'). \end{aligned} \tag{5–2.65}$$

We proceed to convert this into a differential equation by using the following form of the Dirac equation obeyed by $G_+(x - x')$,

$$(-i\,\partial_0 + H)G_+(x - x') = \gamma^0\,\delta(x^0 - x^{0\prime})\,\delta(\mathbf{x} - \mathbf{x}'), \tag{5–2.66}$$

where

$$H = \boldsymbol{\alpha}\cdot\mathbf{p} + \gamma^0 m, \qquad \mathbf{p} = \frac{1}{i}\boldsymbol{\nabla}. \tag{5–2.67}$$

Thus, we have

$$(i\partial_t - H_1 - H_2)G_{1+2}(\mathbf{x}_1\mathbf{x}_2t, \mathbf{x}'_1\mathbf{x}'_2t')_{\text{nonint.}}$$

$$= \delta(t-t')\frac{1}{i}[\delta(\mathbf{x}_1-\mathbf{x}'_1)G_+(\mathbf{x}_2t, \mathbf{x}'_2t)\gamma_2{}^0 + \delta(\mathbf{x}_2-\mathbf{x}'_2)G_+(\mathbf{x}_1t, \mathbf{x}'_1t)\gamma_1{}^0]. \quad (5\text{–}2.68)$$

The value of each $G_+(\mathbf{x}t, \mathbf{x}'t)$ is computed as an average of the two limits, $t - t' \to \pm 0$:

$$G_+(\mathbf{x}t, \mathbf{x}'t) = \frac{1}{2}\left[i\int\frac{(d\mathbf{p})}{(2\pi)^3}\frac{\exp[i\mathbf{p}\cdot(\mathbf{x}-\mathbf{x}')]}{2p^0}(m - \boldsymbol{\gamma}\cdot\mathbf{p} + \gamma^0p^0)\right.$$
$$\left. + i\int\frac{(d\mathbf{p})}{(2\pi)^3}\frac{\exp[i\mathbf{p}\cdot(\mathbf{x}-\mathbf{x}')]}{2p^0}(m - \boldsymbol{\gamma}\cdot\mathbf{p} - \gamma^0p^0)\right], \quad (5\text{–}2.69)$$

or

$$\frac{1}{i}G_+(\mathbf{x}t, \mathbf{x}'t)\gamma^0 = H\int\frac{(d\mathbf{p})}{(2\pi)^3}\frac{\exp[i\mathbf{p}\cdot(\mathbf{x}-\mathbf{x}')]}{2p^0}$$
$$= \frac{1}{2}\frac{H}{W}\delta(\mathbf{x}-\mathbf{x}'). \quad (5\text{–}2.70)$$

The latter is a symbolic way of presenting the result, using the notation

$$W = (\mathbf{p}^2 + m^2)^{1/2}. \quad (5\text{–}2.71)$$

It puts the differential equation of (5–2.68) into the form

$$(i\partial_t - H_1 - H_2)G_{1+2}(\mathbf{x}_1\mathbf{x}_2t, \mathbf{x}'_1\mathbf{x}'_2t')_{\text{nonint.}}$$
$$= \delta(t-t')\frac{1}{2}\left(\frac{H_1}{W_1} + \frac{H_2}{W_2}\right)\delta(\mathbf{x}_1-\mathbf{x}'_1)\,\delta(\mathbf{x}_2-\mathbf{x}'_2), \quad (5\text{–}2.72)$$

which differs from the nonrelativistic version in (5–1.14) by the presence of the factor involving the $H/W$.

We observe that

$$\left(\frac{H}{W}\right)^2 = \frac{(\boldsymbol{\alpha}\cdot\mathbf{p} + \gamma^0m)^2}{\mathbf{p}^2 + m^2} = 1, \quad (5\text{–}2.73)$$

which ascribes the eigenvalues $\pm 1$ to the Hermitian quantity $H/W$. Accordingly, the additional factor or the right side of (5–2.72) has the eigenvalues $1, 0, -1$. That is made explicit on writing

$$\frac{1}{2}\left(\frac{H_1}{W_1} + \frac{H_2}{W_2}\right) = \frac{1}{4}\left[\left(1 + \frac{H_1}{W_1}\right)\left(1 + \frac{H_2}{W_2}\right) - \left(1 - \frac{H_1}{W_1}\right)\left(1 - \frac{H_2}{W_2}\right)\right]. \quad (5\text{–}2.74)$$

The related solution of the differential equation (5–2.72) is

$$G_{1+2}(\mathbf{x}_1\mathbf{x}_2t, \mathbf{x}'_1\mathbf{x}'_2t')_{\text{nonint.}} = \frac{1}{4}\left(1 + \frac{H_1}{W_1}\right)\left(1 + \frac{H_2}{W_2}\right)G(\mathbf{x}_1\mathbf{x}_2, \mathbf{x}'_1\mathbf{x}'_2|t-t')_{\text{nonint.}}$$

$$+\frac{1}{4}\left(1-\frac{H_1}{W_1}\right)\left(1-\frac{H_2}{W_2}\right)G(\mathbf{x}_1\mathbf{x}_2,\mathbf{x}'_1\mathbf{x}'_2|t'-t)_{\text{nonint.}},\qquad(5\text{–}2.75)$$

where $G(\mathbf{x}_1\mathbf{x}_2,\mathbf{x}'_1\mathbf{x}'_2|t-t')$ is the retarded Green's function that obeys

$$(i\partial_t-W_1-W_2)G(\mathbf{x}_1\mathbf{x}_2,\mathbf{x}'_1\mathbf{x}'_2|t-t')_{\text{nonint.}}=\delta(t-t')\,\delta(\mathbf{x}_1-\mathbf{x}'_1)\,\delta(\mathbf{x}_2-\mathbf{x}'_2).\qquad(5\text{–}2.76)$$

Of course, this result is obtained directly by multiplying the equal-time forms of the two single-particle Green's functions,

$$x^0\gtrless x^{0\prime}:\qquad G_+(x-x')\gamma^0$$
$$=i\int\frac{(d\mathbf{p})}{(2\pi)^3}\frac{1}{2p^0}(\boldsymbol{\alpha}\cdot\mathbf{p}+m\gamma^0\pm p^0)\exp[i\mathbf{p}\cdot(\mathbf{x}-\mathbf{x}')-ip^0|x^0-x^{0\prime}|]$$
$$=\frac{i}{2}\left[\left(\frac{H}{W}\right)\pm 1\right]\exp(-iW|x^0-x^{0\prime}|)\,\delta(\mathbf{x}-\mathbf{x}').\qquad(5\text{–}2.77)$$

The structure of the differential equation (5–2.72) can be simplified by noting that

$$H/W=U^{-1}\gamma^0U,\qquad(5\text{–}2.78)$$

where

$$U=\exp[\tfrac{1}{2}(\boldsymbol{\gamma}\cdot\mathbf{p}/|\mathbf{p}|)\phi],\qquad\sin\phi=|\mathbf{p}|/W,\qquad\cos\phi=m/W,\qquad(5\text{–}2.79)$$

is a unitary matrix. Thus, for the noninteraction situation,

$$\bar{G}_{1+2}=U_1U_2G_{1+2}U_1^{-1}U_2^{-1}\qquad(5\text{–}2.80)$$

obeys the equation

$$[i\partial_t-(\gamma^0W)_1-(\gamma^0W)_2]\bar{G}_{1+2}(\mathbf{x}_1\mathbf{x}_2t,\mathbf{x}'_1\mathbf{x}'_2t')_{\text{nonint.}}$$
$$=\delta(t-t')\tfrac{1}{2}(\gamma_1{}^0+\gamma_2{}^0)\,\delta(\mathbf{x}_1-\mathbf{x}'_1)\,\delta(\mathbf{x}_2-\mathbf{x}'_2).\qquad(5\text{–}2.81)$$

In a representation where both $\gamma_1{}^0$ and $\gamma_2{}^0$ are diagonal matrices, with the eigenvalues $\pm 1$, only two possibilities appear,

$$\gamma_1{}^{0\prime}=\gamma_2{}^{0\prime}=+1:\ \bar{G}_{1+2}(\mathbf{x}_1\mathbf{x}_2t,\mathbf{x}'_1\mathbf{x}'_2t')_{\text{nonint.}}=G(\mathbf{x}_1\mathbf{x}_2,\mathbf{x}'_1\mathbf{x}'_2|t-t')_{\text{nonint.}}\qquad(5\text{–}2.82)$$

and

$$\gamma_1{}^{0\prime}=\gamma_2{}^{0\prime}=-1:\ \bar{G}_{1+2}(\mathbf{x}_1\mathbf{x}_2t,\mathbf{x}'_1\mathbf{x}'_2t')_{\text{nonint.}}=G(\mathbf{x}_1\mathbf{x}_2,\mathbf{x}'_1\mathbf{x}'_2|t'-t)_{\text{nonint.}}.\qquad(5\text{–}2.83)$$

They are united in the construction

$$\bar{G}_{1+2}(\mathbf{x}_1\mathbf{x}_2t, \mathbf{x}'_1\mathbf{x}'_2t')_{\text{nonint.}} = \tfrac{1}{4}(1+\gamma_1{}^0)(1+\gamma_2{}^0)G(\mathbf{x}_1\mathbf{x}_2, \mathbf{x}'_1\mathbf{x}'_2|t-t')_{\text{nonint.}}$$
$$+ \tfrac{1}{4}(1-\gamma_1{}^0)(1-\gamma_2{}^0)G(\mathbf{x}_1\mathbf{x}_2, \mathbf{x}'_1\mathbf{x}'_2|t'-t)_{\text{nonint.}}, \qquad (5\text{–}2.84)$$

which, naturally, is the transformed version of Eq. (5–2.75).

The differential equation derived from (5–2.65) is

$$(i\partial_t - H_1 - H_2)G_{1+2}(\mathbf{x}_1\mathbf{x}_2t, \mathbf{x}'_1\mathbf{x}'_2t')$$
$$-\frac{1}{2}\left(\frac{H_1}{W_1}+\frac{H_2}{W_2}\right)V(\mathbf{x}_1-\mathbf{x}_2)G_{1+2}(\mathbf{x}_1\mathbf{x}_2t, \mathbf{x}'_1\mathbf{x}'_2t')$$
$$= \delta(t-t')\frac{1}{2}\left(\frac{H_1}{W_1}+\frac{H_2}{W_2}\right)\delta(\mathbf{x}_1-\mathbf{x}'_1)\,\delta(\mathbf{x}_2-\mathbf{x}'_2). \qquad (5\text{–}2.85)$$

It is transformed, according to (5–2.80), into

$$[i\partial_t - (\gamma^0W)_1 - (\gamma^0W)_2 - \tfrac{1}{2}(\gamma_1{}^0+\gamma_2{}^0)\bar{V}]\bar{G}_{1+2}$$
$$= \delta(t-t')\tfrac{1}{2}(\gamma_1{}^0+\gamma_2{}^0)\,\delta(\mathbf{x}_1-\mathbf{x}'_1)\,\delta(\mathbf{x}_2-\mathbf{x}'_2), \qquad (5\text{–}2.86)$$

where

$$\bar{V} = U_1U_2VU_1{}^{-1}U_2{}^{-1}. \qquad (5\text{–}2.87)$$

The presence of the factor $\tfrac{1}{2}(\gamma_1{}^0+\gamma_2{}^0)$, particularly in the inhomogeneous term of the equation, implies that only $\gamma_1{}^{0\prime} = \gamma_2{}^{0\prime}$ need be considered in the row and column labels of the matrix $\bar{G}_{1+2}$. However, the introduction of the matrix $\bar{V}$ removes the diagonal nature of $\bar{G}_{1+2}$. Using the symbols + and − to indicate the common value of $\gamma_1{}^{0\prime} = \gamma_2{}^{0\prime}$ in row and column indices, and employing three-dimensional coordinate matrix notation, we write out (5–2.86) as the two pairs of equations

$$(i\partial_t - W_1 - W_2 - \bar{V}_{++})\bar{G}_{++} - \bar{V}_{+-}\bar{G}_{-+} = \delta(t-t'),$$
$$(i\partial_t + W_1 + W_2 + \bar{V}_{--})\bar{G}_{-+} + \bar{V}_{-+}\bar{G}_{++} = 0, \qquad (5\text{–}2.88)$$

and

$$(i\partial_t + W_1 + W_2 + \bar{V}_{--})\bar{G}_{--} + \bar{V}_{-+}\bar{G}_{+-} = -\,\delta(t-t'),$$
$$(i\partial_t - W_1 - W_2 - \bar{V}_{++})\bar{G}_{+-} - \bar{V}_{+-}\bar{G}_{--} = 0. \qquad (5\text{–}2.89)$$

Before continuing we must note that

$$\bar{V}_{++} = \bar{V}_{--} \equiv V_0, \qquad (5\text{–}2.90)$$

and that it is possible to arrange matters so that

$$\bar{V}_{+-} = \bar{V}_{-+} \equiv V_1. \tag{5–2.91}$$

These are comments about the structure of $V$ [Eq. (5–2.60)] in its dependence on the matrices $\gamma^0$ and the complementary matrices $\gamma_5$. The submatrices $\bar{V}_{++}$ and $\bar{V}_{--}$ contain no $\gamma_5$ matrices, and an individual term may have no $\gamma^0$ matrices, or the factor $\gamma_1{}^0\gamma_2{}^0$. Single $\gamma^0$ matrices are absent, for the origin of any $\gamma^0$ is in the matrix

$$\boldsymbol{\gamma} = i\gamma^0\gamma_5\boldsymbol{\sigma}, \tag{5–2.92}$$

as distinguished from

$$\boldsymbol{\alpha} = i\gamma_5\boldsymbol{\sigma}, \tag{5–2.93}$$

and no single $\gamma_5$ matrix survives. Since only $\gamma_1{}^0\gamma_2{}^0 \rightarrow 1$ occurs, there is no distinction between $\bar{V}_{++}$ and $\bar{V}_{--}$, as claimed. The submatrices $\bar{V}_{+-}$ and $\bar{V}_{-+}$ come from the part of $\bar{V}$ that is proportional to $\gamma_{51}\gamma_{52}$, where any factor $\gamma_1{}^0\gamma_2{}^0 \rightarrow 1$, as before. The arbitrary phases that can enter the elements of the Hermitian matrix $\gamma_{51}\gamma_{52}$ could always be adjusted to make this matrix symmetrical, which is the content of (5–2.91).

Using the notation

$$H_0 = W_1 + W_2 + V_0, \tag{5–2.94}$$

we now convey the Eqs. (5–2.88, 89) by the sets

$$\begin{aligned} (i\partial_t - H_0)\bar{G}_{++} - V_1\bar{G}_{-+} &= \delta(t - t'), \\ (i\partial_t + H_0)\bar{G}_{-+} + V_1\bar{G}_{++} &= 0, \end{aligned} \tag{5–2.95}$$

and

$$\begin{aligned} (i\partial_t + H_0)\bar{G}_{--} + V_1\bar{G}_{+-} &= -\,\delta(t - t'), \\ (i\partial_t - H_0)\bar{G}_{+-} - V_1\bar{G}_{--} &= 0. \end{aligned} \tag{5–2.96}$$

The non-diagonal elements of $\bar{G}$ can then be found with the aid of the retarded and advanced Green's functions that obey

$$\begin{aligned} (i\partial_t - H_0)G_{\text{ret.}}(t - t') &= \delta(t - t'), \\ (i\partial_t + H_0)G_{\text{adv.}}(t - t') &= -\,\delta(t - t'), \end{aligned} \tag{5–2.97}$$

together with

$$G_{\text{adv.}}(t - t') = G_{\text{ret.}}(t' - t). \tag{5–2.98}$$

This construction is given symbolically by

$$\bar{G}_{-+} = G_{\text{adv.}}V_1\bar{G}_{++}, \qquad \bar{G}_{+-} = G_{\text{ret.}}V_1\bar{G}_{--}. \tag{5–2.99}$$

The retarded and advanced functions are used again in recasting the remaining equations in integral form:

$$\bar{G}_{++} = G_{\text{ret.}} + G_{\text{ret.}} V_1 G_{\text{adv.}} V_1 \bar{G}_{++},$$
$$\bar{G}_{--} = G_{\text{adv.}} + G_{\text{adv.}} V_1 G_{\text{ret.}} V_1 \bar{G}_{--}. \tag{5-2.100}$$

The time symmetry of this system indicates the additional relation,

$$\bar{G}_{--}(t - t') = \bar{G}_{++}(t' - t). \tag{5-2.101}$$

However, these functions individually are not of the retarded or advanced type, but satisfy in a more general way the boundary conditions of the $G_+$ class of Green's functions.

We shall be content with the approximate solution of Eq. (5-2.100) that is produced by one iteration,

$$\bar{G}_{++} = G_{\text{ret.}} + G_{\text{ret.}} V_1 G_{\text{adv.}} V_1 G_{\text{ret.}}. \tag{5-2.102}$$

The time variables are made explicit in

$$\bar{G}_{++}(t - t') = G_{\text{ret.}}(t - t') + \int dt_1\, dt_2\, G_{\text{ret.}}(t - t_1) V_1 G_{\text{adv.}}(t_1 - t_2) V_1 G_{\text{ret.}}(t_2 - t')$$
$$= G_{\text{ret.}}(t - t') + \int_0^\infty d\tau \int_0^\infty d\tau'\, G_{\text{ret.}}(\tau) V_1 G_{\text{adv.}}(t - t' - \tau - \tau') V_1 G_{\text{ret.}}(\tau'). \tag{5-2.103}$$

To see that $G_+$ time boundary conditions are satisfied, it suffices to consider individual time exponentials in the construction of the various Green's functions. Thus, for $t - t' > 0$ we have, say,

$$\int_0^\infty d\tau\, d\tau'\, \eta(\tau + \tau' - (t - t')) \exp(-iE\tau) \exp[-iE'(\tau + \tau' - (t - t'))] \exp(-iE''\tau'), \tag{5-2.104}$$

which, using the variables

$$T = \tau + \tau', \qquad s = \tfrac{1}{2}(\tau - \tau'), \tag{5-2.105}$$

becomes the integral

$$\int_{t-t'}^\infty dT \int_{-(1/2)T}^{(1/2)T} ds \exp[-iE(\tfrac{1}{2}T + s)] \exp[-iE'(T - (t - t'))] \exp[-iE''(\tfrac{1}{2}T - s)]$$
$$= \int_{t-t'}^\infty dT \frac{i}{E - E''} [\exp(-iET) - \exp(-iE''T)] \exp[-iE'(T - (t - t'))]$$

$$= \frac{1}{E - E''}\left[\frac{\exp[-iE(t-t')]}{E+E'} - \frac{\exp[-iE''(t-t')]}{E''+E'}\right]. \qquad (5\text{–}2.106)$$

Here are the required positive frequencies for positive $t - t'$. When $t - t' < 0$, on the other hand, the advanced Green's function imposes no further restriction on $\tau$ and $\tau'$, which immediately supplies the time dependence

$$\exp[iE'(t-t')] = \exp(-iE'|t-t'|); \qquad (5\text{–}2.107)$$

negative frequencies appear for $t - t'$ negative. These general characteristics also apply to $\bar{G}_{--}(t-t')$.

The spectrum of the energy operator $H_0$ [Eq. (5–2.94)] will be discussed in an essentially non-relativistic context. That is to say, only first deviations from non-relativistic behavior will be considered corresponding, for example, to retaining only the indicated terms of the expansion

$$W = (\mathbf{p}^2 + m^2)^{1/2} \cong m + \frac{\mathbf{p}^2}{2m} - \frac{1}{2m}\left(\frac{\mathbf{p}^2}{2m}\right)^2 + \cdots. \qquad (5\text{–}2.108)$$

The unitary matrix [Eq. (5–2.79)]

$$U = \cos \tfrac{1}{2}\phi + \frac{\boldsymbol{\Upsilon}\cdot\mathbf{p}}{|\mathbf{p}|}\sin\tfrac{1}{2}\phi = \left(\frac{W+m}{2W}\right)^{1/2} + \frac{\boldsymbol{\Upsilon}\cdot\mathbf{p}}{|\mathbf{p}|}\left(\frac{W-m}{2W}\right)^{1/2} \qquad (5\text{–}2.109)$$

is therefore simplified to

$$U \cong 1 - \frac{\mathbf{p}^2}{8m^2} + \frac{1}{2m}\boldsymbol{\Upsilon}\cdot\mathbf{p}. \qquad (5\text{–}2.110)$$

The combinations of interest are

$$U_1U_2 \cong 1 - \frac{1}{8}\left(\frac{\mathbf{p}_1^2}{m_1^2} + \frac{\mathbf{p}_2^2}{m_2^2}\right) + \frac{1}{4}\left(\frac{\boldsymbol{\Upsilon}\cdot\mathbf{p}}{m}\right)_1\left(\frac{\boldsymbol{\Upsilon}\cdot\mathbf{p}}{m}\right)_2 + \frac{1}{2}\left[\left(\frac{\boldsymbol{\Upsilon}\cdot\mathbf{p}}{m}\right)_1 + \left(\frac{\boldsymbol{\Upsilon}\cdot\mathbf{p}}{m}\right)_2\right],$$

$$U_1^{-1}U_2^{-1} \cong 1 - \frac{1}{8}\left(\frac{\mathbf{p}_1^2}{m_1^2} + \frac{\mathbf{p}_2^2}{m_2^2}\right) + \frac{1}{4}\left(\frac{\boldsymbol{\Upsilon}\cdot\mathbf{p}}{m}\right)_1\left(\frac{\boldsymbol{\Upsilon}\cdot\mathbf{p}}{m}\right)_2 - \frac{1}{2}\left[\left(\frac{\boldsymbol{\Upsilon}\cdot\mathbf{p}}{m}\right)_1 + \left(\frac{\boldsymbol{\Upsilon}\cdot\mathbf{p}}{m}\right)_2\right].$$

$$(5\text{–}2.111)$$

There are two kinds of terms in $V$. One is proportional to the unit matrix,

$$V_a = \frac{(eq)_1(eq)_2}{4\pi}\frac{1}{|\mathbf{x}_1 - \mathbf{x}_2|}, \qquad (5\text{–}2.112)$$

and the other contains products of Dirac matrices. This we write as $\gamma_{51}\gamma_{52}V_b$, with

$$V_b = \frac{(eq)_1(eq)_2}{4\pi}\frac{1}{2}\left[\frac{\boldsymbol{\sigma}_1\cdot\boldsymbol{\sigma}_2}{|\mathbf{x}_1-\mathbf{x}_2|} + \frac{\boldsymbol{\sigma}_1\cdot(\mathbf{x}_1-\mathbf{x}_2)\boldsymbol{\sigma}_2\cdot(\mathbf{x}_1-\mathbf{x}_2)}{|\mathbf{x}_1-\mathbf{x}_2|^3}\right]. \qquad (5\text{–}2.113)$$

In evaluating the submatrix $V_0$, all terms containing a $\gamma_5$ matrix of either particle are rejected. Accordingly,

$$(U_1U_2V_aU_1^{-1}U_2^{-1})_{++} \cong V_a - \frac{1}{8}\left\{\left(\frac{\mathbf{p}_1^2}{m_1^2}+\frac{\mathbf{p}_2^2}{m_2^2}\right), V_a\right\}$$

$$+ \frac{1}{4m_1^2}\boldsymbol{\sigma}_1\cdot\mathbf{p}_1V_a\boldsymbol{\sigma}_1\cdot\mathbf{p}_1 + \frac{1}{4m_2^2}\boldsymbol{\sigma}_2\cdot\mathbf{p}_2V_a\boldsymbol{\sigma}_2\cdot\mathbf{p}_2$$

$$= V_a + \frac{1}{8}\left(\frac{1}{m_1^2}+\frac{1}{m_2^2}\right)\nabla^2V_a + \frac{1}{4m_1^2}\boldsymbol{\sigma}_1\cdot\boldsymbol{\nabla}_1V_a\times\mathbf{p}_1$$

$$+ \frac{1}{4m_2^2}\boldsymbol{\sigma}_2\cdot\boldsymbol{\nabla}_2V_a\times\mathbf{p}_2, \qquad (5\text{–}2.114)$$

which uses the relation

$$\boldsymbol{\sigma}\cdot\mathbf{p}V_a\boldsymbol{\sigma}\cdot\mathbf{p} = \tfrac{1}{2}\{\mathbf{p}^2, V_a\} + \tfrac{1}{2}\nabla^2V_a + \boldsymbol{\sigma}\cdot\boldsymbol{\nabla}V_a\times\mathbf{p}, \qquad (5\text{–}2.115)$$

and

$$(U_1U_2\gamma_{51}\gamma_{52}V_bU_1^{-1}U_2^{-1})_{++} \cong -\frac{1}{4m_1m_2}\{\boldsymbol{\sigma}_1\cdot\mathbf{p}_1\boldsymbol{\sigma}_2\cdot\mathbf{p}_2, V_b\}$$

$$-\frac{1}{4m_1m_2}(\boldsymbol{\sigma}_1\cdot\mathbf{p}_1V_b\boldsymbol{\sigma}_2\cdot\mathbf{p}_2 + \boldsymbol{\sigma}_2\cdot\mathbf{p}_2V_b\boldsymbol{\sigma}_1\cdot\mathbf{p}_1)$$

$$= -\frac{1}{4m_1m_2}\{\boldsymbol{\sigma}_1\cdot\mathbf{p}_1, \{\boldsymbol{\sigma}_2\cdot\mathbf{p}_2, V_b\}\}. \qquad (5\text{–}2.116)$$

For the latter calculation, where

$$V_b = \boldsymbol{\sigma}_1\cdot\boldsymbol{\Lambda}\cdot\boldsymbol{\sigma}_2,$$

$$\boldsymbol{\Lambda} = \frac{(eq)_1(eq)_2}{4\pi}\frac{1}{2}\left[\frac{1}{|\mathbf{x}_1-\mathbf{x}_2|} + \frac{(\mathbf{x}_1-\mathbf{x}_2)(\mathbf{x}_1-\mathbf{x}_2)}{|\mathbf{x}_1-\mathbf{x}_2|^3}\right], \qquad (5\text{–}2.117)$$

the origin of this combination in the transverse propagation function (5–2.54) makes symmetrized multiplication unnecessary ($\boldsymbol{\nabla}\cdot\boldsymbol{\Lambda} = 0$) in such combinations as $\mathbf{p}_1\cdot\boldsymbol{\Lambda}\cdot\mathbf{p}_2$:

$$\{\boldsymbol{\sigma}_1\cdot\mathbf{p}_1, \{\boldsymbol{\sigma}_2\cdot\mathbf{p}_2, V_b\}\} = 4\mathbf{p}_1\cdot\boldsymbol{\Lambda}\cdot\mathbf{p}_2 + 2\boldsymbol{\sigma}_1\cdot\boldsymbol{\nabla}_1\times\boldsymbol{\Lambda}\cdot\mathbf{p}_2$$

$$+ 2\boldsymbol{\sigma}_2\cdot\boldsymbol{\nabla}_2\times\boldsymbol{\Lambda}\cdot\mathbf{p}_1 + (\boldsymbol{\sigma}_1\times\boldsymbol{\nabla}_1)(\boldsymbol{\sigma}_2\times\boldsymbol{\nabla}_2):\boldsymbol{\Lambda}; \qquad (5\text{–}2.118)$$

the last term is a way of writing the scalar product of the dyadic $\boldsymbol{\Lambda}$ with the two vectors. When the transverse structure is made explicit,

$$\mathbf{\Lambda} = \frac{(eq)_1(eq)_2}{4\pi}\left[1 - \frac{\boldsymbol{\nabla}\boldsymbol{\nabla}}{\nabla^2}\right]\frac{1}{|\mathbf{x}_1 - \mathbf{x}_2|}, \tag{5–2.119}$$

we see that

$$\boldsymbol{\nabla} \times \mathbf{\Lambda} = \boldsymbol{\nabla} V_a \times . \tag{5–2.120}$$

In particular, this gives

$$\begin{aligned}(\boldsymbol{\sigma}_1 \times \boldsymbol{\nabla}_1)(\boldsymbol{\sigma}_2 \times \boldsymbol{\nabla}_2) : \mathbf{\Lambda} &= (\boldsymbol{\sigma}_1 \times \boldsymbol{\nabla}_1)\cdot(\boldsymbol{\sigma}_2 \times \boldsymbol{\nabla}_2) V_a \\ &= -(\boldsymbol{\sigma}_1\cdot\boldsymbol{\nabla}_2\boldsymbol{\sigma}_2\cdot\boldsymbol{\nabla}_1 + \tfrac{1}{3}\boldsymbol{\sigma}_1\cdot\boldsymbol{\sigma}_2\nabla^2)V_a - \tfrac{2}{3}\boldsymbol{\sigma}_1\cdot\boldsymbol{\sigma}_2\nabla^2 V_a,\end{aligned} \tag{5–2.121}$$

where the last term isolates the result of a spatial rotational averaging process.

The outcome of this procedure is the energy operator

$$\begin{aligned}H_0 = m_1 + m_2 &+ \frac{\mathbf{p}_1{}^2}{2m_1} + \frac{\mathbf{p}_2{}^2}{2m_2} - \frac{1}{2m_1}\left(\frac{\mathbf{p}_1{}^2}{2m_1}\right)^2 - \frac{1}{2m_2}\left(\frac{\mathbf{p}_2{}^2}{2m_2}\right)^2 \\ &+ \frac{e_1e_2}{4\pi}\left\{\frac{1}{r} - \frac{1}{m_1m_2}\frac{1}{2}\mathbf{p}_1\cdot\left[\frac{1}{r} + \frac{\mathbf{r}\mathbf{r}}{r^3}\right]\cdot\mathbf{p}_2 - \frac{\pi}{2}\left(\frac{1}{m_1{}^2} + \frac{1}{m_2{}^2}\right)\delta(\mathbf{r})\right. \\ &- \frac{1}{4m_1{}^2}\frac{\boldsymbol{\sigma}_1\cdot\mathbf{r}\times\mathbf{p}_1}{r^3} + \frac{1}{4m_2{}^2}\frac{\boldsymbol{\sigma}_2\cdot\mathbf{r}\times\mathbf{p}_2}{r^3} + \frac{1}{2m_1m_2}\left(\frac{\boldsymbol{\sigma}_1\cdot\mathbf{r}\times\mathbf{p}_2}{r^3} - \frac{\boldsymbol{\sigma}_2\cdot\mathbf{r}\times\mathbf{p}_1}{r^3}\right) \\ &\left.- \frac{1}{4m_1m_2}\left(3\frac{\boldsymbol{\sigma}_1\cdot\mathbf{r}\boldsymbol{\sigma}_2\cdot\mathbf{r}}{r^5} - \frac{\boldsymbol{\sigma}_1\cdot\boldsymbol{\sigma}_2}{r^3}\right) - \frac{1}{m_1m_2}\frac{2\pi}{3}\boldsymbol{\sigma}_1\cdot\boldsymbol{\sigma}_2\,\delta(\mathbf{r})\right\},\end{aligned} \tag{5–2.122}$$

where

$$\mathbf{r} = \mathbf{x}_1 - \mathbf{x}_2, \tag{5–2.123}$$

and, for simplicity, we have written $e_{1,2}$ in place of $(eq)_{1,2}$. Of principal concern is the spectrum of $H_0$ in the rest frame of the two-particle system—the internal energy. Nevertheless, it is of some interest to see how the anticipated dependence of the energy on the total momentum of the system emerges under these circumstances of small relativistic deviations from non-relativistic behavior. Let us insert the momentum relations of (5–1.95) and extract the terms of $H_0$ that involve $\mathbf{P}$:

$$\frac{\mathbf{P}^2}{2M} - \frac{1}{2M}\left(\frac{\mathbf{P}^2}{2M}\right)^2 - \frac{\mathbf{P}^2}{2M^2}\frac{\mathbf{p}^2}{2\mu} - \frac{1}{2\mu}\left(\frac{\mathbf{P}\cdot\mathbf{p}}{M}\right)^2 - \frac{e_1e_2}{4\pi}\frac{1}{2M^2}\mathbf{P}\cdot\left(\frac{1}{r} + \frac{\mathbf{r}\mathbf{r}}{r^3}\right)\cdot\mathbf{P}, \tag{5–2.124}$$

although we have omitted expressions such as $\mathbf{P}\cdot(1/r)\mathbf{p}$ and $\boldsymbol{\sigma}_{1,2}\times\mathbf{r}\cdot\mathbf{P}$, which will not contribute to expectation values in a state of definite internal parity.

We note the appearance in Eq. (5–2.124) of the non-relativistic internal energy operator

$$H_{\text{int.}} = \frac{\mathbf{p}^2}{2\mu} + \frac{e_1 e_2}{4\pi}\frac{1}{r}, \tag{5–2.125}$$

which enters the expectation value of (5–2.124) through its eigenvalue $E_{\text{int.}}$,

$$\frac{\mathbf{P}^2}{2M}\left(1 - \frac{E_{\text{int.}}}{M}\right) - \frac{1}{2M}\left(\frac{\mathbf{P}^2}{2M}\right)^2 - \frac{1}{2M^2}\mathbf{P}\cdot\left\langle\frac{\mathbf{pp}}{\mu} + \frac{e_1 e_2}{4\pi}\frac{\mathbf{rr}}{r^3}\right\rangle\cdot\mathbf{P}. \tag{5–2.126}$$

The first two terms are the expected ones, where the total mass of the system is recognized to be $M + E_{\text{int.}}$, with $E_{\text{int.}} \ll M$. It is necessary, therefore, that the last term vanish:

$$\left\langle\frac{\mathbf{pp}}{\mu} + \frac{e_1 e_2}{4\pi}\frac{\mathbf{rr}}{r^3}\right\rangle = 0. \tag{5–2.127}$$

If we consider the diagonal sum of this dyadic relation,

$$\left\langle\frac{\mathbf{p}^2}{\mu} + \frac{e_1 e_2}{4\pi}\frac{1}{r}\right\rangle = \langle 2T + V\rangle = 0, \tag{5–2.128}$$

we encounter the familiar virial theorem connection between average internal kinetic and potential energies, in the form appropriate to the Coulomb field. To be verified is the dyadic generalization of the virial theorem stated in (5–2.127).

The proof follows directly from a simple generalization of the scale transformation that implies the usual virial theorem. Consider the infinitesimal unitary transformation described by

$$U = 1 + iG, \qquad G = \tfrac{1}{2}(\mathbf{p}\cdot\delta\mathbf{K}\cdot\mathbf{r} + \mathbf{r}\cdot\delta\mathbf{K}\cdot\mathbf{p}), \tag{5–2.129}$$

where $\delta\mathbf{K}$ is an infinitesimal dyadic that is real and symmetrical. The transformation is [cf. Eqs. (1–1.18, 19)]

$$\bar{\mathbf{r}} = U^{-1}\mathbf{r}U = \mathbf{r} - \delta\mathbf{r}, \qquad \bar{\mathbf{p}} = U^{-1}\mathbf{p}U = \mathbf{p} - \delta\mathbf{p}, \tag{5–2.130}$$

with

$$\delta\mathbf{r} = \frac{1}{i}[\mathbf{r}, G] = \delta\mathbf{K}\cdot\mathbf{r}, \qquad \delta\mathbf{p} = \frac{1}{i}[\mathbf{p}, G] = -\,\delta\mathbf{K}\cdot\mathbf{p}. \tag{5–2.131}$$

The induced change in the operator $H$,

$$\delta H = \frac{1}{i}[H, G], \tag{5–2.132}$$

is given by

$$\delta H = -\left[\frac{\mathbf{p}\cdot\delta\mathbf{K}\cdot\mathbf{p}}{\mu} + \frac{e_1e_2}{4\pi}\frac{\mathbf{r}\cdot\delta\mathbf{K}\cdot\mathbf{r}}{r^3}\right]. \tag{5-2.133}$$

But, since the expectation value of $H$ is stationary with respect to variations of the wave function or, more directly, in consequence of the commutator structure of $\delta H$, the expectation value of $\delta H$ vanishes. That is the content of Eq. (5–2.127).

The specialization of the energy operator (5–2.122) to the rest frame is

$$\begin{aligned} H_0 = m_1 + m_2 + \frac{\mathbf{p}^2}{2\mu} - \frac{1}{8}\left(\frac{1}{m_1{}^3} + \frac{1}{m_2{}^3}\right)(\mathbf{p}^2)^2 + \frac{e_1e_2}{4\pi}\bigg\{\frac{1}{r} + \frac{1}{2m_1m_2}\mathbf{p}\cdot\left(\frac{1}{r}+\frac{\mathbf{rr}}{r^3}\right)\cdot\mathbf{p} \\ - \frac{\pi}{2}\left(\frac{1}{m_1{}^2}+\frac{1}{m_2{}^2}\right)\delta(\mathbf{r}) - \frac{1}{4}\left(\frac{1}{m_1{}^2}+\frac{2}{m_1m_2}\right)\frac{\boldsymbol{\sigma}_1\cdot\mathbf{r}\times\mathbf{p}}{r^3} \\ - \frac{1}{4}\left(\frac{1}{m_2{}^2}+\frac{2}{m_1m_2}\right)\frac{\boldsymbol{\sigma}_2\cdot\mathbf{r}\times\mathbf{p}}{r^3} - \frac{1}{4m_1m_2}\left(3\frac{\boldsymbol{\sigma}_1\cdot\mathbf{r}\boldsymbol{\sigma}_2\cdot\mathbf{r}}{r^5} - \frac{\boldsymbol{\sigma}_1\cdot\boldsymbol{\sigma}_2}{r^3}\right) \\ - \frac{1}{m_1m_2}\frac{2\pi}{3}\boldsymbol{\sigma}_1\cdot\boldsymbol{\sigma}_2\,\delta(\mathbf{r})\bigg\}. \end{aligned} \tag{5-2.134}$$

The last terms, which contain both spins, will be recognized as the hyperfine structure interaction for particles with magnetic moments characterized by $\frac{1}{2}g = 1$, as the choice of primitive interaction implies. Let us ignore all terms containing $\boldsymbol{\sigma}_2$, say, thereby omitting hyperfine structure, and consider the situation, appropriate to the hydrogen atom and muonium ($\mu^+ + e^-$), where

$$m_1(\equiv m) \ll m_2(\equiv M). \tag{5-2.135}$$

When second and higher powers of $m/M$ are neglected, the energy operator $H_0$ reduces to $[e_1e_2/4\pi = -\alpha]$

$$\begin{aligned} H_0 = \left\{m + \frac{\mathbf{p}^2}{2m} - \frac{1}{2m}\left(\frac{\mathbf{p}^2}{2m}\right)^2 - \frac{\alpha}{r} + \frac{\pi\alpha}{2m^2}\delta(\mathbf{r}) + \frac{\alpha}{4m^2}\frac{\boldsymbol{\sigma}\cdot\mathbf{r}\times\mathbf{p}}{r^3}\right\} \\ + M + \frac{1}{2M}\left[\mathbf{p}^2 - \frac{\alpha}{m}\mathbf{p}\cdot\left(\frac{1}{r}+\frac{\mathbf{rr}}{r^3}\right)\cdot\mathbf{p} + \frac{\alpha}{m}\frac{\boldsymbol{\sigma}\cdot\mathbf{r}\times\mathbf{p}}{r^3}\right]. \end{aligned} \tag{5-2.136}$$

The combination in braces will be recognized as the approximate transformed version of the Dirac energy operator for a particle in a Coulomb field:

$$\left(U\left(\boldsymbol{\alpha}\cdot\mathbf{p} + \gamma^0 m - \frac{\alpha}{r}\right)U^{-1}\right)_{++} \cong \{\ \}. \tag{5-2.137}$$

That is what one expects to find as $m/M \to 0$. The term with $1/M$ as a factor thus supplies a first correction to the idealized treatment of the more massive body as a fixed source.

Elementary perturbation theory indicates the corrected energy value

$$E = E_{\text{rel.}} + M + \frac{1}{2M}\left\langle \mathbf{p}^2 - \frac{\alpha}{m}\,\mathbf{p}\cdot\left(\frac{1}{r} + \frac{\mathbf{rr}}{r^3}\right)\cdot\mathbf{p} + \frac{\alpha}{m}\,\frac{\boldsymbol{\sigma}\cdot\mathbf{r}\times\mathbf{p}}{r^3}\right\rangle, \quad (5\text{–}2.138)$$

where $E_{\text{rel.}}$ is a typical energy value associated with the transformed Dirac energy operator:

$$E_{\text{rel.}} = \left\langle m + \frac{\mathbf{p}^2}{2m} - \frac{1}{2m}\left(\frac{\mathbf{p}^2}{2m}\right)^2 - \frac{\alpha}{r} + \frac{\pi\alpha}{2m^2}\,\delta(\mathbf{r}) + \frac{\alpha}{4m^2}\,\frac{\boldsymbol{\sigma}\cdot\mathbf{r}\times\mathbf{p}}{r^3}\right\rangle. \quad (5\text{–}2.139)$$

A useful result is obtained by applying an isotropic scale transformation to this operator,

$$0 = \left\langle 2\,\frac{\mathbf{p}^2}{2m} - 4\,\frac{1}{2m}\left(\frac{\mathbf{p}^2}{2m}\right)^2 - \frac{\alpha}{r} + 3\,\frac{\pi\alpha}{2m^2}\,\delta(\mathbf{r}) + 3\,\frac{\alpha}{4m^2}\,\frac{\boldsymbol{\sigma}\cdot\mathbf{r}\times\mathbf{p}}{r^3}\right\rangle. \quad (5\text{–}2.140)$$

The comparison of the last two equations supplies the information that

$$\left\langle \mathbf{p}^2 + \frac{\alpha}{m}\,\frac{\boldsymbol{\sigma}\cdot\mathbf{r}\times\mathbf{p}}{r^3}\right\rangle = 2m(m - E_{\text{rel.}}) + \left\langle 3\left(\frac{\mathbf{p}^2}{2m}\right)^2 - \frac{2\pi\alpha}{m}\,\delta(\mathbf{r})\right\rangle. \quad (5\text{–}2.141)$$

An evaluation for the remaining structure is produced by exploiting a modification of the scale transformation. It is generated by

$$G = \delta\lambda(\mathbf{r}/r)\cdot\mathbf{p}, \quad (5\text{–}2.142)$$

where symmetrized multiplication is understood. This transformation is

$$\delta\mathbf{r} = \frac{1}{i}\,[\mathbf{r}, G] = \delta\lambda\,\frac{\mathbf{r}}{r}, \qquad \delta\mathbf{p} = \frac{1}{i}\,[\mathbf{p}, G] = -\,\delta\lambda\left(\frac{1}{r} - \frac{\mathbf{rr}}{r^3}\right)\cdot\mathbf{p}. \quad (5\text{–}2.143)$$

Since the relevant term in (5–2.138) carries the additional factor of $\alpha$, it suffices to consider the non-relativistic energy operator in applying this transformation:

$$\delta\left(\frac{\mathbf{p}^2}{2m} - \frac{\alpha}{r}\right) = \delta\lambda\left[\frac{1}{m}\,\mathbf{p}\cdot\left(\frac{1}{r} + \frac{\mathbf{rr}}{r^3}\right)\cdot\mathbf{p} - \frac{2}{m}\left(\mathbf{p}\cdot\frac{1}{r}\,\mathbf{p}\right)_{\text{sym.}} + \frac{\alpha}{r^2}\right], \quad (5\text{–}2.144)$$

where the full symmetrization required in $\mathbf{p}\cdot(1/r)\mathbf{p}$ is made explicit in

$$\left(\mathbf{p}\cdot\frac{1}{r}\,\mathbf{p}\right)_{\text{sym.}} = \frac{1}{4}\left[\left\{\mathbf{p}^2, \frac{1}{r}\right\} + 2\mathbf{p}\cdot\frac{1}{r}\,\mathbf{p}\right]. \quad (5\text{–}2.145)$$

Alternatively, one can write

$$\begin{aligned}\mathbf{p}\cdot\frac{1}{r}\,\mathbf{p} &= \frac{1}{2}\left\{\mathbf{p}^2, \frac{1}{r}\right\} - \frac{1}{2}\left[\mathbf{p}\cdot, \left[\mathbf{p}, \frac{1}{r}\right]\right]\\ &= \frac{1}{2}\left\{\mathbf{p}^2, \frac{1}{r}\right\} - 2\pi\,\delta(\mathbf{r}) \end{aligned} \quad (5\text{–}2.146)$$

and get

$$\left(\mathbf{p}\cdot\frac{1}{r}\mathbf{p}\right)_{\text{sym.}} = \frac{1}{2}\left\{\mathbf{p}^2, \frac{1}{r}\right\} - \pi\,\delta(\mathbf{r}). \tag{5–2.147}$$

The vanishing expectation value of (5–2.144) then asserts that

$$\frac{\alpha}{m}\left\langle \mathbf{p}\cdot\left(\frac{1}{r}+\frac{\mathbf{rr}}{r^3}\right)\cdot\mathbf{p}\right\rangle = \left\langle 2\left\{\frac{\mathbf{p}^2}{2m}, \frac{\alpha}{r}\right\} - \frac{2\pi\alpha}{m}\delta(\mathbf{r}) - \left(\frac{\alpha}{r}\right)^2\right\rangle, \tag{5–2.148}$$

which combines with (5–2.141) to give

$$\left\langle \mathbf{p}^2 - \frac{\alpha}{m}\mathbf{p}\cdot\left(\frac{1}{r}+\frac{\mathbf{rr}}{r^3}\right)\cdot\mathbf{p} + \frac{\alpha}{m}\frac{\boldsymbol{\sigma}\cdot\mathbf{r}\times\mathbf{p}}{r^3}\right\rangle$$
$$= 2m(m - E_{\text{rel.}}) + \langle 3T^2 + 2\{T, V\} + V^2\rangle. \tag{5–2.149}$$

In writing the last expression we used the non-relativistic symbols

$$T = \frac{\mathbf{p}^2}{2m}, \qquad V = -\frac{\alpha}{r}. \tag{5–2.150}$$

For a state of non-relativistic energy $E_{\text{non-rel.}}$, one has

$$\langle 3T^2 + 2\{T, V\} + V^2\rangle = E_{\text{non-rel.}}(E_{\text{non-rel.}} + 2\langle T\rangle)$$
$$= -(m - E_{\text{rel.}})^2, \tag{5–2.151}$$

which employs the non-relativistic form of the virial theorem (5–2.141),

$$\langle T\rangle = -E_{\text{non-rel.}}. \tag{5–2.152}$$

Thus, we have the following first indication of the mass dependence in the spectrum of the two-body system, with $M \gg m$,

$$E - M = E_{\text{rel.}} + \frac{m^2 - E_{\text{rel.}}^2}{2M}. \tag{5–2.153}$$

We have not been concerned with fine structure before now [except for the indirect reference to the fact that simple fine structure theory does not remove the degeneracy of certain levels, which accompanied Eq. (4–11.109)], and lack an explicit expression for $E_{\text{rel.}}$. But we have only to introduce the known values for the various terms in the expectation value of Eq. (5–2.139). Thus, for $l \neq 0$,

$$E_{\text{rel.}} = m - \frac{m\alpha^2}{2n^2} - \frac{1}{2m}\left\langle\left(E_{\text{non-rel.}} + \frac{\alpha}{r}\right)^2\right\rangle + \frac{m\alpha^4}{2}\frac{1}{n^3}\frac{1}{2l+1}\begin{cases} j = l + \frac{1}{2}: & \frac{1}{l+1}, \\ j = l - \frac{1}{2}: & -\frac{1}{l} \end{cases} \tag{5–2.154}$$

according to Eqs. (4–11.104, 107), and

$$\left\langle\left(E_{\text{non-rel.}}+\frac{\alpha}{r}\right)^2\right\rangle=\frac{m^2\alpha^4}{n^3}\left(\frac{1}{l+\frac{1}{2}}-\frac{3}{4n}\right), \tag{5–2.155}$$

which uses (4–11.106) and the virial theorem statement, equivalent to (5–2.152),

$$\left\langle\frac{\alpha}{r}\right\rangle=-2E_{\text{non-rel.}}. \tag{5–2.156}$$

On remarking that

$$\frac{1}{l+\frac{1}{2}}-\frac{1}{2l+1}\begin{cases} j=l+\frac{1}{2}: & \frac{1}{l+1} \\ j=l-\frac{1}{2}: & -\frac{1}{l}\end{cases}=\begin{cases} j=l+\frac{1}{2}: & \frac{1}{l+1} \\ j=l-\frac{1}{2}: & \frac{1}{l}\end{cases}=\frac{1}{j+\frac{1}{2}}, \tag{5–2.157}$$

we get the desired result:

$$E_{\text{rel.}}(n,j)=m-\frac{m\alpha^2}{2n^2}-\frac{m\alpha^4}{2n^3}\left(\frac{1}{j+\frac{1}{2}}-\frac{3}{4n}\right)+\cdots, \tag{5–2.158}$$

which continues to hold for $l=0$, $j=\frac{1}{2}$, as one sees by replacing the actually vanishing last term of (5–2.154), $m\alpha^4/2n^3$, with the delta function term

$$\left\langle\frac{\pi\alpha}{2m^2}\delta(\mathbf{r})\right\rangle=\frac{\alpha}{2m^2}\left(\frac{\alpha m}{n}\right)^3=\frac{m\alpha^4}{2n^3}. \tag{5–2.159}$$

The relativistic correction exhibited in Eq. (5–2.158) describes the splitting of the $2n^2$ degenerate levels of quantum number $n$ into $n$ distinct levels that are labeled by the total angular momentum quantum number $j=\frac{1}{2},\frac{3}{2},\ldots,n-\frac{1}{2}$. For each $j\leqslant n-\frac{3}{2}$, two different values of the orbital angular momentum can produce the given $j$ and the multiplicity of the level is, therefore, $2(2j+1)$. The exception is $j=n-\frac{1}{2}$, where the multiplicity is $2j+1=2n$. The first point to notice about the mass dependence given in (5–2.153) is that it is a function of $E_{\text{rel.}}(n,j)$ and hence introduces no new splitting of the still degenerate levels. For a more quantitative expression, let us write

$$E_{\text{rel.}}=m(1-g),\qquad g=\frac{\alpha^2}{2n^2}+\frac{\alpha^4}{2n^3}\left(\frac{1}{j+\frac{1}{2}}-\frac{3}{4n}\right), \tag{5–2.160}$$

and get

$$E-M=m-m\left(1-\frac{m}{M}\right)g-\frac{m^2}{2M}g^2. \tag{5–2.161}$$

To the extent that

$$m\left(1-\frac{m}{M}\right) \cong \frac{m}{1+\dfrac{m}{M}} = \mu, \tag{5–2.162}$$

the reduced mass, the latter is the mass parameter that enters both the gross structure and the fine structure splittings:

$$E - M = m - \frac{\mu\alpha^2}{2n^2} - \frac{\mu\alpha^4}{2n^3}\left(\frac{1}{j+\frac{1}{2}} - \frac{3}{4n}\right) - \frac{m^2\alpha^4}{8Mn^4}. \tag{5–2.163}$$

But we shall see that this simple mass characterization of the fine structure ceases to be valid when the remainder of the theory is consulted.

Before embarking on this task, however, we shall add a comment to the calculation just concluded. Observe that the last term of (5–2.138) is the approximate, transformed version of

$$\frac{1}{2M}\left\langle \mathbf{p}^2 - \alpha\boldsymbol{\alpha}\cdot\left(\frac{1}{r} + \frac{\mathbf{rr}}{r^3}\right)\cdot\mathbf{p}\right\rangle, \tag{5–2.164}$$

which form would be obtained directly by applying the unitary transformation only to the heavy particle. Let us ask what emerges if we evaluate this expectation value as it stands. Now we apply the transformation (5–2.143) to the Dirac energy operator:

$$\delta\left(\boldsymbol{\alpha}\cdot\mathbf{p} + \gamma^0 m - \frac{\alpha}{r}\right) = \delta\lambda\left[\boldsymbol{\alpha}\cdot\frac{\mathbf{rr}}{r^3}\cdot\mathbf{p} - \boldsymbol{\alpha}\cdot\frac{1}{r}\mathbf{p} + \frac{\alpha}{r^2}\right], \tag{5–2.165}$$

and get the expectation value implication

$$\left\langle \boldsymbol{\alpha}\cdot\frac{\mathbf{rr}}{r^3}\cdot\mathbf{p}\right\rangle = \left\langle \frac{1}{r}\boldsymbol{\alpha}\cdot\mathbf{p} - \frac{\alpha}{r^2}\right\rangle. \tag{5–2.166}$$

With this substitution, the expression (5–2.164) becomes

$$\frac{1}{2M}\left\langle\left(\boldsymbol{\alpha}\cdot\mathbf{p} - \frac{\alpha}{r}\right)^2\right\rangle = \frac{1}{2M}\langle(E_{\text{rel.}} - \gamma^0 m)^2\rangle$$

$$= \frac{1}{2M}(E_{\text{rel.}}^2 + m^2 - 2mE_{\text{rel.}}\langle\gamma^0\rangle). \tag{5–2.167}$$

The calculation is completed by applying an ordinary scale transformation to the energy operator, which yields

$$0 = \left\langle \boldsymbol{\alpha}\cdot\mathbf{p} - \frac{\alpha}{r}\right\rangle = E_{\text{rel.}} - m\langle\gamma^0\rangle. \tag{5–2.168}$$

This result

$$\langle\gamma^0\rangle = E_{\text{rel.}}/m, \tag{5–2.169}$$

gives an expression for (5–2.164) that is identical with what is displayed in Eq. (5–2.153).

There is an implicit difference, of course. In the last calculation, $E_{\text{rel.}}$ refers to the exact eigenvalues of the Dirac equation, while $E_{\text{rel.}}$ in (5–2.153) signifies the terms displayed in the expansion of Eq. (5–2.158). It is worth verifying that no discrepancy arises to the order of accuracy that has been retained in the approximate treatment, which is such that the fine structure, or its mass dependence, is regarded as small, of order $\alpha^2$, relative to the gross structure. We shall use the second order form of the homogeneous Dirac equation,

$$[\Pi^2 - eq\sigma F + m^2]\psi = 0. \tag{5–2.170}$$

Specialized to the charge assignment required for a bound state in the Coulomb field of a nuclear charge $Ze$, the equation for the energy eigenvalue $E$ reads

$$\left[\mathbf{p}^2 + m^2 - \left(E + \frac{Z\alpha}{r}\right)^2 + Z\alpha\gamma_5\frac{\boldsymbol{\sigma}\cdot\mathbf{r}}{r^3}\right]\psi = 0. \tag{5–2.171}$$

We write the decomposition of $\mathbf{p}^2$ into radial and angular parts as

$$\mathbf{p}^2 = p_r^{\,2} + \frac{\mathbf{L}^2}{r^2}, \qquad \mathbf{L} = \mathbf{r}\times\mathbf{p}. \tag{5–2.172}$$

That gives

$$\left[p_r^{\,2} + m^2 - E^2 - \frac{2EZ\alpha}{r} + \frac{\mathbf{L}^2 - (Z\alpha)^2 + Z\alpha\gamma_5\boldsymbol{\sigma}\cdot\mathbf{r}/r}{r^2}\right]\psi = 0, \tag{5–2.173}$$

which we compare with the non-relativistic equation appropriate to the quantum numbers $n$ and $l$,

$$\left[p_r^{\,2} + \left(\frac{mZ\alpha}{n}\right)^2 - \frac{2mZ\alpha}{r} + \frac{l(l+1)}{r^2}\right]\psi_{nl} = 0, \tag{5–2.174}$$

where

$$n - l - 1 = n_r = 0, 1, 2, \ldots . \tag{5–2.175}$$

Immediately evident are the correspondences

$$2mZ\alpha \leftrightarrow 2EZ\alpha, \qquad \left(\frac{mZ\alpha}{n}\right)^2 \leftrightarrow m^2 - E^2, \tag{5–2.176}$$

but what corresponds to the non-relativistic $l(l+1)$ still appears as an operator. We need its eigenvalues.

Concerning the angular momentum properties of spin $\frac{1}{2}$ particles, we know [Section 2–7] that a given total angular momentum quantum number $j$ can be realized with $l = j + \frac{1}{2}$ or $j - \frac{1}{2}$, which are states of opposite orbital parity. We also know that the effect of multiplication by $i\gamma_5 \boldsymbol{\sigma} \cdot (\mathbf{r}/r)$, where $i\gamma_5$ reverses intrinsic parity ($\gamma^0$), is to interchange the two kinds of spin-orbit functions associated with a given $j$. Accordingly, the coefficient of $1/r^2$ in (5–2.173) can be represented by a two-dimensional matrix, with row and column labeled by $l = j + \frac{1}{2}, j - \frac{1}{2}$:

$$\begin{pmatrix} (j+\frac{1}{2})(j+\frac{3}{2}) - (Z\alpha)^2 & -iZ\alpha \\ -iZ\alpha & (j-\frac{1}{2})(j+\frac{1}{2}) - (Z\alpha)^2 \end{pmatrix}. \qquad (5\text{–}2.177)$$

The eigenvalues of this matrix are of the form $l'(l'+1)$, with

$$l' = [(j+\tfrac{1}{2})^2 - (Z\alpha)^2]^{1/2},\ [(j+\tfrac{1}{2})^2 - (Z\alpha)^2]^{1/2} - 1, \qquad (5\text{–}2.178)$$

which, as $Z\alpha \to 0$, reduce to $j + \frac{1}{2}, j - \frac{1}{2}$, respectively. The correspondences (5–2.176) are thus completed by

$$l(l+1) \leftrightarrow l'(l'+1). \qquad (5\text{–}2.179)$$

The result obtained for the relativistic energy spectrum by applying these relations is

$$m^2 - E^2 = \frac{(EZ\alpha)^2}{(n_r + l' + 1)^2}, \qquad (5\text{–}2.180)$$

or, using either value of $l'$ and identifying $n$ accordingly,

$$\left(\frac{m}{E}\right)^2 = 1 + \frac{(Z\alpha)^2}{(n - j - \frac{1}{2} + [(j+\frac{1}{2})^2 - (Z\alpha)^2]^{1/2})^2}. \qquad (5\text{–}2.181)$$

As a by-product, we observe that the analogous equation for a spin 0 particle lacks the spin term of (5–2.171). Hence, the coefficient of $1/r^2$ in Eq. (5–2.173) becomes

$$l(l+1) - (Z\alpha)^2 = l'(l'+1), \qquad (5\text{–}2.182)$$

and

$$l' = [(l+\tfrac{1}{2})^2 - (Z\alpha)^2]^{1/2} - \tfrac{1}{2}. \qquad (5\text{–}2.183)$$

The corresponding energy formula is again (5–2.181), but with the integer $+\frac{1}{2}$ quantity, $j$ replaced by the integer $l$. In the special circumstance $n = 1, j = \frac{1}{2}$, the formula (5–2.181) reduces to

$$\left(\frac{m}{E}\right)^2 = 1 + \frac{(Z\alpha)^2}{1 - (Z\alpha)^2} = \frac{1}{1 - (Z\alpha)^2}, \qquad (5\text{–}2.184)$$

or

$$E = m[1 - (Z\alpha)^2]^{1/2}, \tag{5-2.185}$$

which has already been encountered in Eq. (4-17.31). A similar result holds for all $n = j + \frac{1}{2}$,

$$\begin{aligned} E &= m\left[1 - \left(\frac{Z\alpha}{n}\right)^2\right]^{1/2} \\ &\cong m\left[1 - \frac{1}{2}\left(\frac{Z\alpha}{n}\right)^2 - \frac{1}{8}\left(\frac{Z\alpha}{n}\right)^4 - \frac{1}{16}\left(\frac{Z\alpha}{n}\right)^6 - \cdots\right], \end{aligned} \tag{5-2.186}$$

and the expanded version agrees with Eq. (5-2.158), where $Z = 1$, as far as the latter is stated. The same agreement appears when the general formula of (5-2.181) is expanded.

With regard to the mass dependence exhibited in Eq. (5-2.153), when the result of (5-2.181) is used for $E_{\text{rel.}}$, the fact that the latter is an even function of $\alpha$ implies that no term appears of the form

$$\alpha^5 \frac{m^2}{M} \sim \frac{m}{M} \alpha \text{ (fine structure)}. \tag{5-2.187}$$

The next section is devoted to evaluating effects of this type, which are contained in the full theory.

## 5-3 TWO-PARTICLE INTERACTIONS. RELATIVISTIC THEORY II

We begin this section by demonstrating that there are energy shifts of the magnitude indicated in Eq. (5-2.187); effects of order $\alpha$, rather than $\alpha^2$, relative to the mass dependence $(m/M)$ of the fine structure. Although it is not the most important one numerically, the simplest example of such an effect appears in the comparison between the calculation just concluded, employing the Dirac equation with the instantaneous interaction, and the one based on the equation system of (5-2.94, 95). Since the entire transverse part of the interaction is advantageously handled in another way, only the instantaneous Coulomb interaction will now be considered. To put the two approaches on the same footing, the Eqs. (5-2.94, 95) should be simplified, in the sense of retaining only the first power of $m/M$, which characterizes our present limited objective. Thus we have

$$W_2 \cong M + \frac{\mathbf{p}^2}{2M}, \qquad V_0 \cong -\left(U\frac{\alpha}{r}U^{-1}\right)_{(+)(+)} \tag{5-3.1}$$

where $U$ refers only to the particle of mass $m$, and the $(+)$ subscripts correspondingly indicate only the $\gamma^0$ eigenvalue of this particle. The latter notation is used to distinguish such labels from the ones in (5–2.95) where +, for example, indicates the common eigenvalue of $\gamma_1{}^0$ and $\gamma_2{}^0$.

First, let us recognize that in the equation of (5–2.95), combined as

$$\left[i\partial_t - H_0 + V_1 \frac{1}{i\partial_t + H_0} V_1\right] \bar{G}_{++} = \delta(t-t'), \tag{5–3.2}$$

the term involving $V_1$ is of relative order $(m/M)^2$, and will not be retained. A rough indication of the magnitude of $V_1$ is given by

$$V_1 \sim \frac{1}{M}\left|\nabla \frac{\alpha}{r}\right|, \tag{5–3.3}$$

and the denominator appearing in the $V_1$ term can be approximated by $2M$. The inference that the term in question has three powers of $M$ in the denominator is incorrect, however, as one recognizes from a momentum transcription of its expectation value in which the momentum associated with the wave function is neglected (this is, equivalently, the use of the wave function at the origin of the relative coordinates)

$$\left\langle V_1 \frac{1}{i\partial_t + H_0} V_1 \right\rangle \sim |\psi(0)|^2 \int \frac{(d\mathbf{p})}{(2\pi)^3} \frac{1}{2M}\left(\frac{\mathbf{p}}{M}\right)^2 \left(\frac{\alpha}{\mathbf{p}^2}\right)^2$$
$$\sim |\psi(0)|^2 \frac{\alpha^2}{M^3}\int dp \sim |\psi(0)|^2 \frac{\alpha^2}{M^2}. \tag{5–3.4}$$

The final step recognizes that the non-relativistic evaluations, such as (5–3.3), become incorrect and give an overestimate at momentum transfers of magnitude $\sim M$. Since $|\psi(0)|^2 \sim (\alpha m)^3$, this effect is of the order of $(m/M)^2\alpha(\alpha^4 m)$, and is omitted.

The comparison of interest is that between the eigenvalues of the energy operator

$$H_0 = M + \frac{\mathbf{p}^2}{2M} + W(\mathbf{p}) - \left(U\frac{\alpha}{r}U^{-1}\right)_{(+)(+)},$$
$$W(\mathbf{p}) = (\mathbf{p}^2 + m^2)^{1/2}, \tag{5–3.5}$$

and those of

$$H = M + \frac{\mathbf{p}^2}{2M} + \boldsymbol{\alpha}\cdot\mathbf{p} + \gamma^0 m - \frac{\alpha}{r}. \tag{5–3.6}$$

Consider, then, the Green's function equation associated with $H$ or, rather, with the transformed version of $H$,

$$(i\partial_t - \bar{H})\bar{G} = \delta(t-t'),$$

$$\mathcal{H} = M + \frac{\mathbf{p}^2}{2M} + \gamma^0 W(\mathbf{p}) - U\frac{\alpha}{r}U^{-1}. \qquad (5\text{–}3.7)$$

This equation is now decomposed in accordance with the eigenvalues of $\gamma^0$:

$$\begin{aligned}(i\partial_t - H_0)\bar{G}_{(+)(+)} - \mathcal{H}_{(+)(-)}\bar{G}_{(-)(+)} &= \delta(t-t'),\\ (i\partial_t - \mathcal{H}_{(-)(-)})\bar{G}_{(-)(+)} - \mathcal{H}_{(-)(+)}\bar{G}_{(+)(+)} &= 0,\end{aligned} \qquad (5\text{–}3.8)$$

giving

$$\left[i\partial_t - H_0 - \mathcal{H}_{(+)(-)}\frac{1}{i\partial_t - \mathcal{H}_{(-)(-)}}\mathcal{H}_{(-)(+)}\right]\bar{G}_{(+)(+)} = \delta(t-t'). \qquad (5\text{–}3.9)$$

It will suffice to ignore the interaction term in the denominator of (5–3.9),

$$\mathcal{H}_{(-)(-)} \cong M + \frac{\mathbf{p}^2}{2M} - W(\mathbf{p}). \qquad (5\text{–}3.10)$$

As another simplification, the energy of the state of interest, which replaces $i\partial_t$ in the denominator, can be approximated by $M + m$, omitting the binding energy. The expectation value, in an eigenstate of $H_0$, of the effective energy operator obtained in this way, gives an evaluation for the corresponding eigenvalue of $H$,

$$E = E_0 + \left\langle \mathcal{H}_{(+)(-)}\frac{1}{m + W(\mathbf{p}) - [\mathbf{p}^2/2M]}\mathcal{H}_{(-)(+)}\right\rangle. \qquad (5\text{–}3.11)$$

This determines the $E_0$ spectrum from the known $E$ spectrum.

The eigenfunctions of $H_0$ [cf. Eq. (5–2.136)] are approximately those of the non-relativistic system with the reduced mass $\mu$ [Eq. (5–2.162)]. In the application to (5–3.11), which is dominated by relativistic energies $\sim m$ [rather than $\sim M$, as in (5–3.4)], the momentum associated with the wave function is negligible. That reduces (5–3.11) to

$$E - E_0 = |\psi(0)|^2 \int \frac{(d\mathbf{p})}{(2\pi)^3}\mathcal{H}_{(+)(-)}\frac{1}{m + W(\mathbf{p}) - [\mathbf{p}^2/2M]}\mathcal{H}_{(-)(+)}, \qquad (5\text{–}3.12)$$

in which the matrix elements of $\mathcal{H}$ are now momentum transforms, which take into account that $U$ reduces to unity for $\mathbf{p} = 0$. Thus, extracting the coefficient of $i\gamma_5$, we have

$$\mathcal{H}_{(+)(-)} = \mathcal{H}_{(-)(+)} = \frac{\boldsymbol{\sigma}\cdot\mathbf{p}}{|\mathbf{p}|}\left(\frac{W - m}{2W}\right)^{1/2}\frac{4\pi\alpha}{\mathbf{p}^2} \qquad (5\text{–}3.13)$$

where, illustrating the discussion that accompanied Eq. (5–3.4), the factor $[(W - m)/2W]^{1/2}$ behaves as $|\mathbf{p}|/2m$ non-relativistically and approaches a constant for $W \gg m$. This gives

$$E_0 - E = -|\psi(0)|^2(4\pi\alpha)^2 \int \frac{(d\mathbf{p})}{(2\pi)^3}\left(\frac{1}{\mathbf{p}^2}\right)^2 \frac{W-m}{2W}\frac{1}{W+m-[\mathbf{p}^2/2M]}, \quad (5\text{–}3.14)$$

or, simplified by $m/M \ll 1$,

$$E_0 - E = -4\alpha^2|\psi(0)|^2 \int_0^\infty dp\, \frac{1}{W(W+m)^2}\left[1 + \frac{W-m}{2M}\right]. \quad (5\text{–}3.15)$$

Introducing a new integration variable $\theta$,

$$p = m \sinh\theta, \qquad W = m\cosh\theta, \quad (5\text{–}3.16)$$

we evaluate the two integrals:

$$\int_0^\infty dp\, \frac{1}{W(W+m)^2} = \frac{1}{m^2}\int_0^\infty d\theta\, \frac{1}{(\cosh\theta + 1)^2} = \frac{1}{3m^2},$$

$$\int_0^\infty dp\, \frac{W-m}{W(W+m)^2} = \frac{1}{m}\int_0^\infty d\theta\, \frac{\cosh\theta - 1}{(\cosh\theta+1)^2} = \frac{1}{3m}. \quad (5\text{–}3.17)$$

The result, stated for an $s$-state of principal quantum number $n$, is

$$E_0 - E = -\frac{4}{3}\frac{\alpha^2}{m^2}|\psi(0)|^2\left(1 + \frac{m}{2M}\right) = -\frac{4}{3\pi}\frac{\alpha^5}{n^3}\frac{\mu^3}{m^2}\left(1+\frac{m}{2M}\right). \quad (5\text{–}3.18)$$

Here is an example of an energy shift of order $\alpha(m/M)(\alpha^4 m)$.

The other implication of this calculation is somewhat disconcerting, however. In the limit $m/M \to 0$, the difference of the two energies does not vanish. That is, the spectrum of $H_0$ in this limit is not that of a charge in the Coulomb field of a static source. The apparent discrepancy is a reminder that another aspect of the instantaneous interaction remains to be considered—the interaction $I^{(2)}$ of Eqs. (5–2.22—24). When only the instantaneous Coulomb part is retained, the latter becomes

$$I^{(2)}_{\text{Coul.}} = -\alpha^2 \frac{\delta(x_1{}^0 - \bar{x}_2{}^0)}{|\mathbf{x}_1 - \bar{\mathbf{x}}_2|}\frac{\delta(x_2{}^0 - \bar{x}_1{}^0)}{|\mathbf{x}_2 - \bar{\mathbf{x}}_1|}\gamma_1{}^0 G_+(x_1 - \bar{x}_1)\gamma_1{}^0\gamma_2{}^0 G_+(x_2 - \bar{x}_2)\gamma_2{}^0. \quad (5\text{–}3.19)$$

To draw the consequences of this non-local interaction, we need a further development of the Green's function equation for a local interaction, since the description of the equal-time situation no longer suffices.

Consider, then, the Green's function equation

$$[(\gamma p + m)_1(\gamma p + m)_2 - i\delta(x_1{}^0 - x_2{}^0)\gamma_1{}^0\gamma_2{}^0 V]G = 1, \quad (5\text{–}3.20)$$

and its equivalent integral form

$$G = G_1 G_2 + G_1 G_2 i\delta(x_1{}^0 - x_2{}^0)\gamma_1{}^0\gamma_2{}^0 V G. \quad (5\text{–}3.21)$$

Let us set $x_1{}^0 = x_2{}^0 = t$ in $G(x_1x_2, x'_1x'_2)$, as the interaction term requires, but leave $x_1{}^{0\prime}$ and $x_2{}^{0\prime}$ free. We need the differential equation obeyed by $G_1G_2$ under these circumstances. It is conveyed by

$$\begin{aligned}(i\partial_t - H_1 - H_2)G_+(x_1 - x'_1)\gamma_1{}^0G_+(x_2 - x'_2)\gamma_2{}^0 \\ = -\,[\delta(x_1 - x'_1)G_+(x_2 - x'_2)\gamma_2{}^0 + \delta(x_2 - x'_2)G_+(x_1 - x'_1)\gamma_1{}^0].\end{aligned} \tag{5-3.22}$$

The implied differential equation for $G$, presented in symbolic notation, is

$$\left[i\partial_t - H_1 - H_2 - \frac{1}{2}\left(\frac{H_1}{W_1} + \frac{H_2}{W_2}\right)V\right] iG\gamma_1{}^0\gamma_2{}^0 = -\,i(G_1\gamma_1{}^0 + G_2\gamma_2{}^0), \tag{5-3.23}$$

which reduces to (5–2.85) when one sets $x_1{}^{0\prime} = x_2{}^{0\prime} = t'$. In the transformed version of this equation one has

$$H \to \gamma^0 W, \qquad V \to \bar{V}, \tag{5-3.24}$$

and

$$-\,iG_+\gamma^0 \to \left[\eta(x^0 - x^{0\prime})\,\frac{1 + \gamma^0}{2} - \eta(x^{0\prime} - x^0)\,\frac{1 - \gamma^0}{2}\right]\exp(-\,iW|x^0 - x^{0\prime}|), \tag{5-3.25}$$

where the factor in brackets can also be written as

$$\tfrac{1}{2}(\gamma^0 + \varepsilon(x^0 - x^{0\prime})), \tag{5-3.26}$$

in which

$$\varepsilon(x^0 - x^{0\prime}) = \begin{cases} x^0 > x^{0\prime}: & +1 \\ x^0 = x^{0\prime}: & 0\,. \\ x^0 < x^{0\prime}: & -1 \end{cases} \tag{5-3.27}$$

As with any Green's function equation, there is another, transposed, form for which the second set of variables is involved in the differential aspects of the equation. This counterpart of (5–3.23) is

$$iG\gamma_1{}^0\gamma_2{}^0\left[i\partial'_t{}^T - H_1 - H_2 - \tfrac{1}{2}V\left(\frac{H_1}{W_1} + \frac{H_2}{W_2}\right)\right] = -\,i(G_1\gamma_1{}^0 + G_2\gamma_2{}^0), \tag{5-3.28}$$

where we have written $\partial^T$ to indicate differentiation to the left, with an additional minus sign.

Now let us add to the instantaneous interaction of (5–3.20) a non-local interaction, written as

$$I^{(2)} = \gamma_1{}^0\gamma_2{}^0 v, \tag{5-3.29}$$

which replaces the integral equation (5–3.21) with

$$G = G_1G_2 + G_1G_2\gamma_1^0\gamma_2^0[i\delta(x_1^0 - x_2^0)V + v]G. \tag{5–3.30}$$

Then the differential equation (5–3.23) becomes

$$\left[i\partial_t - H_1 - H_2 - \frac{1}{2}\left(\frac{H_1}{W_1} + \frac{H_2}{W_2}\right)V\right]iG\gamma_1^0\gamma_2^0$$
$$= -i(G_1\gamma_1^0 + G_2\gamma_2^0) - (G_1\gamma_1^0 + G_2\gamma_2^0)viG\gamma_1^0\gamma_2^0. \tag{5–3.31}$$

In virtue of the non-local nature of $v$, the Green's function $G$ in the last term does not have equal time values for its two left-hand coordinates. An approximate description of the needed relative time dependence can be obtained from (5–3.28), where it is the right-hand time variables that are equal. The equal-time Green's function associated with the instantaneous interaction obeys the equations

$$\left[i\partial_t - H_1 - H_2 - \frac{1}{2}\left(\frac{H_1}{W_1} + \frac{H_2}{W_2}\right)V\right]G_{1+2}$$
$$= G_{1+2}\left[i\partial'^{\,T}_t - H_1 - H_2 - V\frac{1}{2}\left(\frac{H_1}{W_1} + \frac{H_2}{W_2}\right)\right] = \delta(t - t')\frac{1}{2}\left(\frac{H_1}{W_1} + \frac{H_2}{W_2}\right). \tag{5–3.32}$$

With its aid, a sufficiently accurate solution of (5–3.28) is exhibited as

$$iG\gamma_1^0\gamma_2^0 = -i(G_1\gamma_1^0 + G_2\gamma_2^0)G_{1+2}, \tag{5–3.33}$$

which we use in the approximate conversion of (5–3.31) into the equal-time equation

$$\left[i\partial_t - H_1 - H_2 - \frac{1}{2}\left(\frac{H_1}{W_1} + \frac{H_2}{W_2}\right)V - i(G_1\gamma_1^0 + G_2\gamma_2^0)v(G_1\gamma_1^0 + G_2\gamma_2^0)\right]G_{1+2}$$
$$= \delta(t - t')\frac{1}{2}\left(\frac{H_1}{W_1} + \frac{H_2}{W_2}\right). \tag{5–3.34}$$

The transformed version of this equation, in the anticipated approximation that retains only the + components, reads [hopefully, the transitional mixture of three- and four-dimensional matrix notations for coordinates is not too unsettling]

$$(i\partial_t - W_1 - W_2 - V_0 - \delta V_0)\bar{G}_{++} = \delta(t - t'), \tag{5–3.35}$$

where

$$\delta V_0 = i(G_{1\,\text{ret.}} + G_{2\,\text{ret.}})v_0(G_{1\,\text{ret.}} + G_{2\,\text{ret.}}), \qquad v_0 = \bar{v}_{++}, \tag{5–3.36}$$

and

$$G_{\text{ret.}} = (1/i)\eta(t - t') \exp[-iW(t - t')]. \tag{5-3.37}$$

The time variables that have been suppressed in (5-3.36) are written out as

$$\begin{aligned}\delta V_0(t, t') = {} & i \int dt_1\, dt'_1\, G_{1\,\text{ret.}}(t - t_1)v_0(t_1 t, t'_1 t')G_{1\,\text{ret.}}(t'_1 - t') \\ & + i \int dt_1\, dt'_2\, G_{1\,\text{ret.}}(t - t_1)v_0(t_1 t, t't'_2)G_{2\,\text{ret.}}(t'_2 - t') \\ & + i \int dt_2\, dt'_1\, G_{2\,\text{ret.}}(t - t_2)v_0(tt_2, t'_1 t')G_{1\,\text{ret.}}(t'_1 - t') \\ & + i \int dt_2\, dt'_2\, G_{2\,\text{ret.}}(t - t_2)v_0(tt_2, t't'_2)G_{2\,\text{ret.}}(t'_2 - t').\end{aligned} \tag{5-3.38}$$

In the application to the non-local interaction (5-3.19), which has the form

$$v_0(t_1 t_2, t'_1 t'_2) = \delta(t_1 - t'_2)\, \delta(t_2 - t'_1)\, v(t_1 - t_2), \tag{5-3.39}$$

the terms in (5-3.38) that involve the Green's functions of different particles do not survive, as illustrated by

$$G_{1\,\text{ret.}}(t - t_1)G_{2\,\text{ret.}}(t_1 - t) = 0. \tag{5-3.40}$$

This reduces (5-3.38) to

$$\begin{aligned}\delta V_0(t, t') = {} & iG_{1\,\text{ret.}}(t - t')v(t' - t)G_{1\,\text{ret.}}(t - t') \\ & + iG_{2\,\text{ret.}}(t - t')v(t - t')G_{2\,\text{ret.}}(t - t').\end{aligned} \tag{5-3.41}$$

If $\delta V_0$ were an instantaneous interaction, its expectation value, a three-dimensional coordinate integration, would estimate the energy displacement that it induces. But the actual structure involves a time integration,

$$\int d\bar{t}\, \delta V_0(t, \bar{t})\bar{G}_{++}(\bar{t}, t'), \tag{5-3.42}$$

and one must recognize that, when a particular state of energy $E$ is considered, $\bar{G}_{++}(\bar{t}, t')$ effectively differs from $\bar{G}_{++}(t, t')$ by the phase factor

$$\exp[-iE(\bar{t} - t)]. \tag{5-3.43}$$

Accordingly, the energy shift is

$$\delta E = \left\langle \int d\bar{t}\, \delta V_0(t, \bar{t}) \exp[-iE(\bar{t} - t)] \right\rangle. \tag{5-3.44}$$

When the structure (5-3.41) is introduced, with

$$t - \bar{t} = \tau > 0, \tag{5-3.45}$$

this becomes

$$\delta E = i\left\langle\int_0^\infty d\tau\,[G_{1\,\mathrm{ret.}}(\tau)v(-\tau)G_{1\,\mathrm{ret.}}(\tau)\exp(iE\tau) + G_{2\,\mathrm{ret.}}(\tau)v(\tau)G_{2\,\mathrm{ret.}}(\tau)\exp(iE\tau)]\right\rangle. \tag{5-3.46}$$

As we shall see, only relativistic energies contribute to the effect under consideration. That enables the energy of the state to be approximated, $E \simeq M + m$; the internal wave function is replaced by $\psi(0)$, which reduces the unitary transformation $U_1U_2$ to unity, and the retarded Green's functions refer to the rest mass of the associated particle. The energetic simplifications reduce (5–3.46) to

$$\delta E = -i\left\langle\int_0^\infty d\tau\,\{\exp[i(M-m)\tau]\,v(-\tau) + \exp[i(m-M)\tau]\,v(\tau)\}\right\rangle. \tag{5-3.47}$$

In view of the $+$ character of the wave function ($\gamma_1^{0\prime} = \gamma_2^{0\prime} = +1$), the functions $v(\pm\tau)$ obtained from Eqs. (5–3.19, 29, 39) are

$$v(\pm\tau) = \alpha^2\frac{1}{|\mathbf{x}_1-\bar{\mathbf{x}}_2|}\frac{1}{|\mathbf{x}_2-\bar{\mathbf{x}}_1|}\int\frac{(d\mathbf{p}_1)}{(2\pi)^3}\frac{1}{2W_1}(m\pm W_1)\exp[i\mathbf{p}_1\cdot(\mathbf{x}_1-\bar{\mathbf{x}}_1)-iW_1\tau] \times\int\frac{(d\mathbf{p}_2)}{(2\pi)^3}\frac{1}{2W_2}(M\mp W_2)\exp[i\mathbf{p}_2\cdot(\mathbf{x}_2-\bar{\mathbf{x}}_2)-iW_2\tau]. \tag{5-3.48}$$

Let us recognize immediately that the energy shift associated with the $v(\tau)$ term is of relative order $(m/M)^2$, and is therefore omitted. In a non-relativistic treatment of the heavy particle, the $\tau$ dependence of that term is $\sim\exp[-i2M\tau]$, which contributes a factor $\sim 1/M$ on performing the $\tau$ integration. We also have in $(W_2 - M)/W_2$ a factor $\sim \mathbf{p}_2^{\,2}/M^2$. As in the discussion of Eq. (5–3.4), the apparent $1/M^3$ dependence is reduced to $1/M^2$ on examining the momentum dependence of the integral.

The expectation value of (5–3.47) involves integrations over both particle coordinates, as restricted by the wave function [cf. Eq. (5–1.113)]

$$\psi(\mathbf{x}_1\mathbf{x}_2) = \psi(\mathbf{r})\psi_{\mathbf{P}}(\mathbf{R}), \tag{5-3.49}$$

which describes the factorization of the essentially non-relativistic motion into relative and center of mass motion. Since we are concerned with the system in its center of mass frame, and neglect relative momentum, the first set of integrations is

$$\int(d\mathbf{x}_1)(d\mathbf{x}_2)\psi^*(\mathbf{x}_1\mathbf{x}_2)v(-\tau) \simeq \psi^*(0)\psi^*_{\mathbf{P}}(\bar{\mathbf{R}})\int(d\mathbf{x}_1)(d\mathbf{x}_2)v(-\tau), \tag{5-3.50}$$

where

$$\int (d\mathbf{x}_1)(d\mathbf{x}_2)v(-\tau) = (4\pi\alpha)^2 \int \frac{(d\mathbf{p}_1)}{(2\pi)^3}\frac{(d\mathbf{p}_2)}{(2\pi)^3}\frac{m - W_1}{2W_1}\frac{\exp[i\mathbf{p}_1\cdot(\bar{\mathbf{x}}_2 - \bar{\mathbf{x}}_1)]}{\mathbf{p}_1^{\,2}}$$

$$\times \frac{\exp[i\mathbf{p}_2\cdot(\bar{\mathbf{x}}_1 - \bar{\mathbf{x}}_2)]}{\mathbf{p}_2^{\,2}}\exp[-i(W_1 + W_2)\tau], \quad (5\text{–}3.51)$$

and we have proceeded cautiously with the center of mass wave function to avoid obscuring the application of the normalization condition

$$\int (d\mathbf{R})|\psi_{\mathbf{P}}(\mathbf{R})|^2 = 1. \quad (5\text{–}3.52)$$

Since (5–3.51) only involves relative coordinates, the remaining integrations directly give

$$\langle v(-\tau)\rangle = -|\psi(0)|^2(4\pi\alpha)^2\int\frac{(d\mathbf{p})}{(2\pi)^3}\left(\frac{1}{\mathbf{p}^2}\right)^2\frac{W - m}{2W}\exp\{-i[W + M + (\mathbf{p}^2/2M)]\tau\}, \quad (5\text{–}3.53)$$

where we have used the non-relativistic form for the heavy particle energy, and omitted the subscript on $W_1$. Carrying out the relative time integration in (5–3.47) now gives

$$\delta E = |\psi(0)|^2(4\pi\alpha)^2\int\frac{(d\mathbf{p})}{(2\pi)^3}\left(\frac{1}{\mathbf{p}^2}\right)^2\frac{W - m}{2W}\frac{1}{W + m + (\mathbf{p}^2/2M)}, \quad (5\text{–}3.54)$$

which is very closely related to (5–3.14). Indeed, on neglecting $\mathbf{p}^2/2M$ the two expressions cancel, which is as it should be. Also, the terms of relative order $m/M$ in (5–3.14) and (5–3.54) are equal. Hence, the net energy shift obtained from the instantaneous Coulomb interaction is

$$(\delta E)_{\text{Coul.}} = -\frac{4}{3}\frac{\alpha^2}{m^2}\frac{m}{M}|\psi(0)|^2 = -\frac{4}{3\pi}\frac{m}{M}\frac{\alpha^5}{n^3}m; \quad (5\text{–}3.55)$$

the last form refers to an $ns$ state, in which the distinction between $\mu$ and $m$ has not been retained. For the important example of $n = 2$, this reads (the numerical value refers to hydrogen)

$$2s_{1/2}\colon\quad (\delta E)_{\text{Coul.}} = -\frac{1}{3\pi}\frac{m}{M}\alpha^3\,\text{Ry} = -0.07\,\text{MHz}. \quad (5\text{–}3.56)$$

Before embarking on the relativistic discussion of the transverse interaction, let us return to the non-relativistic considerations of Section 4–11. The causal vacuum amplitude for the exchange of a photon and a particle is written in Eq.

(4–11.33), where the choice of particle propagation function describes the dynamical nature of the system. For a composite of two particles with charge and mass assignments given by $-e, m$ and $e, M$, considered in the center of mass frame, one has only to make the substitution

$$-\frac{e}{m}\mathbf{p} \to -\frac{e}{m}\mathbf{p} + \frac{e}{M}(-\mathbf{p}), \tag{5–3.57}$$

while indicating, through the coordinate dependence of the photon propagation function, which particles are involved in the emission and absorption acts. [See the related discussion in Section 3–15, pp. 357–358 where, however, the symbol $M$ indicates the mass of the composite particle.] The additional coupling between different particles has an explicit $1/M$ factor. Accordingly, the remaining calculation is performed as though $m/M = 0$. This enables the heavy particle to be stationed at the origin, while the light particle is assigned the coordinate vector $\mathbf{r}$. The two vacuum amplitude terms obtained in this way are

$$-\frac{e^2}{mM}\int (d\mathbf{r})\,dt\,(d\mathbf{r}')\,dt'\,\psi^*{}_1(\mathbf{r}t)\mathbf{p}\cdot[\mathbf{D}(\mathbf{r},t-t')G(\mathbf{r}\mathbf{r}',t-t')$$
$$+ G(\mathbf{r}\mathbf{r}',t-t')\mathbf{D}(\mathbf{r}',t-t')]\cdot\mathbf{p}\psi_2(\mathbf{r}'t'). \tag{5–3.58}$$

The presence of the photon propagation function [Eq. (4–11.25)],

$$\mathbf{D}(\mathbf{r},t-t') = i\int d\omega_k \exp[i\mathbf{k}\cdot\mathbf{r} - ik^0(t-t')]\left(\mathbf{1} - \frac{\mathbf{k}\mathbf{k}}{\mathbf{k}^2}\right), \tag{5–3.59}$$

adds the energy $k^0$ to the composite particle energy, as before. When the space-time extrapolation is performed, the resulting addition to $\delta V$ [Eq. (4–11.58)] is

$$\delta V' = \frac{e^2}{mM}\int d\omega_k\left(\mathbf{1} - \frac{\mathbf{k}\mathbf{k}}{\mathbf{k}^2}\right) : \mathbf{p}\left[\exp(i\mathbf{k}\cdot\mathbf{r})\frac{1}{E + i\varepsilon - H - k^0}\right.$$
$$\left. + \frac{1}{E + i\varepsilon - H - k^0}\exp(i\mathbf{k}\cdot\mathbf{r})\right]\mathbf{p}, \tag{5–3.60}$$

or

$$\delta V' = -2\frac{e^2}{mM}\int d\omega_k\left(\mathbf{1} - \frac{\mathbf{k}\mathbf{k}}{\mathbf{k}^2}\right) : \mathbf{p}\frac{\exp(i\mathbf{k}\cdot\mathbf{r})}{k^0}\mathbf{p}$$
$$+ \frac{e^2}{mM}\int d\omega_k\left(\mathbf{1} - \frac{\mathbf{k}\mathbf{k}}{\mathbf{k}^2}\right) : \mathbf{p}\left[\frac{\exp(i\mathbf{k}\cdot\mathbf{r})}{k^0}\frac{E - H}{E + i\varepsilon - H - k^0}\right.$$
$$\left. + \frac{E - H}{E + i\varepsilon - H - k^0}\frac{\exp(i\mathbf{k}\cdot\mathbf{r})}{k^0}\right]\mathbf{p}. \tag{5–3.61}$$

The point of the latter decomposition appears on noting that

$$2\int d\omega_k\left(1-\frac{\mathbf{kk}}{\mathbf{k}^2}\right)\frac{\exp(i\mathbf{k}\cdot\mathbf{r})}{k^0}=\left(1-\frac{\nabla\nabla}{\nabla^2}\right)\int\frac{(d\mathbf{k})}{(2\pi)^3}\frac{\exp(i\mathbf{k}\cdot\mathbf{r})}{\mathbf{k}^2}=\left(1-\frac{\nabla\nabla}{\nabla^2}\right)\mathscr{D}(\mathbf{r});\tag{5–3.62}$$

the corresponding term is the non-relativistic version of the instantaneous part of the transverse interaction. It has already been considered, in its relativistic form, and should be removed from (5–3.61).

The presence of the exponential factor $\exp(i\mathbf{k}\cdot\mathbf{r})$ in the significant term of (5–3.61) introduces a characteristic photon energy $K$, a small numerical multiple of $(a_0)^{-1} = \alpha m$, say

$$K \sim \alpha^{3/2}m.\tag{5–3.63}$$

For $k^0 \lesssim K$, the exponential factor is effectively unity, and the performance of the integration over all photon directions gives

$$\delta V'_{<K} = \frac{4}{3}\frac{\alpha}{\pi}\frac{1}{mM}\int_0^K dk^0\,\mathbf{p}\cdot\frac{E-H}{E+i\varepsilon-H-k^0}\mathbf{p}.\tag{5–3.64}$$

One will recognize, multiplied by $2m/M$, a more compact form of the structure of Eq. (4–11.59). Underlying this simple result is the observation, based on (5–3.57), that the two particles emit and absorb long wavelength photons in the same way as the single particle, but with the coupling constant alteration $e/m \to e/\mu$. The additional energy displacement for a $2s$ level is obtained from (4–11.91, 92), with $Z = 1$, as

$$\langle\delta V'_{<K}\rangle_{2s} = \frac{2}{3\pi}\frac{m}{M}\alpha^3\,\mathrm{Ry}\left[\log\frac{K}{\mathrm{Ry}} - 2.8118\right].\tag{5–3.65}$$

For the $2p$ level, according to (4–11.97), we have

$$\langle\delta V'_{<K}\rangle_{2p} = \frac{2}{3\pi}\frac{m}{M}\alpha^3\,\mathrm{Ry}\,[0.0300].\tag{5–3.66}$$

When $k^0 > K$, which is an order of magnitude larger than the atomic binding energies, one can neglect $E - H$ relative to $k^0$ in the denominator of (5–3.61):

$$\delta V'_{>K} = -\frac{e^2}{mM}\int_K^\infty d\omega_k\frac{1}{\mathbf{k}^2}\left(1-\frac{\mathbf{kk}}{\mathbf{k}^2}\right):\mathbf{p}[\exp(i\mathbf{k}\cdot\mathbf{r})(E-H) + (E-H)\exp(i\mathbf{k}\cdot\mathbf{r})]\mathbf{p}.\tag{5–3.67}$$

For the expectation value of this operator in an eigenstate of $H$ with eigenvalue $E$, one can modify (5–3.67) according to

$$(E-H)\mathbf{p} \to [E-H, \mathbf{p}] = -i\nabla V,$$
$$\mathbf{p}(E-H) \to [\mathbf{p}, E-H] = i\nabla V, \tag{5–3.68}$$

which converts it into

$$\delta V'_{>K} = -\frac{e^2}{mM}\int_K^\infty d\omega_k \frac{1}{\mathbf{k}^2}\left(1-\frac{\mathbf{kk}}{\mathbf{k}^2}\right) : i(\nabla V \exp(i\mathbf{k}\cdot\mathbf{r})\,\mathbf{p} - \mathbf{p}\exp(i\mathbf{k}\cdot\mathbf{r})\,\nabla V), \tag{5–3.69}$$

or

$$\delta V'_{>K} = \frac{e^2}{mM}\int_K^\infty d\omega_k \frac{1}{\mathbf{k}^2}\left(1-\frac{\mathbf{kk}}{\mathbf{k}^2}\right) : \exp(i\mathbf{k}\cdot\mathbf{r})\,\nabla\nabla V. \tag{5–3.70}$$

It is simpler to use (5–3.69) with a real wave function choice, giving

$$\langle\delta V'_{>K}\rangle = -\frac{e^2}{mM}\int_K^\infty d\omega_k \frac{1}{\mathbf{k}^2}\left(1-\frac{\mathbf{kk}}{\mathbf{k}^2}\right) : \int (d\mathbf{r})\,\nabla V \exp(i\mathbf{k}\cdot\mathbf{r})\,\nabla(\psi(\mathbf{r}))^2. \tag{5–3.71}$$

As in Eqs. (4–11.75, 76), we shall employ dimensionless variables,

$$\mathbf{r} = na_0\mathbf{x}, \qquad \psi(\mathbf{r}) = \left[\frac{1}{\pi n^3 a_0{}^3}\right]^{1/2}\psi(\mathbf{x}), \tag{5–3.72}$$

together with

$$\mathbf{k} = \frac{1}{na_0}\boldsymbol{\kappa}. \tag{5–3.73}$$

On introducing the cosine of the angle between $\mathbf{x}$ and $\boldsymbol{\kappa}$,

$$\mu = \frac{\mathbf{x}\cdot\boldsymbol{\kappa}}{x\kappa}, \tag{5–3.74}$$

we find that

$$\langle\delta V'_{>K}\rangle = -\frac{4}{\pi}\frac{m}{M}\frac{\alpha^3}{n^3}\,\mathrm{Ry}\int_{-1}^{1} d\mu(1-\mu^2)\int_{na_0K}^\infty \frac{d\kappa}{\kappa}\int_0^\infty dx\,\exp(i\kappa x\mu)\,\frac{d}{dx}(\psi(x))^2. \tag{5–3.75}$$

According to the generating function [Eq. (4–11.84)]

$$\frac{1}{(1-t)^2}\exp\{-[(1+t)/(1-t)]x\} = \sum_{n=1}^\infty t^{n-1}n\psi_{ns}(x), \tag{5–3.76}$$

which, incidentally, is verified as a solution of the differential equation

$$\left[\frac{\partial^2}{\partial x^2} + \frac{2}{x}\frac{\partial}{\partial x} - 1 + \frac{2}{x}\frac{\partial}{\partial t}t\right]\frac{1}{(1-t)^2}\exp\{-[(1+t)/(1-t)]x\} = 0, \tag{5–3.77}$$

we have

$$\frac{1}{(1-t)^2(1-t')^2}\exp(-\lambda x) = \sum_{n,n'=1}^{\infty} t^{n-1}t'^{n'-1}nn'\psi_{ns}(x)\psi_{n's}(x),$$

$$\lambda = \frac{1+t}{1-t} + \frac{1+t'}{1-t'} = 2\frac{1-tt'}{(1-t)(1-t')}, \quad (5\text{-}3.78)$$

and therefore

$$(\psi_{ns}(x))^2 = \frac{1}{n!}\left(\frac{d}{dt}\right)^{n-1}\frac{1}{n!}\left(\frac{d}{dt'}\right)^{n-1}\frac{1}{(1-t)^2(1-t')^2}\exp(-\lambda x)\bigg|_{t=t'=0} \quad (5\text{-}3.79)$$

One now encounters the successive integrals

$$\lambda\int_0^\infty dx\exp(i\kappa x\mu)\exp(-\lambda x) = \frac{\lambda}{\lambda - i\kappa\mu} \to \frac{\lambda^2}{\lambda^2+\kappa^2\mu^2}, \quad (5\text{-}3.80)$$

where the last step refers to the nature of the subsequent $\mu$ integration; then,

$$\int_{na_0K}^\infty \frac{d\kappa}{\kappa}\frac{\lambda^2}{\lambda^2+\kappa^2\mu^2} \simeq \log\frac{\lambda}{na_0K\mu}, \qquad na_0K \ll 1; \quad (5\text{-}3.81)$$

and finally,

$$\int_{-1}^1 d\mu(1-\mu^2)\log\frac{\lambda}{na_0K\mu} = \frac{4}{3}\left(\log\frac{\lambda}{na_0K} + \frac{4}{3}\right). \quad (5\text{-}3.82)$$

The resulting generating function is

$$\frac{16}{3\pi}\frac{m}{M}\frac{\alpha^3}{n^3}\,\mathrm{Ry}\frac{1}{(1-t)^2(1-t')^2}\left(\log\frac{\lambda}{na_0K} + \frac{4}{3}\right), \quad (5\text{-}3.83)$$

in which the coefficient of $t^{n-1}t'^{n-1}$, divided by $n^2$, is the desired expectation value in the $ns$ state. For the important example of $n = 2$, this recipe gives

$$\langle\delta V'_{>K}\rangle_{2s} = \frac{2}{3\pi}\frac{m}{M}\alpha^3\,\mathrm{Ry}\left(\log\frac{1}{a_0K} + \frac{25}{12}\right). \quad (5\text{-}3.84)$$

When (5-3.65) is added, we get

$$\langle\delta V'\rangle_{2s} = \frac{2}{3\pi}\frac{m}{M}\alpha^3\,\mathrm{Ry}\left[\log\frac{2}{\alpha} - 2.8118 + \frac{25}{12}\right]. \quad (5\text{-}3.85)$$

We have only had occasion to exhibit the wave functions of $s$-states, based on the generating function (5-3.76). The relation between the states of common $n$ but different $l$ quantum numbers can be described with the aid of the axial vector

$$\mathbf{A} = \frac{\mathbf{r}}{r} + a_0\tfrac{1}{2}(\mathbf{L}\times\mathbf{p} - \mathbf{p}\times\mathbf{L}). \quad (5\text{-}3.86)$$

It is directly verified to be a constant of the motion as a consequence of the equations of motion,

$$m\frac{d\mathbf{r}}{dt} = \mathbf{p}, \qquad \frac{d\mathbf{p}}{dt} = -\alpha\frac{\mathbf{r}}{r^3}, \tag{5–3.87}$$

on employing the symmetrized multiplication that is justified by the quadratic dependence of the energy operator on the momentum [cf. Eq. (1–2.51)]. As a constant of the motion, the action of **A** on a wave function of given energy must produce another such wave function. According to a commutation relation that characterizes **p** as a vector,

$$\mathbf{L}\times\mathbf{p} + \mathbf{p}\times\mathbf{L} = 2i\mathbf{p}, \tag{5–3.88}$$

the effect on an *s*-state wave function $[\mathbf{L}\psi_{ns} = 0]$ is

$$\mathbf{A}\psi_{ns}(r) = \frac{\mathbf{r}}{r}\left(1 + a_0\frac{d}{dr}\right)\psi_{ns}(r) \tag{5–3.89}$$

which, as a vector, constitutes a *p*-state wave function. For $n = 1$, where only an *s*-state exists, we conclude that

$$\left(1 + a_0\frac{d}{dr}\right)\psi_{1s}(r) = 0, \tag{5–3.90}$$

which implies the simple exponential function known to represent this state. The radial dependence of the 2*s* wave function [Eq. (4–11.90)],

$$\psi_{2s}(r) \sim \left(1 - \frac{r}{2a_0}\right)\exp(-r/2a_0), \tag{5–3.91}$$

produces the three 2*p* wave functions, which are exhibited as components of a vector:

$$\psi_{2pk}(\mathbf{r}) = \left[\frac{1}{32\pi a_0^3}\right]^{1/2}\frac{\mathbf{r}_k}{a_0}\exp(-r/2a_0); \tag{5–3.92}$$

it has been supplied with the proper normalization constant.

Instead of introducing one of the *p* wave functions into (5–3.71), we shall employ their average,

$$\frac{1}{3}\sum_{k=1}^{3}(\psi_{2pk}(\mathbf{r}))^2 = \frac{1}{96\pi a_0^3}\left(\frac{r}{a_0}\right)^2\exp(-r/a_0) = \frac{1}{\pi(2a_0)^3}\frac{x^2}{3}\exp(-2x), \tag{5–3.93}$$

where the latter form uses the dimensionless variables for $n = 2$. The application of (5–3.75) then gives

$$\langle \delta V'_{>K} \rangle_{2p} = -\frac{4}{\pi}\frac{m}{M}\frac{\alpha^3}{8}\,\mathrm{Ry}\int_{-1}^{1} d\mu(1-\mu^2)\int_{2a_0K}^{\infty}\frac{d\kappa}{\kappa}\int_0^\infty dx$$

$$\times \exp(i\kappa x\mu)\,\frac{d}{dx}\left(\frac{x^2}{3}\exp(-2x)\right). \tag{5–3.94}$$

If we write

$$x^2\exp(-2x) = \left(\frac{d}{d\lambda}\right)^2\exp(-\lambda x)|_{\lambda=2}, \tag{5–3.95}$$

we can make use of the generating function (5–3.83), without the factor $(1-t)^{-2} \times (1-t')^{-2}$. This yields

$$\langle \delta V'_{>K} \rangle_{2p} = \frac{16}{3\pi}\frac{m}{M}\frac{\alpha^3}{8}\,\mathrm{Ry}\,\frac{1}{3}\left(\frac{d}{d\lambda}\right)^2\left(\log\frac{\lambda}{2a_0K}+\frac{4}{3}\right)\Bigg|_{\lambda=2}$$

$$= -\frac{1}{18\pi}\frac{m}{M}\,\alpha^3\,\mathrm{Ry}, \tag{5–3.96}$$

which is combined with (5–3.66) in

$$\langle \delta V' \rangle_{2p} = \frac{2}{3\pi}\frac{m}{M}\,\alpha^3\,\mathrm{Ry}\left[0.0300-\frac{1}{12}\right]. \tag{5–3.97}$$

The magnitude of the effects we have been discussing are indicated by

$$\delta E \sim \frac{\alpha}{m}\frac{\alpha}{M}\,|\psi(0)|^2, \tag{5–3.98}$$

which can also be realized through the exchange of two photons between the two particles. Indeed, we recognize in $\alpha/m$ and $\alpha/M$ the amplitudes for low energy photon scattering by the respective particles [recall the Thomson cross section], which also govern the emission and absorption of two low energy photons. We shall evaluate this effect for the non-relativistic regime. Supplementing the action term of (4–11.30) is

$$-\int (d\mathbf{r})\,dt\,\psi^*(\mathbf{r}t)\,\frac{e^2}{2m}\,(\mathbf{A}(\mathbf{r}t))^2\psi(\mathbf{r}t), \tag{5–3.99}$$

which is extended to the two-particle situation in

$$-\int (d\mathbf{r}_1)(d\mathbf{r}_2)\,dt\,\psi^*(\mathbf{r}_1\mathbf{r}_2t)\left[\frac{e^2}{2m}(\mathbf{A}(\mathbf{r}_1t))^2+\frac{e^2}{2M}(\mathbf{A}(\mathbf{r}_2t))^2\right]\psi(\mathbf{r}_1\mathbf{r}_2t). \tag{5–3.100}$$

It is the latter process, rather than two successive acts governed by the two-particle extension of (4–11.30), that dominates the non-relativistic situation.

Thus, the effective source for the emission of two photons and the composite particle by an extended source of composite particles is

$$J_k(x)J_l(x')\eta(\mathbf{r}_1\mathbf{r}_2t)|_{\text{eff.}}$$

$$= -\,\delta(x-x')\,\delta(x^0-t)\delta_{kl}\left[\frac{e^2}{m}\,\delta(\mathbf{x}-\mathbf{r}_1)+\frac{e^2}{M}\,\delta(\mathbf{x}-\mathbf{r}_2)\right]\psi(\mathbf{r}_1\mathbf{r}_2t), \quad (5\text{–}3.101)$$

with a similar expression involving $\psi^*(\mathbf{r}_1\mathbf{r}_2t)$ for the three-particle absorption process. The vacuum amplitude describing the three-particle exchange between different component particles is

$$i\,\frac{e^2}{m}\,\frac{e^2}{M}\int(d\mathbf{r}_1)\cdots dt'\,\psi^*(\mathbf{r}_1\mathbf{r}_2t)\,\delta_{kl}\tfrac{1}{2}[D(\mathbf{r}_1-\mathbf{r}'_2,t-t')_{km}D(\mathbf{r}_1-\mathbf{r}'_2,t-t')_{ln}$$

$$+\,D(\mathbf{r}_2-\mathbf{r}'_1,t-t')_{km}D(\mathbf{r}_2-\mathbf{r}'_1,t-t')_{ln}]G(\mathbf{r}_1\mathbf{r}_2t,\mathbf{r}'_1\mathbf{r}'_2t')\delta_{mn}\psi(\mathbf{r}'_1\mathbf{r}'_2t'). \quad (5\text{–}3.102)$$

Note that the tensors describing the transverse character of the photon propagation function are combined in

$$\delta_{kl}\left(\delta_{km}-\frac{\mathbf{k}_k\mathbf{k}_m}{\mathbf{k}^2}\right)\left(\delta_{ln}-\frac{\mathbf{k}'_l\mathbf{k}'_n}{\mathbf{k}'^2}\right)\delta_{mn}=1+\frac{(\mathbf{k}\cdot\mathbf{k}')^2}{\mathbf{k}^2\mathbf{k}'^2}, \quad (5\text{–}3.103)$$

which is reminiscent of the angle dependence in low energy photon scattering [Eq. (3–13.118), for example]. We also observe that the presence of the factor $1/M$ enables us to station the heavy particle at the origin, and then infer an addition to the action that has the form (4–11.58), with the interaction

$$\delta V''=\frac{e^2}{m}\,\frac{e^2}{M}\int d\omega_k\,d\omega_{k'}\left(1+\frac{(\mathbf{k}\cdot\mathbf{k}')^2}{\mathbf{k}^2\mathbf{k}'^2}\right)\frac{1}{2}\left[\exp[i(\mathbf{k}+\mathbf{k}')\cdot\mathbf{r}]\,\frac{1}{E-H-k^0-k^{0\prime}}\right.$$

$$\left.+\,\frac{1}{E-H-k^0-k^{0\prime}}\exp[i(\mathbf{k}+\mathbf{k}')\cdot\mathbf{r}]\right]. \quad (5\text{–}3.104)$$

The expectation value of $\delta V''$ in a null eigenstate of $E-H$ is

$$\langle\delta V''\rangle=-\,\frac{e^2}{m}\,\frac{e^2}{M}\int d\omega_k\,d\omega_{k'}\left(1+\frac{(\mathbf{k}\cdot\mathbf{k}')^2}{\mathbf{k}^2\mathbf{k}'^2}\right)\frac{1}{k^0+k^{0\prime}}\int(d\mathbf{r})\,\exp[i(\mathbf{k}+\mathbf{k}')\cdot\mathbf{r}]\,|\psi(\mathbf{r})|^2. \quad (5\text{–}3.105)$$

Consider first an approximation in which the particle momentum is neglected, corresponding to the replacement of $\psi(\mathbf{r})$ by $\psi(0)$. This gives

$$\langle\delta V''\rangle\sim-\,\frac{e^2}{m}\,\frac{e^2}{M}\,|\psi(0)|^2\int d\omega_k\,d\omega_{k'}\left(1+\frac{(\mathbf{k}\cdot\mathbf{k}')^2}{\mathbf{k}^2\mathbf{k}'^2}\right)\frac{1}{k^0+k^{0\prime}}\,(2\pi)^3\,\delta(\mathbf{k}+\mathbf{k}')$$

$$= -2\frac{\alpha^2}{mM}|\psi(0)|^2\int\frac{dk^0}{k^0}. \tag{5–3.106}$$

The photon energy integration appearing here will be terminated by relativistic effects at $k^0 \sim m$, and by the finite momentum associated with the bound state at $k^0 \sim \alpha m$. Hence,

$$\int\frac{dk^0}{k^0} \sim \log\frac{1}{\alpha}, \tag{5–3.107}$$

and we get the rough estimate, for $n = 2$,

$$\langle\delta V''\rangle_{2s} \sim -\frac{1}{2\pi}\frac{m}{M}\alpha^3\log\frac{1}{\alpha}\,\mathrm{Ry}. \tag{5–3.108}$$

The effect is approximately $\frac{3}{4}$ of (5–3.85), and of opposite sign.

Now let us improve this estimate at the low energy end. According to the generating function for $s$-states, Eq. (5–3.76), the coordinate integration of (5–3.105), expressed in dimensionless variables, involves

$$\int(d\mathbf{x})\exp[i(\boldsymbol{\kappa}+\boldsymbol{\kappa}')\cdot\mathbf{x}]\exp(-\lambda x) = 4\pi\left(-\frac{d}{d\lambda}\right)\int_0^\infty dx\exp(-\lambda x)\frac{\sin|\boldsymbol{\kappa}+\boldsymbol{\kappa}'|x}{|\boldsymbol{\kappa}+\boldsymbol{\kappa}'|}$$
$$= 4\pi\left(-\frac{d}{d\lambda}\right)\frac{1}{(\boldsymbol{\kappa}+\boldsymbol{\kappa}')^2+\lambda^2}. \tag{5–3.109}$$

The representation

$$\frac{1}{(\boldsymbol{\kappa}+\boldsymbol{\kappa}')^2+\lambda^2} = \int_0^\infty ds\exp\{-s[(\boldsymbol{\kappa}+\boldsymbol{\kappa}')^2+\lambda^2]\} \tag{5–3.110}$$

then presents (5–3.109) as

$$\int(d\mathbf{x})\exp[i(\boldsymbol{\kappa}+\boldsymbol{\kappa}')\cdot\mathbf{x}]\exp(-\lambda x) = 8\pi\lambda\int_0^\infty ds\,s\exp\{-s[(\boldsymbol{\kappa}+\boldsymbol{\kappa}')^2+\lambda^2]\}. \tag{5–3.111}$$

Using the term in the same sense as in the context of Eq. (5–3.83), the generating function associated with (5–3.105) is given by

$$-\frac{8}{\pi^2}\frac{m}{M}\frac{\alpha^3}{n^3}\mathrm{Ry}\frac{\lambda}{(1-t)^2(1-t')^2}\int_0^\infty d\kappa\,d\kappa'\frac{\kappa\kappa'}{\kappa+\kappa'}\int_{-1}^1 d\mu(1+\mu^2)\int_0^\infty ds\,s$$
$$\times\exp\{-s[\kappa^2+\kappa'^2+2\kappa\kappa'\mu+\lambda^2]\}, \tag{5–3.112}$$

where $\kappa$ and $\kappa'$ indicate the magnitudes of the respective vectors and $\mu$ is the cosine of the angle between them. It is convenient to introduce a change of variables:

$$\kappa = y\,2^{-1/2}(1+v), \qquad \kappa' = y\,2^{-1/2}(1-v),$$
$$-1 < v < 1, \qquad 0 < y < Y. \tag{5-3.113}$$

The upper limit $Y$ expresses the necessity of stopping the photon energy integration before one enters the relativistic region. With these variables, (5-3.112) reads

$$-\frac{2}{\pi}\frac{m}{M}\frac{\alpha^3}{n^3}\,\mathrm{Ry}\,\frac{1}{(1-t)^2(1-t')^2}\left[\frac{2^{3/2}}{\pi}\lambda I\right], \tag{5-3.114}$$

where

$$I = \int_0^Y dy\,y^2 \int_0^1 dv(1-v^2)\int_{-1}^1 d\mu(1+\mu^2)\int_0^\infty ds\,s$$
$$\times \exp\{-s[(1+v^2)y^2 + \mu(1-v^2)y^2 + \lambda^2]\}. \tag{5-3.115}$$

We shall treat the two terms appearing in the rearrangement

$$1+\mu^2 = 2-(1-\mu^2) \tag{5-3.116}$$

in different ways. For the constant term, the integrations over $\mu$, and then $s$, are performed first:

$$I' = 2\int_0^Y dy\,y^2\int_0^1 dv(1-v^2)\int_{-1}^1 d\mu\int_0^\infty ds\,s\exp\{-s[(1+v^2)y^2+\mu(1-v^2)y^2+\lambda^2]\}$$
$$= 2\int_0^Y dy\int_0^1 dv\int_0^\infty ds\exp(-s\lambda^2)\,[\exp(-s2v^2y^2)-\exp(-s2y^2)]$$
$$= 2\int_0^Y dy\int_0^1 dv\left[\frac{1}{2v^2y^2+\lambda^2}-\frac{1}{2y^2+\lambda^2}\right]. \tag{5-3.117}$$

Then the $v$ integration gives

$$I' = 2\int_0^Y dy\left[\frac{1}{2^{1/2}\lambda y}\tan^{-1}\frac{2^{1/2}y}{\lambda}-\frac{1}{2y^2+\lambda^2}\right]$$
$$= \frac{2^{1/2}}{\lambda}\log\left(\frac{2^{1/2}Y}{\lambda}\right)\tan^{-1}\frac{2^{1/2}Y}{\lambda} - 2\int_0^Y dy\,\frac{\log(2^{1/2}y/\lambda)+1}{2y^2+\lambda^2}, \tag{5-3.118}$$

where the latter form is produced by partial integration. The logarithmic dependence on $Y$ is a consequence of the $vy$ combination in (5-3.117) for small values of $v$. Thus, it refers to the situation $\kappa \sim \kappa' \sim 2^{-1/2}y$. In consequence,

$$2^{-1/2}Y = Lna_0 = \frac{nL}{\alpha m}, \tag{5-3.119}$$

where $L$ is an upper limit to photon energies which, in anticipation of a junction with a relativistic calculation, is chosen intermediate between nonrelativistic momenta ($\sim \alpha m$) and relativistic ones ($\sim m$), say,

$$L \sim \alpha^{1/2} m. \tag{5-3.120}$$

The fairly large value of $Y \sim \alpha^{-1/2}$ permits the simplification of (5-3.118) to

$$I' \cong \frac{2^{-1/2}\pi}{\lambda}\left(\log \frac{2^{1/2} Y}{\lambda} - 1\right), \tag{5-3.121}$$

which employs the integral

$$\int_0^\infty dt \, \frac{\log t}{1 + t^2} = 0. \tag{5-3.122}$$

In the second piece of $I$,

$$\begin{aligned} I'' = -\int_0^Y dy\, y^2 \int_0^1 dv (1 - v^2) \int_{-1}^1 d\mu (1 - \mu^2) \int_0^\infty ds\, s \\ \times \exp\{-s[(1 + v^2) y^2 + \mu(1 - v^2) y^2 + \lambda^2]\}, \end{aligned} \tag{5-3.123}$$

the vanishing of $1 - \mu^2$ for $\mu = -1$, which is the high energy situation according to (5-3.106), enables one to replace $Y$ by $\infty$. Here we first perform the $y$ integration, and then the $s$ and $\mu$ integrals:

$$\begin{aligned} I'' &= -\frac{\pi^{1/2}}{4} \int_0^1 dv (1 - v^2) \int_{-1}^1 d\mu (1 - \mu^2) \int_0^\infty ds\, s^{-1/2} \\ &\quad \times \exp(-\lambda^2 s)\, [1 + v^2 + \mu(1 - v^2)]^{-3/2} \\ &= -\frac{\pi}{4\lambda} \int_0^1 dv (1 - v^2) \int_{-1}^1 d\mu (1 - \mu^2) [1 + v^2 + \mu(1 - v^2)]^{-3/2}. \end{aligned} \tag{5-3.124}$$

To aid in the $\mu$ integration, a partial integration is used,

$$\begin{aligned} I'' &= \frac{\pi}{\lambda} \int_0^1 dv \int_{-1}^1 d\mu\, \mu [1 + v^2 + \mu(1 - v^2)]^{-1/2} \\ &= -\frac{2^{3/2}\pi}{\lambda} \frac{1}{3} \int_0^1 dv \, \frac{1 - v}{(1 + v)^2}, \end{aligned} \tag{5-3.125}$$

and, finally,

$$I'' = -\frac{2^{3/2}\pi}{\lambda} \frac{1}{3} (1 - \log 2). \tag{5-3.126}$$

The two parts of $I$ are united in

$$\frac{2^{3/2}}{\pi}\lambda I = 2\left(\log\frac{2^{1/2}Y}{\lambda} - 1\right) - \frac{8}{3}(1 - \log 2). \qquad (5\text{–}3.127)$$

The generating function (5–3.114) is thus found to be

$$-\frac{4}{\pi}\frac{m}{M}\frac{\alpha^3}{n^3}\mathrm{Ry}\frac{1}{(1-t)^2(1-t')^2}\left[\log\frac{2nL}{\lambda\alpha m} - \frac{7}{3} + \frac{4}{3}\log 2\right], \qquad (5\text{–}3.128)$$

and its consequence for $n = 2$ is

$$\langle\delta V''\rangle_{2s} = -\frac{1}{2\pi}\frac{m}{M}\alpha^3\,\mathrm{Ry}\left[\log\frac{2L}{\alpha m} - \frac{37}{12} + \frac{4}{3}\log 2\right]. \qquad (5\text{–}3.129)$$

As in the discussion of (5–3.96), the result for the $2p$ level is obtained by applying the operator $\frac{1}{3}(d/d\lambda)^2$ to the generating function (5–3.128), without the factor $(1-t)^{-2}(1-t')^{-2}$. This gives

$$\begin{aligned}\langle\delta V''\rangle_{2p} &= -\frac{1}{2\pi}\frac{m}{M}\alpha^3\,\mathrm{Ry}\frac{1}{3}\frac{d^2}{d\lambda^2}\log\frac{1}{\lambda}\bigg|_{\lambda=2}\\ &= -\frac{1}{24\pi}\frac{m}{M}\alpha^3\,\mathrm{Ry}. \end{aligned} \qquad (5\text{–}3.130)$$

The principal remaining task is the calculation of the relativistic process that joins with (5–3.129) to remove the dependence upon the parameter $L$. Let us return to the construction of Eq. (5–2.12) where, in the radiation gauge, $D_+(x-x')_{\mu\nu}$ has the two types of components restated by

$$D_+(x-x')_{00} = -\delta(x^0 - x^{0'})\mathscr{D}(\mathbf{x} - \mathbf{x}') \qquad (5\text{–}3.131)$$

and

$$D_+(x-x')_{kl} = \left(\delta_{kl} - \frac{\partial_k\partial_l}{\nabla^2}\right)D_+(x-x'). \qquad (5\text{–}3.132)$$

We have dealt with the instantaneous interaction by devising a differential equation for the two-particle Green's function. The transverse interaction will be handled in a more elementary way, by using a power series expansion for that part of the functional differential operator:

$$\begin{aligned}&\exp\left[-i\int(d\xi)(d\xi')\frac{\delta}{\delta A_{1k}(\xi)}D_+(\xi-\xi')_{kl}\frac{\delta}{\delta A_{2l}(\xi')}\right]\\ &= 1 - i\int(d\xi)(d\xi')\frac{\delta}{\delta A_{1k}(\xi)}D_+(\xi-\xi')_{kl}\frac{\delta}{\delta A_{2l}(\xi')} - \frac{1}{2}\int(d\xi)\cdots(d\xi''')\frac{\delta}{\delta A_{1k}(\xi)}\\ &\quad\times\frac{\delta}{\delta A_{1l}(\xi')}D_+(\xi-\xi'')_{km}D_+(\xi'-\xi''')_{ln}\frac{\delta}{\delta A_{2m}(\xi'')}\frac{\delta}{\delta A_{2n}(\xi''')} + \cdots.\end{aligned} \qquad (5\text{–}3.133)$$

These functional derivatives can be applied directly to the individual particle Green's functions, in accordance with

$$\frac{\delta}{\delta A_k(\xi)} G_+{}^A(x, x') = G_+{}^A(x, \xi) eq\gamma_k G_+{}^A(\xi, x') \qquad (5\text{-}3.134)$$

and

$$\frac{\delta}{\delta A_k(\xi)} \frac{\delta}{\delta A_l(\xi')} G_+{}^A(x, x') = G_+{}^A(x, \xi) eq\gamma_k G_+{}^A(\xi, \xi') eq\gamma_l G_+{}^A(\xi', x')$$
$$+ G_+{}^A(x, \xi') eq\gamma_l G_+{}^A(\xi', \xi) eq\gamma_k G_+{}^A(\xi, x'). \qquad (5\text{-}3.135)$$

This is unnecessarily complicated for the heavy particle, however. Since only effects of order $1/M$ will be retained, it is simpler to replace the non-relativistic reduction of these forms by a direct derivation from the non-relativistic version of the Green's function. A transitional form of the latter that follows from the construction

$$G_+{}^A = \frac{1}{\gamma\Pi + M} = \frac{M - \gamma\Pi}{-(\gamma\Pi)^2 + M^2} \qquad (5\text{-}3.136)$$

is ($p^0 \to M + p^0$)

$$x^0 > x^{0\prime}: \quad G_+{}^A(x, x') \simeq \tfrac{1}{2}(1 + \gamma^0) \exp[-iM(x^0 - x^{0\prime})]$$
$$\times \frac{1}{[(\mathbf{p} - e\mathbf{A})^2/2M] - (p^0 - eA^0)} \delta(\mathbf{x} - \mathbf{x}'), \quad (5\text{-}3.137)$$

referring to a particle of charge $e$. Inasmuch as we are interested in the dependence on $\mathbf{A}$, which already involves the smallness parameter $1/M$, the kinetic energy $\mathbf{p}^2/2M$ can be neglected to give

$$x^0 > x^{0\prime}: \quad G_+{}^A(x, x') \simeq \frac{i}{2}(1 + \gamma^0) \exp[-iM(x^0 - x^{0\prime})] \cdot$$
$$\times \exp\left[-i\int_{x^{0\prime}}^{x^0} dt \left(eA^0 - \frac{e}{M}\mathbf{p}\cdot\mathbf{A} + \frac{e^2\mathbf{A}^2}{2M}\right)\right] \delta(\mathbf{x} - \mathbf{x}'),$$
$$(5\text{-}3.138)$$

which is a generalization of Eq. (5-2.30). The functional derivatives obtained in this way are

$$\frac{\delta}{\delta A_k(\xi)} G_+{}^A(x, x') \simeq i\frac{e}{M} p_k \, \delta(\boldsymbol{\xi} - \mathbf{x}) G_+{}^A(x, x') \qquad (5\text{-}3.139)$$

and

$$\frac{\delta}{\delta A_k(\xi)}\frac{\delta}{\delta A_l(\xi')}G_+{}^A(x,x') \simeq -i\frac{e^2}{M}\delta_{kl}\,\delta(\boldsymbol{\xi}-\boldsymbol{\xi}')\,\delta(\xi^0-\xi^{0\prime})\,\delta(\boldsymbol{\xi}-\mathbf{x})\,G_+{}^A(x-x'), \tag{5-3.140}$$

for the situation where the times $\xi^0$ and $\xi^{0\prime}$ are intermediate between $x^0$ and $x^{0\prime}$; otherwise, the functional derivatives vanish.

It may be helpful to interpose here an illustration of the technique for extracting an energy shift when the Green's function is given by such an expansion. For simplicity we consider a single particle, the Green's function of which has an expansion with the leading terms

$$G_+(x,x') - \int (dy)(dy')G_+(x,y)V(\mathbf{y}\mathbf{y}',y^0-y^{0\prime})G_+(y',x') + \cdots. \tag{5-3.141}$$

The initial Green's function has the eigenfunction representation

$$x^0 > x^{0\prime}:\quad G_+(x,x') = i\sum \psi(\mathbf{x})\exp[-iE(x^0-x^{0\prime})]\,\psi^*(\mathbf{x}')\gamma^0 \tag{5-3.142}$$

with positive frequencies, under the indicated time circumstances, while negative frequencies occur for $x^0 < x^{0\prime}$. Associated with a particular eigenfunction in (5-3.141), as the factor of $i\psi(\mathbf{x})\psi^*(\mathbf{x}')\gamma^0$, is $(x^0 - x^{0\prime} = T)$

$$\exp(-iET) - i\exp(-iET)\int dy^0\,dy^{0\prime}\,V(y^0-y^{0\prime})\exp[iE(y^0-y^{0\prime})], \tag{5-3.143}$$

where

$$V(y^0-y^{0\prime}) = \int (d\mathbf{y})(d\mathbf{y}')\psi^*(\mathbf{y})\gamma^0 V(\mathbf{y}\mathbf{y}',y^0-y^{0\prime})\psi(\mathbf{y}'), \tag{5-3.144}$$

and we have only exhibited the contribution associated with the time domain

$$x^0 > y^0,\qquad y^{0\prime} > x^{0\prime}. \tag{5-3.145}$$

The reason for that becomes apparent on giving the integral of (5-3.143) an asymptotic evaluation in which the microscopic time variable $t = y^0 - y^{0\prime}$ effectively ranges from $-\infty$ to $\infty$, while the remaining time variable $y^0 \simeq y^{0\prime}$ covers the interval of duration $T$. This gives for (5-3.143):

$$\exp(-iET)\,[1 - i\,\delta ET] \cong \exp[-i(E+\delta E)T], \tag{5-3.146}$$

with

$$\delta E = \int_{-\infty}^{\infty} dt\, V(t)\exp(iEt). \tag{5-3.147}$$

In contrast to the secular variation exhibited in (5-3.146), the significant time intervals in the regions $y^0 > x^0$ and $y^{0\prime} < x^{0\prime}$ are microscopic and do not contribute to the energy displacement formula (5–3.147), which is the counterpart of Eq. (5–3.44).

Another useful observation follows from the remark that

$$\frac{1}{i}\,\partial_\mu[G_+^A(x,\bar{x})\gamma^\mu G_+^A(\bar{x},x')] = [\delta(\bar{x}-x')-\delta(x-\bar{x})]G_+^A(x,x'), \qquad (5\text{–}3.148)$$

which is a version of the divergence equation [cf. Eq. (3–6.48)]

$$\partial_\mu[\tfrac{1}{2}\psi(x)\gamma^0\gamma^\mu eq\psi(x)] = i\psi(x)\gamma^0 eq\eta(x). \qquad (5\text{–}3.149)$$

Consider the situation with $x^0 > x^{0\prime}$, and let $\bar{x}^0$ be such that

$$x^0 > \bar{x}^0 > x^{0\prime}. \qquad (5\text{–}3.150)$$

Now perform a space-time integration over the semi-infinite region with time values less than $\bar{x}^0$. This gives

$$\frac{1}{i}\int(d\bar{\mathbf{x}})G_+^A(x,\bar{x})\gamma^0 G_+^A(\bar{x},x') = G_+^A(x,x'), \qquad (5\text{–}3.151)$$

which is a multiplicative composition property for the function $(1/i)G_+^A\gamma^0$. This is an elementary statement for free particles; here is a generalization to arbitrary electromagnetic fields. Note, incidentally, that if both inequalities in (5–3.150) are reversed, a minus sign appears on the right side of (5–3.151). If only one inequality is reversed, the integral of (5–3.151) vanishes.

The factorization of the heavy particle Green's function that is exhibited in Eqs. (5–3.139, 140) enables one to consider an effective single particle Green's function for the light particle. This elimination of the two-particle aspect also involves the restriction to relative motion, according to which the heavy particle momentum in (5–3.139) is replaced by the negative of the light particle momentum at the same time. To do that we make explicit the heavy particle emission or absorption time, using the analysis of Eq. (5–3.151). With the position of the heavy particle adopted as spatial origin, the effective change of the light particle Green's function that is associated with single photon exchange is

$$\begin{aligned}-i\frac{e^2}{M}\int(d\xi)(d\xi')D_+(\boldsymbol{\xi},\xi^0-\xi^{0\prime})_{kl}\exp[\ ]\{&\eta(\xi^0-\xi^{0\prime})G_+^A(x,\xi)\gamma_k G_+^A(\xi,\xi')\gamma^0\\ &\times p_l G_+^A(\xi',x')+\eta(\xi^{0\prime}-\xi^0)G_+^A(x,\xi')\gamma^0 p_l G_+^A(\xi',\xi)\\ &\times\gamma_k G_+^A(\xi,x')\}\exp\left(-i\int dt\,eA^0\right)\Big|_{A^0=0},\end{aligned} \qquad (5\text{–}3.152)$$

where exp[ ] indicates the instantaneous interaction part of the functional operator in (5–2.12). The effect of the latter is most simply described if one lets the time span of the heavy particle, the range of time integration in the last factor of (5–3.152), be large compared to $x^0 - x^{0\prime} = T$ while completely including this interval. With this elimination of end effects, the functional operator simply replaces $eqA^0$ in $G_+$ by the static Coulomb potential. Comparison with the structure (5–3.141) then gives the energy shift formula (somewhat different notation is used, and $D_{+kl}$ is made explicit)

$$\delta E = -\frac{e^2}{M}\int d\omega_k\left(\delta_{lm} - \frac{k_l k_m}{\mathbf{k}^2}\right)\int_0^\infty dt\, \exp[i(E-k^0)t]\, \langle \exp(i\mathbf{k}\cdot\mathbf{x})\, \alpha_l G_+(t)\gamma^0 p_m + p_m G_+(t)\gamma^0\alpha_l \exp(i\mathbf{k}\cdot\mathbf{x})\rangle. \quad (5\text{–}3.153)$$

Now if we introduce the unitary transformation of (5–2.79) and exploit the essentially non-relativistic nature of the system, according to

$$\alpha_l \to p_l/m \quad (5\text{–}3.154)$$

and [cf. Eqs. (5–3.137, 138)]

$$\int_0^\infty dt\, \exp[i(E-k^0)t]\, G_+(t)\gamma^0 \to \int_0^\infty dt\, \exp[i(E_{\text{non-rel.}} - k^0)t]\, i\exp(-iHt) = -\frac{1}{E_{\text{non-rel.}} - k^0 - H}, \quad (5\text{–}3.155)$$

the resulting energy shift formula is just that implied by the consequence (5–3.60) of the completely non-relativistic discussion.

Two-photon exchange produces the following change in the propagation function:

$$i\frac{(e^2)^2}{M}\int (d\xi)\cdots(d\xi''')D_+(\xi-\xi'')_{km}D_+(\xi'-\xi''')_{ln}\exp[\;]\,G_+^A(x,\xi)\gamma_k G_+^A(\xi,\xi') \times \gamma_l G_+^A(\xi',x')\,\delta_{mn}\,\delta(\xi''-\xi''')\,\delta(\boldsymbol{\xi}'')\exp\left(-i\int dt\, eA^0\right)\Bigg|_{A^0=0}. \quad (5\text{–}3.156)$$

There is an integration here over the common time variable $\xi^{0\prime\prime} = \xi^{0\prime\prime\prime}$. If we confine our attention to the relativistic domain where the momentum associated with the wave function is negligible, and the function $G_+(\xi, \xi')$ can be approximated as the translationally invariant free particle function, we then effectively encounter the four-dimensional integral

$$\int (d\xi'')D_+(\xi'-\xi'')_{km}D_+(\xi'-\xi'')_{lm} = \int \frac{(dk)}{(2\pi)^4}\frac{1}{(k^2-i\varepsilon)^2}\left(\delta_{kl} - \frac{\mathbf{k}_k\mathbf{k}_l}{\mathbf{k}^2}\right). \quad (5\text{–}3.157)$$

The energy shift formula is ($E \simeq m$)

$$\frac{(4\pi\alpha)^2}{M}\psi(0)^*\frac{1}{i}\int\frac{(dk)}{(2\pi)^4}\frac{1}{(k^2-i\varepsilon)^2}\left(\delta_{kl}-\frac{\mathbf{k}_k\mathbf{k}_l}{\mathbf{k}^2}\right)\gamma_k\frac{1}{-\boldsymbol{\gamma}\cdot\mathbf{k}+m-\gamma^0(m-k^0)}\gamma_l\psi(0), \tag{5-3.158}$$

where

$$\frac{1}{-\boldsymbol{\gamma}\cdot\mathbf{k}+m-\gamma^0(m-k^0)}=\frac{m+\boldsymbol{\gamma}\cdot\mathbf{k}+\gamma^0(m-k^0)}{k^2+2mk^0-i\varepsilon}. \tag{5-3.159}$$

Since $\psi(0)$ is an eigenvector of $\gamma^0$ with $\gamma^{0\prime} = +1$, the expectation value of any odd power of $\boldsymbol{\gamma}$ vanishes, so that $\boldsymbol{\gamma}\cdot\mathbf{k}$ can be omitted, while $\gamma^0$ in (5-3.159), which acts on $\gamma_l\psi(0)$, can be set equal to $-1$. That reduces (5-3.158) to

$$-\frac{(4\pi\alpha)^2}{mM}|\psi(0)|^2\frac{1}{i}\int\frac{(dk)}{(2\pi)^4}\frac{1}{(k^2-i\varepsilon)^2}\frac{2mk^0}{k^2+2mk^0-i\varepsilon}. \tag{5-3.160}$$

The frequency integral appearing here can be evaluated by contour integration,

$$\frac{1}{i}\int_{-\infty}^{\infty}\frac{dk_0}{2\pi}\left[\frac{1}{(k^2-i\varepsilon)^2}-\frac{1}{k^2-i\varepsilon}\frac{1}{k^2+2mk^0-i\varepsilon}\right]=\frac{1}{4}\left[\frac{1}{|\mathbf{k}|^3}-\frac{1}{\mathbf{k}^2(\mathbf{k}^2+m^2)^{1/2}}\right]. \tag{5-3.161}$$

In order to join with the non-relativistic calculation, the subsequent three-dimensional momentum integral will be stopped at the lower limit $|\mathbf{k}| = L \ll m$. This gives the partial energy shift

$$-\frac{2\alpha^2}{mM}|\psi(0)|^2\log\frac{m}{2L}=-\frac{4}{\pi}\frac{m}{M}\frac{\alpha^3}{n^3}\,\mathrm{Ry}\log\frac{m}{2L}, \tag{5-3.162}$$

which unites with the generating function (5-3.128) or with the explicit result (5-3.129), for $n = 2$, to effectively replace $L$ by $\frac{1}{2}m$. Thus,

$$\delta E''_{2s}=-\frac{1}{2\pi}\frac{m}{M}\alpha^3\,\mathrm{Ry}\left[\log\frac{1}{\alpha}-\frac{37}{12}+\frac{4}{3}\log 2\right]. \tag{5-3.163}$$

The several contributions to the energy shift of the $2s$ level, listed in Eqs. (5-3.56, 85, 163) are combined in

$$\delta E_{2s}=\frac{2}{3\pi}\frac{m}{M}\alpha^3\,\mathrm{Ry}\left[\frac{1}{4}\log\frac{1}{\alpha}-2.8118-\frac{1}{2}+\frac{25}{12}+\frac{37}{16}\right], \tag{5-3.164}$$

while the $2p$ displacements listed in Eqs. (5-3.97, 130) give

$$\delta E_{2p}=\frac{2}{3\pi}\frac{m}{M}\alpha^3\,\mathrm{Ry}\left[0.0300-\frac{1}{12}-\frac{1}{16}\right]. \tag{5-3.165}$$

The unit that appears in these results is, for hydrogen,

$$\text{H:}\quad \frac{2}{3\pi}\frac{m}{M}\alpha^3\,\text{Ry} = 2\,\frac{135.64}{1836.1}\,\text{MHz} = 0.148\ \text{MHz}, \tag{5–3.166}$$

and

$$\delta E_{2s} = 0.342\ \text{MHz}, \qquad \delta E_{2p} = -\,0.017\ \text{MHz}. \tag{5–3.167}$$

The additional relative displacement of 0.36 MHz adds to the last theoretical estimate of (4–17.132) to produce the new value:

$$\text{H:}\quad E_{2s_{1/2}} - E_{2p_{1/2}} = 1058.17\ \text{MHz}. \tag{5–3.168}$$

This time we have somewhat overshot the nominal experimental result.

## 5-4 PHOTON PROPAGATION FUNCTION II

The treatment of the modified photon propagation function given in Section 4–3 involved the exchange of a non-interacting pair of oppositely charged particles between extended photon sources that are in causal array. We shall now consider the next dynamical level for this process which, for example, recognizes the possibility of interaction between the particles as produced by the exchange of a virtual photon. Other mechanisms must also be taken into account at this dynamical level, as one can see from several points of view. Thus, any process employing a virtual photon can have a counterpart employing a real photon. In this causal arrangement it arises through the possibility of emitting, not a pair of real particles, but a real and a virtual particle. The latter radiates a photon to produce a three-particle emission act. The corresponding absorption process takes place by having the photon combine with a charged particle to produce a virtual particle, where the latter, together with the other real particle, is subsequently detected by the extended photon source. There are two possibilities here, however, since the photon can either be absorbed by the same particle that earlier emitted it, or by the other particle. It is the second possibility, involving a photon exchanged between the oppositely charged particles, that is the counterpart of the scattering, or virtual photon exchange, process. Alternatively, we can remark that mechanisms must exist to introduce the appropriate modifications in the various elements of the two-particle exchange act. These are: the primitive interaction that interconverts a virtual photon with a particle pair, and the particle propagation functions. We recognize, in the particular three-particle process where one of the particles does not share in the exchange of the photon, just the mechanism for the modification in the particle propagation functions. And the scattering process is the one that introduces the form factor modification of the primitive interaction. Since the latter occurs at both the emission and absorption ends of the process, the scattering mechanism, with appropriate causal controls, must serve both functions.

The three-particle exchange process is a new feature of such problems, and we discuss it first. As an introduction, let us consider the three-particle kinematical integral that is the generalization of (4–1.23),

$$I(M, m_a, m_b, m_c) = (2\pi)^3 \int d\omega_a\, d\omega_b\, d\omega_c\, \delta(P - p_a - p_b - p_c). \qquad (5\text{–}4.1)$$

One elementary procedure first performs the integrations that unite two particles into a composite system of variable mass $M'$, and then deals with the effective two-particle system that remains. This point of view is conveyed by writing

$$\delta(P - p_a - p_b - p_c) = \int \delta(P - P' - p_c)\, d\omega_{P'}\, dM'^2\, (2\pi)^3\, \delta(P' - p_a - p_b). \qquad (5\text{–}4.2)$$

Carrying out the successive integrations for two particle systems gives

$$\begin{aligned} I(M, m_a, m_b, m_c) &= \int dM'^2\, I(M', m_a, m_b)(2\pi)^3 \int d\omega_{P'}\, d\omega_c\, \delta(P - P' - p_c) \\ &= \int dM'^2\, I(M', m_a, m_b) I(M, M', m_c). \end{aligned} \qquad (5\text{–}4.3)$$

The simplest example of the final integration over $M'$ occurs when the individual particle masses are zero or, equivalently, under the circumstances $M \gg m_a, m_b, m_c$. Then (5–4.3) reduces to

$$I(M, 0, 0, 0) = \int_0^{M^2} dM'^2 \frac{1}{(4\pi)^2} \frac{1}{(4\pi)^2} \left(1 - \frac{M'^2}{M^2}\right) = \frac{1}{(4\pi)^4} \frac{M^2}{2}. \qquad (5\text{–}4.4)$$

Next in simplicity is the situation where only one mass differs from zero:

$$\begin{aligned} I(M, m, 0, 0) &= \int_{m^2}^{M^2} dM'^2 \frac{1}{(4\pi)^2} \left(1 - \frac{m^2}{M'^2}\right) \frac{1}{(4\pi)^2} \left(1 - \frac{M'^2}{M^2}\right) \\ &= \frac{1}{(4\pi)^4} \left[\frac{(M^2 + m^2)(M^2 - m^2)}{2M^2} - m^2 \log \frac{M^2}{m^2}\right]. \end{aligned} \qquad (5\text{–}4.5)$$

For the system of present interest, where $m_a = m_b = m$, $m_c = 0$, we have

$$\begin{aligned} I(M, m, m, 0) &= \int_{(2m)^2}^{M^2} dM'^2 \frac{1}{(4\pi)^2} \left(1 - \frac{4m^2}{M'^2}\right)^{1/2} \frac{1}{(4\pi)^2} \left(1 - \frac{M'^2}{M^2}\right) \\ &= \frac{8m^2}{(4\pi)^4} \int_0^{v_0} dv\, \frac{v^2(v_0^2 - v^2)}{(1 - v^2)^3} \\ &= \frac{2m^2}{(4\pi)^4} \left[\frac{\frac{3}{2}v_0 - \frac{1}{2}v_0^3}{1 - v_0^2} - \frac{3 + v_0^2}{4} \log \frac{1 + v_0}{1 - v_0}\right], \end{aligned} \qquad (5\text{–}4.6)$$

which evaluation uses the variable

$$v = \left(1 - \frac{4m^2}{M'^2}\right)^{1/2}, \tag{5–4.7}$$

with

$$v_0 = \left(1 - \frac{4m^2}{M^2}\right)^{1/2}. \tag{5–4.8}$$

The asymptotic behavior for $M^2 \gg (2m)^2$, $v_0 \sim 1$, is indeed given by (5–4.4), while near the threshold, $M^2 \gtrsim (2m)^2$, $v_0 \ll 1$, we find

$$M^2 \gtrsim (2m)^2:\quad I(M, m, m, 0) \simeq \frac{1}{(4\pi)^4}\frac{4}{15}\frac{(M^2 - 4m^2)^{5/2}}{(2m)^3}. \tag{5–4.9}$$

A treatment that is more symmetrical among the particles can be supplied by using an infinite momentum frame, generalizing the discussion that led to Eq. (4–1.32). We note the invariant momentum space element (4–1.28), and the delta function expression that is the three-particle counterpart of (4–1.30):

$$\delta(P - p_a - p_b - p_c) = \delta(\mathbf{p}_{aT} + \mathbf{p}_{bT} + \mathbf{p}_{cT})\,\delta(1 - u_a - u_b - u_c)\,\eta(u_a)\eta(u_b)\eta(u_c)$$
$$\times\, 2\delta\left(\frac{\mathbf{p}_{aT}^2 + m_a^{\,2}}{u_a} + \frac{\mathbf{p}_{bT}^2 + m_b^{\,2}}{u_b} + \frac{\mathbf{p}_{cT}^2 + m_c^{\,2}}{u_c} - M^2\right). \tag{5–4.10}$$

In a temporarily unsymmetrical performance of the transverse momenta integrations, we use the relevant delta function of (5–4.10) to eliminate $\mathbf{p}_{cT}$, thereby producing in the energy delta function the quadratic form

$$\left(\frac{1}{u_a} + \frac{1}{u_c}\right)\mathbf{p}_{aT}^2 + \left(\frac{1}{u_b} + \frac{1}{u_c}\right)\mathbf{p}_{bT}^2 + \frac{2}{u_c}\mathbf{p}_{aT}\cdot\mathbf{p}_{bT} = \lambda_1\mathbf{p}_{1T}^2 + \lambda_2\mathbf{p}_{2T}^2. \tag{5–4.11}$$

We have also indicated here the possibility of diagonalizing the quadratic form by an orthogonal transformation, which is such that

$$(d\mathbf{p}_{aT})(d\mathbf{p}_{bT}) = (d\mathbf{p}_{1T})(d\mathbf{p}_{2T}) \to \pi\, d\mathbf{p}_{1T}^2\, \pi\, d\mathbf{p}_{2T}^2. \tag{5–4.12}$$

The latter form exploits the rotational symmetries of the integrand, which now appears as

$$\frac{1}{4}\frac{1}{(2\pi)^6}\pi^2 \int du_a\, du_b\, du_c\, d\mathbf{p}_{1T}^2\, d\mathbf{p}_{2T}^2\, \frac{1}{u_a u_b u_c}\,\delta(1 - u_a - u_b - u_c)$$
$$\times\, \delta\left(\lambda_1\mathbf{p}_{1T}^2 + \lambda_2\mathbf{p}_{2T}^2 - \left(M^2 - \frac{m_a^{\,2}}{u_a} - \frac{m_b^{\,2}}{u_b} - \frac{m_c^{\,2}}{u_c}\right)\right), \tag{5–4.13}$$

where it is understood that the $u$-parameters are positive. The spectral restrictions are evident here, in parametric form:

$$M^2 > \sum \frac{m_\kappa^2}{u_\kappa}, \qquad \sum u_\kappa = 1. \tag{5–4.14}$$

The anticipated threshold mass

$$M_0 = \sum m_\kappa \tag{5–4.15}$$

can be used to rewrite (5–4.14) as

$$M^2 - M_0^2 > \sum u_\kappa \left(\frac{m_\kappa}{u_\kappa} - M_0\right)^2 \geqslant 0, \tag{5–4.16}$$

and the possibility of attaining the indicated lower limit, with

$$u_\kappa = \frac{m_\kappa}{M_0}, \qquad \sum u_\kappa = 1, \tag{5–4.17}$$

confirms the significance of $M_0$. As the generality of the notation indicates, these considerations hold for any number of particles.

The remaining momentum integrations in (5–4.13) are performed with appropriate variable changes,

$$\begin{aligned}
&\int d\mathbf{p}_{1T}^2\, d\mathbf{p}_{2T}^2\, \delta\left(\lambda_1 \mathbf{p}_{1T}^2 + \lambda_2 \mathbf{p}_{2T}^2 - \left(M^2 - \sum \frac{m_\kappa^2}{u_\kappa}\right)\right) \\
&= \frac{1}{\lambda_1\lambda_2} \int dx_1\, dx_2\, \delta\left(x_1 + x_2 - \left(M^2 - \sum \frac{m_\kappa^2}{u_\kappa}\right)\right) \\
&= \frac{1}{\lambda_1\lambda_2} \int_0^\infty dx_1\, \eta\left(M^2 - \sum \frac{m_\kappa^2}{u_\kappa} - x_1\right) = \frac{1}{\lambda_1\lambda_2}\left(M^2 - \sum \frac{m_\kappa^2}{u_\kappa}\right),
\end{aligned} \tag{5–4.18}$$

under the circumstances of (5–4.14). The product $\lambda_1\lambda_2$ is the determinant of the quadratic form (5–4.11),

$$\lambda_1\lambda_2 = \left(\frac{1}{u_a} + \frac{1}{u_c}\right)\left(\frac{1}{u_b} + \frac{1}{u_c}\right) - \frac{1}{u_c^2} = \frac{1}{u_a u_b} + \frac{1}{u_a u_c} + \frac{1}{u_b u_c} = \frac{1}{u_a u_b u_c}. \tag{5–4.19}$$

This gives

$$\begin{aligned}
&I(M, m_a, m_b, m_c) \\
&= \frac{1}{(4\pi)^4} \int du_a\, du_b\, du_c\, \delta(1 - u_a - u_b - u_c) \left[M^2 - \frac{m_a^2}{u_a} - \frac{m_b^2}{u_b} - \frac{m_c^2}{u_c}\right],
\end{aligned} \tag{5–4.20}$$

where the integration domain is restricted by the delta function and the requirement of (5–4.14). The high energy limit (5–4.4) is obtained directly:

$$\int du_a\,du_b\,du_c\,\delta(1-u_a-u_b-u_c) = \int du_a\,du_b\,\eta(1-u_a-u_b) = \int_0^1 du_a\,(1-u_a) = \tfrac{1}{2}. \tag{5-4.21}$$

For the situation of two vanishing masses we omit the final integration of (5–4.21),

$$\int du_b\,du_c\,\delta(1-u_a-u_b-u_c) = 1-u_a. \tag{5-4.22}$$

The resulting single parameter integral,

$$I(M,m,0,0) = \frac{1}{(4\pi)^4}\int_{m^2/M^2}^1 du(1-u)\left(M^2-\frac{m^2}{u}\right), \tag{5-4.23}$$

is equivalent to (5–4.5). Turning to the mass assignments $m_a = m_b = m$, $m_c = 0$, we eliminate one parameter to get

$$I(M,m,m,0) = \frac{1}{(4\pi)^4}\int du_a\,du_b\,\eta(1-u_a-u_b)\left[M^2-m^2\left(\frac{1}{u_a}+\frac{1}{u_b}\right)\right]. \tag{5-4.24}$$

It is convenient now to introduce new variables:

$$u_a = u\tfrac{1}{2}(1+v), \qquad u_b = u\tfrac{1}{2}(1-v), \qquad du_a\,du_b = du\,u\,\tfrac{1}{2}dv, \tag{5-4.25}$$

which present $(4\pi)^4 I(M,m,m,0)$ as

$$\int du\,u\,dv\left(M^2-4m^2\frac{1}{u}\frac{1}{1-v^2}\right). \tag{5-4.26}$$

Both variables here range between 0 and 1, subject to the restriction

$$\frac{M^2}{4m^2} > \frac{1}{u}\frac{1}{1-v^2}, \tag{5-4.27}$$

or

$$u > \frac{1-v_0^2}{1-v^2}, \tag{5-4.28}$$

on introducing the definition (5–4.8). Since $u$ cannot exceed unity, it is necessary that

$$v < v_0. \tag{5-4.29}$$

Carrying out the $u$ integration then yields

$$I(M,m,m,0) = \frac{1}{(4\pi)^4}\frac{M^2}{2}\int_0^{v_0} dv\left(\frac{v_0^2-v^2}{1-v^2}\right)^2, \tag{5-4.30}$$

which is reminiscent of the parametric integral in (5–4.6), but differs from it in detail. In fact, the two versions are equivalent, but (5–4.30) leads somewhat more directly to the high and low energy limiting forms.

The process of emitting two particles and a photon from an extended photon source is represented by a coupling among two particle sources and two photon sources. As such, it is another application of the interaction $W_{22}$ which also describes photon-particle scattering [cf. Eq. (3–12.29)]:

$$W_{22} = \tfrac{1}{2}\int (dx)(dx')\phi(x)[2eqp.A(x)\,\Delta_+(x-x')\,2eqp.A(x') - \delta(x-x')\,e^2A^2(x)]\phi(x'). \tag{5–4.31}$$

The part of the vacuum amplitude $iW_{22}$ that is linear in the field of the extended source, $A_2$, and in the field of the emitted photon, is compared with the equivalent three-particle vacuum amplitude

$$\tfrac{1}{2}\left[i\int (dx)K(x)\phi(x)\right]^2 i\int (d\xi)J^\lambda(\xi)A_\lambda(\xi) \tag{5–4.32}$$

to give the effective emission source

$$\begin{aligned}-K_2(x)K_2(x')J_2{}^\lambda(\xi)|_{\text{eff.}} &= 2eqp^\lambda.\delta(x-\xi)\,\Delta_+(x-x')\,2eqp.A_2(x') \\ &\quad + 2eqp.A_2(x)\,\Delta_+(x-x')\,2eqp^\lambda.\delta(x'-\xi) \\ &\quad - \delta(x-x')\,\delta(x-\xi)\,2e^2A_2{}^\lambda(x).\end{aligned} \tag{5–4.33}$$

The momentum version is

$$-K_2(p)K_2(p')J_2{}^\lambda(k)|_{\text{eff.}} = 2e^2V_2{}^{\lambda\nu}A_{2\nu}(K), \tag{5–4.34}$$

where

$$V_2^{\lambda\nu} = \frac{1}{4}\frac{(2p+k)^\lambda(p+k-p')^\nu}{pk} + \frac{1}{4}\frac{(2p'+k)^\lambda(p'+k-p)^\nu}{p'k} - g^{\lambda\nu} \tag{5–4.35}$$

has been simplified by introducing the real particle properties

$$p^2 + m^2 = p'^2 + m^2 = k^2 = 0, \tag{5–4.36}$$

and

$$K = p + p' + k \tag{5–4.37}$$

is the total momentum emitted by the source. We note the conservation and gauge invariance statements

$$k_\lambda V_2^{\lambda\nu} = 0, \qquad V_2^{\lambda\nu}K_\nu = 0. \tag{5–4.38}$$

The corresponding absorption source is

$$-K_1(-p')K_1(-p)J_1{}^\lambda(-k)|_{\text{eff.}} = 2e^2A_{1\mu}(-K)V_1^{\mu\lambda}, \tag{5-4.39}$$

in which

$$V_1^{\mu\lambda} = V_2^{\lambda\mu}. \tag{5-4.40}$$

Incidentally, the unit matrix in the charge space is implicit in the latter objects.

The vacuum amplitude for the three-particle exchange process is

$$-\tfrac{1}{2}\int d\omega_p\, d\omega_{p'}\, d\omega_k\, \text{tr}[K_1(-p')K_1(-p)J_1{}^\lambda(-k)|_{\text{eff.}}K_2(p)K_2(p')J_{2\lambda}(k)|_{\text{eff.}}] \tag{5-4.41}$$

where the trace acts in charge space, and supplies a factor of 2 in the resulting expression

$$-(2e^2)^2\int dM^2\, d\omega_K\, A_1{}^\mu(-K)I_{\mu\nu}(K)A_2{}^\nu(K), \tag{5-4.42}$$

which collects all the internal workings in the tensor

$$I_{\mu\nu}(K) = \int d\omega_p\, d\omega_{p'}\, d\omega_k\, (2\pi)^3\, \delta(K-p-p'-k)V_{1\mu\lambda}V_2{}^\lambda{}_\nu. \tag{5-4.43}$$

According to the relations (5-4.38, 40), $I_{\mu\nu}$ is symmetrical in $\mu$ and $\nu$, and obeys the condition of gauge invariance

$$K^\mu I_{\mu\nu}(K) = 0. \tag{5-4.44}$$

This specifies the tensor structure:

$$I_{\mu\nu}(K) = \left(g_{\mu\nu} + \frac{K_\mu K_\nu}{M^2}\right)I(M^2), \tag{5-4.45}$$

where the scalar $I(M^2)$ is

$$I(M^2) = \tfrac{1}{3}\int d\omega_p\, d\omega_{p'}\, d\omega_k\, (2\pi)^3\, \delta(K-p-p'-k)V_2^{\lambda\nu}V_{2\lambda\nu}. \tag{5-4.46}$$

The tensor $V_2^{\lambda\nu}$ can be exhibited as follows:

$$V_2^{\lambda\nu} = \frac{1}{2}\left(\frac{p}{pk} - \frac{p'}{p'k}\right)^\lambda (p-p')^\nu + \frac{1}{2}\left(\frac{p}{pk} + \frac{p'}{p'k}\right)^\lambda k^\nu - g^{\lambda\nu}$$
$$+ \frac{1}{4}k^\lambda\left[\frac{(p+k-p')^\nu}{pk} + \frac{(p'+k-p)^\nu}{p'k}\right]. \tag{5-4.47}$$

There are three sets of terms here, each of which vanishes on multiplication by $k_\lambda$.

Consequently, the one having $k^\lambda$ as a factor does not contribute to the requi[red] product,

$$V_2^{\lambda\nu}V_{2\lambda\nu} = \frac{1}{4}\left(\frac{p}{pk} - \frac{p'}{p'k}\right)^2 (p-p')^2 - \frac{1}{2}\left(\frac{1}{pk} + \frac{1}{p'k}\right)(p-p')^2$$
$$- \frac{1}{2}m^2\left(\frac{1}{(pk)^2} - \frac{1}{(p'k)^2}\right)k(p-p') + 2. \qquad (5\text{–}4.48)$$

In

$$\left(\frac{p}{pk} - \frac{p'}{p'k}\right)^2 = -\frac{m^2}{(pk)^2} - \frac{m^2}{(p'k)^2} - \frac{2pp'}{pkp'k} \qquad (5\text{–}4.49)$$

we recognize the multiplicative structure that dominates soft photon emission. It has often been encountered in describing the deflection of a particle. Here it refers to the creation of an oppositely charged pair of particles. One can use the relation

$$M^2 = -(p + p' + k)^2 = 2m^2 - 2(pp' + pk + p'k) \qquad (5\text{–}4.50)$$

to combine the first two terms on the right side of (5–4.48),

$$V_2^{\lambda\nu}V_{2\lambda\nu} = \frac{1}{4}\left[-\frac{m^2}{(pk)^2} - \frac{m^2}{(p'k)^2} + \frac{M^2 - 2m^2}{pkp'k}\right](p-p')^2$$
$$- \frac{1}{2}m^2\left(\frac{1}{(pk)^2} - \frac{1}{(p'k)^2}\right)k(p-p') + 2. \qquad (5\text{–}4.51)$$

We shall carry out the integration in the manner of Eq. (5–4.3), first grouping the particles into a composite of mass $M'$:

$$\delta(K - p - p' - k) = \int \delta(K - P - k)\, d\omega_P\, dM'^2\, (2\pi)^3\, \delta(P - p - p'). \qquad (5\text{–}4.52)$$

The integration over the particles, in (5–4.46), now produces the scalar function

$$S(M'^2, M^2) = \int d\omega_p\, d\omega_{p'}\, (2\pi)^3\, \delta(P - p - p')V_2^{\lambda\nu}V_{2\lambda\nu} \qquad (5\text{–}4.53)$$

and the remaining kinematical integral then gives

$$I(M^2) = \frac{1}{3}\int dM'^2 \frac{1}{(4\pi)^2}\left(1 - \frac{M'^2}{M^2}\right)S(M'^2, M^2), \qquad (5\text{–}4.54)$$

although, as $M' \to M$, we must modify the indicated kinematical factor to take account of the fictitious photon mass $\mu$,

$$-\frac{M'^2}{M^2} \to \frac{1}{M^2}[(M^2 - M'^2)^2 - 4\mu^2 M'^2]^{1/2} \cong \frac{2}{M}[(M - M')^2 - \mu^2]^{1/2}. \tag{5-4.55}$$

The integration of (5-4.53), as expressed by

$$S(M'^2, M^2) = \frac{1}{(4\pi)^2}\left(1 - \frac{4m^2}{M'^2}\right)^{1/2} \langle V_2^{\lambda\nu} V_{2\lambda\nu}\rangle, \tag{5-4.56}$$

is performed in the rest frame of $P$. Some invariant expressions for quantities in this coordinate system are:

$$k^0 = \frac{M^2 - M'^2}{2M'}, \qquad |\mathbf{k}| = \frac{1}{2M'}[(M^2 - M'^2)^2 - 4\mu^2 M'^2]^{1/2},$$

$$p^0 = p^{0\prime} = \tfrac{1}{2}M', \qquad |\mathbf{p}| = |\mathbf{p}'| = \tfrac{1}{2}(M'^2 - 4m^2)^{1/2}, \tag{5-4.57}$$

and, there is also the invariant

$$(p - p')^2 = M'^2 - 4m^2. \tag{5-4.58}$$

The infra-red sensitive integrals are

$$\left\langle\frac{1}{(pk)^2}\right\rangle = \left\langle\frac{1}{(p'k)^2}\right\rangle = \int_{-1}^{1}\frac{1}{2}dz\,\frac{1}{(p^0k^0 - |\mathbf{p}|\,|\mathbf{k}|z)^2}$$

$$= \frac{1}{m^2[(M^2 - M'^2)/2M']^2 + \mu^2[(M'^2/4) - m^2]}$$

$$= \begin{cases} M - M' \gg \mu: & \dfrac{4}{(M^2 - M'^2)^2}\dfrac{M'^2}{m^2} \\ M - M' \sim \mu: & \dfrac{1}{m^2}\dfrac{1}{(M - M')^2 + \mu^2[(M^2/4m^2) - 1]} \end{cases}, \tag{5-4.59}$$

and

$$\left\langle\frac{1}{pkp'k}\right\rangle = \int_{-1}^{1}\frac{1}{2}dz\,\frac{1}{(p^0k^0)^2 - (|\mathbf{p}|\,|\mathbf{k}|z)^2}. \tag{5-4.60}$$

We shall only exhibit the latter integral in the two domains that were finally introduced in (5-4.59). For the first one, with $M - M' \gg \mu$, the integral is variously expressed as

$$\frac{4}{(M^2 - M'^2)^2}\frac{M'^2}{m^2}\int_0^1 dz\,\frac{1}{1 + [(M'^2/4m^2) - 1](1 - z^2)}$$

$$= \frac{8}{(M^2 - M'^2)^2}\frac{M'}{m}\int_0^1 dv\,\frac{1}{1 + \frac{1}{2}[(M'/2m) - 1](1 - v^2)}$$

$$= \frac{8}{(M^2 - M'^2)^2}\left(1 - \frac{4m^2}{M'^2}\right)^{-1/2}\log\frac{1 + [1 - (4m^2/M'^2)]^{1/2}}{1 - [1 - (4m^2/M'^2)]^{1/2}}. \tag{5-4.61}$$

In the region $M - M' \sim \mu$ it becomes

$$\frac{4}{M^2}\int_0^1 dz \frac{1}{(M-M')^2 - z^2[1-(4m^2/M^2)][(M-M')^2-\mu^2]}$$

$$= \frac{2}{M^2}\left(1-\frac{4m^2}{M^2}\right)^{-1/2} \frac{1}{(M-M')[(M-M')^2-\mu^2]^{1/2}}$$

$$\times \log\frac{M-M'+[1-(4m^2/M^2)]^{1/2}[(M-M')^2-\mu^2]^{1/2}}{M-M'-[1-(4m^2/M^2)]^{1/2}[(M-M')^2-\mu^2]^{1/2}}. \quad (5\text{–}4.62)$$

The remaining integral is

$$\left\langle\left(\frac{1}{(pk)^2}-\frac{1}{(p'k)^2}\right)k(p-p')\right\rangle = \int_{-1}^1 \frac{1}{2}\,dz \frac{4|\mathbf{k}|\,|\mathbf{p}|z}{(k^0p^0-|\mathbf{k}|\,|\mathbf{p}|z)^2}$$

$$= \frac{16}{M^2-M'^2}\left(1-\frac{4m^2}{M'^2}\right)^{1/2}\int_{-1}^1\frac{1}{2}\,dz\frac{z}{[1-z(1-(4m^2/M'^2))^{1/2}]^2}, \quad (5\text{–}4.63)$$

which is evaluated as

$$\frac{4}{M^2-M'^2}\frac{M'^2}{m^2} - \frac{8}{M^2-M'^2}\left(1-\frac{4m^2}{M'^2}\right)^{-1/2}\log\frac{1+[1-(4m^2/M'^2)]^{1/2}}{1-[1-(4m^2/M'^2)]^{1/2}}. \quad (5\text{–}4.64)$$

Before combining these structures, let us be explicit about the quantity of actual interest. It is the weight function $a(M^2)$, displayed in the action expression of Eq. (4–3.70) and in the implied propagation function formula (4–3.81). The calculation of Section 4–3 gave the two-particle exchange contribution to $a(M^2)$:

$$\text{spin } 0: \quad M^2a^{(2)}(M^2) = \frac{\alpha}{12\pi}\left(1-\frac{4m^2}{M^2}\right)^{3/2}. \quad (5\text{–}4.65)$$

We are now finding the three-particle exchange contribution $a^{(3)}(M^2)$, as another term in the action, of similar structure, and thus appearing additively in $a(M^2)$. The value inferred by comparing the coupling (5–4.42, 54, 56) with (4–3.34, 37) is

$$M^2a^{(3)}(M^2) = \frac{\alpha^2}{12\pi^2}\int\frac{dM'^2}{M^2}\,`\left(1-\frac{M'^2}{M^2}\right)'\left(1-\frac{4m^2}{M'^2}\right)^{1/2}\langle V_2^{\lambda\nu}V_{2\lambda\nu}\rangle, \quad (5\text{–}4.66)$$

where the 'quotation marks' recall the necessity of using the version in (5–4.55) for $M - M' \sim \mu$.

A function of frequent occurrence in these integrations is

$$\chi(v) = \frac{1}{v}\int_0^v dv'\frac{1}{1-v'^2} = \frac{1}{2v}\log\frac{1+v}{1-v}. \quad (5\text{–}4.67)$$

With its aid, we convey the form of $\langle V_2^{\lambda\nu}V_{2\lambda\nu}\rangle$ that is appropriate to $M - M' \gg \mu$ as

$$\langle V_2^{\lambda\nu} V_{2\lambda\nu}\rangle = \left(\frac{4m^2}{M^2 - M'^2}\right)^2 \frac{2v'^2}{1-v'^2}\left[\frac{1+v^2}{1-v^2}\chi(v') - \frac{1}{1-v'^2}\right] + \frac{8m^2}{M^2-M'^2}\left[\chi(v') - \frac{1}{1-v'^2}\right] + 2, \tag{5–4.68}$$

which uses the variables

$$v^2 = 1 - \frac{4m^2}{M^2}, \qquad v'^2 = 1 - \frac{4m^2}{M'^2}. \tag{5–4.69}$$

The latter will also be employed to write

$$\frac{4m^2}{M^2 - M'^2} = \frac{(1-v^2)(1-v'^2)}{v^2 - v'^2}, \tag{5–4.70}$$

and then the coefficient of $\alpha^2/12\pi^2$ in (5–4.66) reads

$$4(1-v^2)^3 \int dv' \, \frac{v'^4}{(1-v'^2)^2} \frac{\dfrac{1+v^2}{1-v^2}\chi(v') - \dfrac{1}{1-v'^2}}{v^2 - v'^2}. \tag{5–4.71}$$

The integration here ranges from $M' = 2m$ to $M' = M - \delta M$, where

$$\mu \ll \delta M \ll m, \tag{5–4.72}$$

or, from $v' = 0$ to $v' = v - \delta v$, with

$$\delta v = \frac{\delta M}{2m} \frac{(1-v^2)^{3/2}}{v}. \tag{5–4.73}$$

Other terms do appear in addition to (5–4.71), namely

$$4(1-v^2)^2 \left[\int_0^v dv' \, \frac{v'^2}{(1-v'^2)^2}\left(\chi(v') - \frac{2}{1-v'^2}\right) + \frac{1}{1-v^2}\int_0^v dv' \, \frac{v'^2}{(1-v'^2)^2}\right], \tag{5–4.74}$$

but this combination vanishes. That is verified by explicit integration, of

$$\int_0^v dv' \, \frac{v'^2}{(1-v'^2)^2}\left(\chi(v') - \frac{2}{1-v'^2}\right) = \frac{1}{2}\int_0^v d\left[\frac{v'\chi(v')}{1-v'^2} - \frac{v'}{(1-v'^2)^2}\right] = \frac{1}{2}\frac{v}{1-v^2}\left(\chi(v) - \frac{1}{1-v^2}\right), \tag{5–4.75}$$

and

$$\int_0^v dv' \, \frac{v'^2}{(1-v'^2)^2} = \frac{1}{2} v \left(\frac{1}{1-v^2} - \chi(v)\right). \tag{5–4.76}$$

In the region $M - M' \sim \mu$, we have

$$\langle V_2^{\lambda\nu} V_{2\lambda\nu}\rangle = 2m^2 \frac{v^2}{1-v^2}\left[\frac{1+v^2}{v}\int_0^v dv' \frac{1}{1-v'^2}\frac{1}{(M-M')^2+\mu^2[v'^2/(1-v'^2)]} - \frac{1}{(M-M')^2+\mu^2[v^2/(1-v^2)]}\right]; \qquad (5\text{–}4.77)$$

only the infra-red singular terms are retained. For this region, the integral of (5–4.66) becomes

$$\frac{(1-v^2)v}{m^2}\int d(M'-M)[(M-M')^2-\mu^2]^{1/2}\langle V_2^{\lambda\nu}V_{2\lambda\nu}\rangle, \qquad (5\text{–}4.78)$$

where the integration over $M - M'$ ranges from $\mu$ to $\delta M$. The basic integral encountered here is [cf. Eq. (4–4.97)]

$$\int_\mu^{\delta M} d(M-M')[(M-M')^2-\mu^2]^{1/2}\frac{1}{(M-M')^2+\mu^2[v^2/(1-v^2)]}$$
$$= \log\left(\frac{2\delta M}{\mu}\right) - \chi(v), \qquad (5\text{–}4.79)$$

and the resulting form of (5–4.78) is

$$2v^3[(1+v^2)\chi(v)-1]\log\left(\frac{2\delta M}{\mu}\right) + 2v^3\chi(v) - 2v^2(1+v^2)\int_0^v dv'\,\frac{\chi(v')}{1-v'^2}. \qquad (5\text{–}4.80)$$

The dependence on $\delta M$ will disappear when (5–4.71) is added. To remove the fictitious photon mass, we must consider the second effect, the modification in the two-particle exchange process.

We know that the modifications in the individual particle-pair emission and absorption acts are described by the form factor

$$F(k) = 1 - \frac{\alpha}{2\pi}\frac{k^2}{4m^2}\int_0^1 dv'(1+v'^2)\frac{\log\left(\frac{4m^2}{\mu^2}\frac{v'^2}{1-v'^2}\right)-2}{1+\frac{k^2}{4m^2}(1-v'^2)-i\varepsilon}. \qquad (5\text{–}4.81)$$

But we must appreciate the causal situation before applying (5–4.81). In the initial two-particle exchange process, there is a causal control over the emission and absorption regions. That ceases to be true, in general, when the form factor is introduced into the description of the individual emission and absorption acts, since there is complete non-locality (propagation) under the energetic circumstances expressed by the vanishing of the denominator in (5–4.81). The situation is similar to that encountered in describing unstable particles where simple sources cannot be considered if a causal control is to be exerted. As the analogue of the extended source employed in the latter discussion, we must exclude, for each choice of $v'$ in (5–4.81), those sources for which $v$ is in the immediate neighborhood of $v'$, where

$$-k^2 = M^2 = \frac{4m^2}{1-v^2}. \tag{5–4.82}$$

And, as in the unstable particle considerations, a final limiting process introduces a principal value integral. Thus, (5–4.81) is effectively replaced by

$$F(v) = 1 - \frac{\alpha}{2\pi} f(v), \tag{5–4.83}$$

with

$$f(v) = P\int_0^1 dv'(1+v'^2)\frac{\log\left(\dfrac{4m^2}{\mu^2}\dfrac{v'^2}{1-v'^2}\right)-2}{v^2-v'^2}. \tag{5–4.84}$$

This represents the modifying effect of those particle interactions that are suitably localized near the emitting or absorbing source. The net effect on the causal two-particle exchange process is conveyed by the factor

$$(F(v))^2 \simeq 1 - \frac{\alpha}{\pi} f(v). \tag{5–4.85}$$

The resulting change in $a^{(2)}(M^2)$ is then given by

$$M^2\,\delta a^{(2)}(M^2) = -\frac{\alpha^2}{12\pi^2} v^3 f(v). \tag{5–4.86}$$

To exhibit the photon mass dependence in the above equation, we decompose $f(v)$,

$$f(v) = 2\left[\log\left(\frac{2m}{\mu}\right)-1\right] P\int_0^1 dv'\,\frac{1+v'^2}{v^2-v'^2} + P\int_0^1 dv'\,\frac{(1+v'^2)\log\dfrac{v'^2}{1-v'^2}}{v^2-v'^2}. \tag{5–4.87}$$

Then, using the fact that

$$P\int_0^1 dv'\,\frac{1}{v^2-v'^2} = \mathrm{Re}\int_0^1 dv'\,\frac{1}{2v}\left(\frac{1}{v+v'}+\frac{1}{v-v'}\right) = \chi(v), \tag{5–4.88}$$

we have

$$P\int_0^1 dv'\,\frac{1+v'^2}{v^2-v'^2} = (1+v^2)\chi(v) - 1. \tag{5–4.89}$$

The resulting coefficient of $\alpha^2/12\pi^2$ in (5-4.86) cancels the photon mass term Eq. (5-4.80).

The remaining integrals in Eqs. (5-4.71, 80, 87) can be performed in terms of one type of transcendental function, which will be described later, but the resulting expression is not very illuminating. Rather, we now propose to use these integrals, as they appear, to extract a numerical consequence of the process under consideration. It is the modification in the vacuum polarization calculation of Section 4-3, where it was recognized that the significant quantity is the zero momentum limit of $\delta D_+(k)$. According to the construction [Eq. (4-3.81)]

$$\bar{D}_+(k) = \frac{1}{k^2}\frac{1}{1 - k^2\displaystyle\int dM^2 \frac{a(M^2)}{k^2 + M^2}}, \tag{5-4.90}$$

this quantity is

$$\delta D_+(0) = \int dM^2 \frac{a(M^2)}{M^2} = \frac{1}{(2m)^2}\int_0^1 dv^2\, M^2 a(M^2). \tag{5-4.91}$$

The two-particle exchange contribution to the integral is

$$\text{spin } 0: \quad \int dv^2\, M^2 a^{(2)}(M^2) = \frac{\alpha}{12\pi}\int_0^1 dv\, 2v^4 = \frac{\alpha}{12\pi}\frac{2}{5}. \tag{5-4.92}$$

The desired supplement to it is given by the $v^2$ integral of the sum of (5-4.71) and (5-4.80), multiplied by $\alpha^2/12\pi^2$, and of (5-4.86).

Let us begin with (5-4.71), first integrating over $v^2$ from $v'^2 + \delta v'^2$ to 1. The basic integral here is

$$\int_{v'^2+\delta v'^2}^1 dv^2 \frac{1}{v^2 - v'^2} = \log\frac{1 - v'^2}{\delta v'^2} = \log\left[\frac{m}{\delta M}\frac{1}{(1 - v'^2)^{1/2}}\right]. \tag{5-4.93}$$

One then verifies inductively, by differentiation with respect to $v'^2$, that

$$\int_{v'^2+\delta v'^2}^1 dv^2 \frac{(1 - v^2)^n}{v^2 - v'^2} = (1 - v'^2)^n\left\{\log\left[\frac{m}{\delta M}\frac{1}{(1 - v'^2)^{1/2}}\right] - \sum_{k=1}^{n}\frac{1}{k}\right\}. \tag{5-4.94}$$

Using these results, we find that the integral of Eq. (5-4.71) becomes (dropping the prime on the remaining integration variable)

$$4\int_0^1 dv\, v^4\left\{[(1 + v^2)\chi(v) - 1]\left[\log\left(\frac{m}{\delta M}\frac{1}{(1 - v^2)^{1/2}}\right) - \frac{3}{2} - \frac{1}{3}\right] + \frac{2}{3}\chi(v)\right\}. \tag{5-4.95}$$

Turning to the integral of (5-4.80), we first observe, through partial integration, that

$$\int_0^1 dv^2\, v^2(1+v^2)\int_0^v dv'\,\frac{\chi(v')}{1-v'^2} = \frac{5}{6}\int_0^1 dv\,(1+v^2)\chi(v) + \frac{1}{3}\int_0^1 dv\, v^4\chi(v). \quad (5\text{–}4.96)$$

The sum of these two contributions, which cancels $\delta M$, is

$$4\int_0^1 dv\, v^4[(1+v^2)\chi(v)-1]\left[\log\left(\frac{2m}{\mu}\frac{1}{(1-v^2)^{1/2}}\right)-\frac{3}{2}-\frac{1}{3}\right] - \frac{5}{3}\int_0^1 dv\,(1+v^2)\chi(v) + 6\int_0^1 dv\, v^4\chi(v). \quad (5\text{–}4.97)$$

To this is added [Eqs. (5–4.86, 87)]

$$-4\int_0^1 dv\, v^4[(1+v^2)\chi(v)-1]\left[\log\left(\frac{2m}{\mu}\right)-1\right] - 2\int_0^1 dv\,(\tfrac{1}{3}+v^2-v^4\chi(v))(1+v^2)\log\frac{v^2}{1-v^2}, \quad (5\text{–}4.98)$$

where we have used the principal value integral

$$P\int_0^1 dv\,\frac{v^4}{v^2-v'^2} = \tfrac{1}{3}+v'^2-v'^4\chi(v'). \quad (5\text{–}4.99)$$

When we add (5–4.97) and (5–4.98), all non-physical parameters disappear, to give

$$2\int_0^1 dv\, v^4[(1+v^2)\chi(v)-1]\left[\log\left(\frac{1}{1-v^2}\right)-\frac{5}{3}\right] - \frac{5}{3}\int_0^1 dv(1+v^2)\chi(v) + 6\int_0^1 dv\, v^4\chi(v) - 2\int_0^1 dv\,(\tfrac{1}{3}+v^2-v^4\chi(v))(1+v^2)\log\frac{v^2}{1-v^2}. \quad (5\text{–}4.100)$$

For the remaining integrations, we perform various partial integrations, as illustrated by

$$2\int_0^1 dv\, v^4(1+v^2)\chi(v)\log v^2 = \int_0^1 dv\left(\frac{5}{3}+\frac{5}{3}v^2+\frac{2}{3}v^4\right)\log v + \int_0^1 dv\left(\frac{5}{3}-v^4-\frac{2}{3}v^6\right)\chi(v), \quad (5\text{–}4.101)$$

use examples of the integrals

$$n\geqslant 1:\quad \int_0^1 dv\, v^{2n}\chi(v) = \frac{1}{2n}\sum_{k=0}^{n-1}\frac{1}{2k+1}, \quad (5\text{–}4.102)$$

and note the specific result

$$\int_0^1 dv\left(1 - v^2 - \frac{10}{3}v^4\right)\log\frac{1}{1-v^2} = -\frac{14}{15}. \tag{5-4.103}$$

The outcome is expressed by

$$\text{spin } 0: \quad \int dv^2\, M^2 a(M^2) = \frac{\alpha}{12\pi}\frac{2}{5} + \frac{\alpha^2}{12\pi^2}\frac{95}{54}$$
$$= \frac{\alpha}{30\pi}\left[1 + \frac{5\alpha}{4\pi}\frac{95}{27}\right]. \tag{5-4.104}$$

This increase in the vacuum polarization effect is roughly one percent. A quantitative statement will be reserved for the more experimentally relevant spin $\frac{1}{2}$ discussion.

The integrals that must be performed in order to exhibit $a(M^2)$ have structures containing a denominator and a logarithm which are different linear functions of one variable. A standard function of this type is

$$0 < x < 1: \quad l(x) = \int_0^x \frac{dt}{t}\log\frac{1}{1-t} = \sum_{n=1}^{\infty}\frac{x^n}{n^2}, \tag{5-4.105}$$

variously called Euler's dilogarithm and the Spence function. As one recognizes through the substitution $t \to 1 - t$, followed by partial integration, this function obeys

$$l(x) + l(1-x) = \frac{\pi^2}{6} - \log\left(\frac{1}{x}\right)\log\left(\frac{1}{1-x}\right), \tag{5-4.106}$$

which incorporates the fact that

$$l(1) = \sum_{n=1}^{\infty}\frac{1}{n^2} = \frac{\pi^2}{6}. \tag{5-4.107}$$

Analogous functions defined for other ranges of $x$ are simply related to $l(x)$. Thus, for $x > 1$ we consider

$$x > 1: \quad \bar{l}(x) = \int_1^x \frac{dt}{t}\log(t-1), \tag{5-4.108}$$

and note that $t \to 1/t$ produces

$$\bar{l}(x) = \int_{1/x}^1 \frac{dt}{t}\log\frac{1-t}{t}, \tag{5-4.109}$$

or

$$\bar{l}(x) = -\frac{\pi^2}{6} + \tfrac{1}{2}(\log x)^2 + l(1/x). \tag{5-4.110}$$

The function that effectively appears on changing the sign of $x$ is

$$0<x<1:\quad \int_0^x \frac{dt}{t}\log(1+t) = \int_0^x \frac{dt}{t}\log\frac{1}{1-t} - \int_0^x \frac{dt}{t}\log\frac{1}{1-t^2}$$
$$= l(x) - \tfrac{1}{2}l(x^2), \tag{5–4.111}$$

and the analogous relation for $x > 1$ reads

$$\int_1^x \frac{dt}{t}\log(1+t) = \tfrac{1}{2}l(x^2) - l(x)$$
$$= \frac{\pi^2}{12} + \tfrac{1}{2}(\log x)^2 + \tfrac{1}{2}l\left(\frac{1}{x^2}\right) - l\left(\frac{1}{x}\right). \tag{5–4.112}$$

These are all aspects of one function, of course, but for numerical purposes we prefer to use $l(x)$ as the standard function.

Other relations appear on making the substitution $t \to 1 + t$ in (5–4.108), yielding

$$l(x) = \int_0^{x-1} \frac{dt}{1+t}\log t = \log x \log(x-1) - \int_0^{x-1} \frac{dt}{t}\log(1+t). \tag{5–4.113}$$

This is given different forms depending upon whether $x - 1$ is greater or less than 1. In the latter situation we can apply (5–4.111) to get

$$1 < x < 2:\quad l(x) = \log x \log(x-1) - l(x-1) + \tfrac{1}{2}l((x-1)^2), \tag{5–4.114}$$

whereas (5–4.112), together with the integral

$$\int_0^1 \frac{dt}{t}\log(1+t) = \tfrac{1}{2}l(1) = \frac{\pi^2}{12}, \tag{5–4.115}$$

is used to produce

$$x > 2:\quad l(x) = -\frac{\pi^2}{6} + \log x \log(x-1) - \tfrac{1}{2}[\log(x-1)]^2$$
$$+ l\left(\frac{1}{x-1}\right) - \tfrac{1}{2}l\left(\frac{1}{(x-1)^2}\right). \tag{5–4.116}$$

With the aid of (5–4.110), these results are transformed into

$$\tfrac{1}{2} < x < 1:\quad l(x) = \frac{\pi^2}{6} + \frac{1}{2}\left(\log\frac{1}{x}\right)^2 - \log\frac{1}{x}\log\frac{1}{1-x}$$
$$- l\left(\frac{1-x}{x}\right) + \tfrac{1}{2}l\left(\left(\frac{1-x}{x}\right)^2\right) \tag{5–4.117}$$

and

$$0<x<\tfrac{1}{2}:\quad l(x)=-\tfrac{1}{2}\left(\log\frac{1}{1-x}\right)^2+l\left(\frac{x}{1-x}\right)-\tfrac{1}{2}l\left(\left(\frac{x}{1-x}\right)^2\right), \qquad (5\text{–}4.118)$$

which are interconnected by the statement of Eq. (5–4.106). A particular consequence is reached by parametrizing $x$ as $\frac{1}{2}(1 \pm v)$ in the respective domains of Eqs. (5–4.117, 118), and subtracting the latter:

$$l\left(\frac{1+v}{2}\right)-l\left(\frac{1-v}{2}\right)=\frac{\pi^2}{6}-\log\frac{2}{1+v}\log\frac{1+v}{1-v}-2l\left(\frac{1-v}{1+v}\right)+l\left(\left(\frac{1-v}{1+v}\right)^2\right). \qquad (5\text{–}4.119)$$

To illustrate the use of the dilogarithmic function, consider the second integral of Eq. (5–4.87), which can be resolved into individual integrals containing either $v - v'$ or $v + v'$ as denominator, and a logarithm of $v'$, $1 + v'$, or $1 - v'$. We first observe that

$$P\int_0^1 \frac{dv'}{v-v'}\log v' = -\operatorname{Re}\int_0^1 d\log\frac{v-v'}{v}\log v'$$

$$=\int_0^v \frac{dv'}{v'}\log\frac{v-v'}{v}+\int_v^1\frac{dv'}{v'}\log\frac{v'-v}{v}. \qquad (5\text{–}4.120)$$

The substitution $v' = vt$ then brings these integrals to the form

$$-\int_0^1\frac{dt}{t}\log\frac{1}{1-t}+\int_1^{1/v}\frac{dt}{t}\log(t-1)=-\frac{\pi^2}{6}+l\left(\frac{1}{v}\right), \qquad (5\text{–}4.121)$$

or

$$P\int_0^1\frac{dv'}{v-v'}\log v' = -\frac{\pi^2}{3}+\frac{1}{2}\left(\log\frac{1}{v}\right)^2+l(v). \qquad (5\text{–}4.122)$$

Another consequence, produced by the substitution $v \to 1 - v$, $v' \to 1 - v'$, is

$$P\int_0^1\frac{dv'}{v-v'}\log(1-v')=\frac{\pi^2}{3}-\frac{1}{2}\left(\log\frac{1}{1-v}\right)^2-l(1-v), \qquad (5\text{–}4.123)$$

and, incidentally,

$$P\int_0^1\frac{dv'}{v-v'}\log\frac{v'}{1-v'}=-\frac{2\pi^2}{3}+\frac{1}{2}\left(\log\frac{1}{v}\right)^2+\frac{1}{2}\left(\log\frac{1}{1-v}\right)^2+l(v)+l(1-v)$$

$$=-\frac{\pi^2}{2}+\frac{1}{2}\left(\log\frac{v}{1-v}\right)^2, \qquad (5\text{–}4.124)$$

according to (5–4.106). Changing the denominator in (5–4.120), we have

$$\int_0^1 \frac{dv'}{v+v'} \log v' = -\int_0^1 \frac{dv'}{v'} \log \frac{v+v'}{v} = -\int_0^{1/v} \frac{dt}{t} \log(1+t)$$

$$= -\frac{\pi^2}{6} - \frac{1}{2}\left(\log \frac{1}{v}\right)^2 + l(v) - \tfrac{1}{2}l(v^2), \qquad (5\text{-}4.125)$$

which is added to (5-4.122) in

$$P\int_0^1 \frac{dv'}{v^2 - v'^2} \log v'^2 = \frac{1}{v}\left[-\frac{\pi^2}{2} + 2l(v) - \tfrac{1}{2}l(v^2)\right]. \qquad (5\text{-}4.126)$$

Also required is

$$P\int_0^1 \frac{dv'}{v-v'} \log(1+v') = \operatorname{Re}\int_0^1 \frac{dv'}{1+v'} \log \frac{v-v'}{1-v}$$

$$= \int_0^v \frac{dv'}{1+v'} \log \frac{v-v'}{1-v} + \int_v^1 \frac{dv'}{1+v'} \log \frac{v'-v}{1-v}. \qquad (5\text{-}4.127)$$

Now it is the transformation $1 + v' = (1 + v)t$ that produces the form

$$\log 2 \log \frac{1+v}{1-v} - \int_{1/(1+v)}^1 \frac{dt}{t} \log \frac{1}{1-t} + \int_1^{2/(1+v)} \frac{dt}{t} \log(t-1)$$

$$= -\frac{\pi^2}{3} + \log 2 \log \frac{1+v}{1-v} + \frac{1}{2}\left(\log \frac{1+v}{2}\right)^2 + l\left(\frac{1}{1+v}\right) + l\left(\frac{1+v}{2}\right). \qquad (5\text{-}4.128)$$

Furthermore, we have

$$\int_0^1 \frac{dv'}{v+v'} \log(1-v') = \int_0^1 \frac{dv'}{1-v'} \log \frac{v+v'}{1+v} = -\int_0^{1/(1+v)} \frac{dt}{t} \log \frac{1}{1-t}$$

$$= -l\left(\frac{1}{1+v}\right), \qquad (5\text{-}4.129)$$

which uses the transformation $v' = 1 - (1 + v)t$, and, with the transformation $v' = -1 + (1 - v)t$,

$$\int_0^1 \frac{dv'}{v+v'} \log(1+v') = -\int_0^1 \frac{dv'}{1+v'} \log \frac{v+v'}{1+v}$$

$$= \log 2 \log \frac{1+v}{1-v} - \int_{1/(1-v)}^{2/(1-v)} \frac{dt}{t} \log(t-1)$$

$$= \log 2 \log \frac{1+v}{2} + \tfrac{1}{2}(\log 2)^2 + l(1-v) - l\left(\frac{1-v}{2}\right). \qquad (5\text{-}4.130)$$

The outcome for the integral in question is

$$P\int_0^1 dv'\,\frac{(1+v'^2)\log[v'^2/(1-v'^2)]}{v^2-v'^2}$$
$$=\frac{1+v^2}{2v}\left[-\pi^2+\frac{1}{2}\left(\log\frac{1-v}{2}\right)^2-\frac{1}{2}\left(\log\frac{1+v}{2}\right)^2\right.$$
$$\left.-2\log 2\log\frac{1+v}{1-v}+4l(v)-l(v^2)-l\left(\frac{1+v}{2}\right)+l\left(\frac{1-v}{2}\right)\right]+2\log 2. \tag{5-4.131}$$

Note that the relation (5-4.119) could be used to give this another form.

Without going into further details about the integration, we state the result for $a(M^2)$:

$$M^2a(M^2)=\frac{\alpha}{12\pi}v^3+\frac{\alpha^2}{12\pi^2}\left\{v^2(1+v^2)\left[\frac{\pi^2}{6}+\log\frac{1+v}{2}\log\frac{1+v}{1-v}\right.\right.$$
$$\left.+2l\left(\frac{1-v}{1+v}\right)+2l\left(\frac{1+v}{2}\right)-2l\left(\frac{1-v}{2}\right)-4l(v)+l(v^2)\right]$$
$$+\left[5\left(\frac{1+v^2}{2}\right)^2-2-3v^2\right]\log\frac{1+v}{1-v}+6v^3\log\frac{1+v}{2}$$
$$\left.-4v^3\log v+\tfrac{3}{2}v(1+v^2)\right\}. \tag{5-4.132}$$

This elaborate structure can better be comprehended in the high energy ($v\to 1$) and low energy ($v\to 0$) limits. Thus,

$$M^2\gg(2m)^2:\quad M^2a(M^2)=\frac{\alpha}{12\pi}+\frac{\alpha^2}{4\pi^2}=\frac{\alpha}{12\pi}\left(1+\frac{3\alpha}{\pi}\right), \tag{5-4.133}$$

where the contribution of order $\alpha^2$ comes entirely from the last term in the braces of (5-4.132), and

$$M^2\sim(2m)^2:\quad M^2a(M^2)=\frac{\alpha}{12\pi}v^3+\frac{\alpha^2}{24}v^2=\frac{\alpha}{12\pi}v^3\left(1+\frac{\pi}{2}\frac{\alpha}{v}\right); \tag{5-4.134}$$

here the $\alpha^2$ term arises from the first bracket in the braces of (5-4.132) and can be traced back to the partial form factor integral (5-4.126). The latter result is particularly interesting since the threshold behavior has been changed. This can be understood from familiar non-relativistic considerations. The effect of the Coulomb attraction between charges that are produced with relative speed $v_{\text{rel.}}$ increases the probability of establishing the state by the factor

$$\frac{(2\pi\alpha/v_{\text{rel.}})}{1-\exp[-(2\pi\alpha/v_{\text{rel.}})]}\simeq 1+\frac{\pi\alpha}{v_{\text{rel.}}}, \tag{5-4.135}$$

where the approximation refers to the circumstances $1 \gg v_{\text{rel.}} \gg \alpha$, which validate the treatment of the Coulomb interaction as a weak, non-relativistic effect. According to the non-relativistic relation

$$M \cong 2m + \tfrac{1}{4}mv_{\text{rel.}}^2, \tag{5–4.136}$$

we have

$$\left(\frac{M}{2m}\right)^2 - 1 = \frac{v^2}{1-v^2} \cong \tfrac{1}{4}v_{\text{rel.}}^2 \tag{5–4.137}$$

or

$$v_{\text{rel.}} \cong 2v, \tag{5–4.138}$$

which indeed identifies (5–4.135) with the modification factor of (5–4.134). Incidentally, the elastic form factor itself, in the non-relativistic limit, is essentially identical with the wave function for relative motion in the Coulomb field, evaluated at the origin, with the normalization set by the unit amplitude of the asymptotic plane wave.

We could repeat the vacuum polarization computation given in Eq. (5–4.104), using the explicit expression for $M^2a(M^2)$. Instead of doing that, let us make the following approximate observation. A simple, but slightly contrived, formula that interpolates between the two limiting forms of Eqs. (5–4.133, 134) is

$$M^2a(M^2) \sim \frac{\alpha}{12\pi}v^3\left[1 + \frac{\pi\alpha}{2v} - \frac{1+v}{2}\left(\frac{\pi}{2} - \frac{3}{\pi}\right)\alpha\right]. \tag{5–4.139}$$

The result of performing the $v^2$ integration of this function approximates the numerical coefficient exhibited in (5–4.104),

$$\frac{95}{54} = 1.759, \tag{5–4.140}$$

by

$$\frac{11}{10} + \frac{\pi^2}{15} = 1.758. \tag{5–4.141}$$

Now let us go through the analogous calculations for spin $\frac{1}{2}$ charged particles. To describe the three-particle exchange process, we begin with [cf. Eq. (3–12.24)]

$$W_{22} = \tfrac{1}{2}\int (dx)(dx')\psi(x)\gamma^0 eq\gamma A(x)G_+(x-x')eq\gamma A(x')\psi(x'). \tag{5–4.142}$$

Comparing the appropriate part of the vacuum amplitude $iW_{22}$ with the equivalent amplitude

$$\tfrac{1}{2}\left[i\int(dx)\psi(x)\gamma^0\eta(x)\right]^2 i\int(d\xi)J^\lambda(\xi)A_\lambda(\xi), \tag{5-4.143}$$

we infer the effective source

$$-\eta_2(x)\eta_2(x')\gamma^0 J_2{}^\lambda(\xi)|_{\text{eff.}} = e^2[\delta(x-\xi)\gamma^\lambda G_+(x-x')\gamma A_2(x') + \gamma A_2(x)G_+(x-x')\gamma^\lambda\,\delta(x'-\xi)]. \tag{5-4.144}$$

The momentum version is

$$-\eta_2(p)\eta_2(p')\gamma^0 J_2{}^\lambda(k)|_{\text{eff.}} = e^2\left[\gamma^\lambda\frac{1}{\gamma(p+k)+m}\gamma A_2(K) + \gamma A_2(K)\frac{1}{-\gamma(p'+k)+m}\gamma^\lambda\right], \tag{5-4.145}$$

and the analogous absorption process is represented by

$$-\eta_1(-p')\eta_1(-p)\gamma^0 J_1{}^\lambda(-k)|_{\text{eff.}} = e^2\left[\gamma^\lambda\frac{1}{-\gamma(p'+k)+m}\gamma A_1(-K) + \gamma A_1(-K)\frac{1}{\gamma(p+k)+m}\gamma^\lambda\right]. \tag{5-4.146}$$

The three-particle exchange vacuum amplitude is then derived from

$$-\tfrac{1}{2}\int d\omega_p\,d\omega_{p'}\,d\omega_k\,\text{tr}[\eta_1(-p')\eta_1(-p)\gamma^0 J_1{}^\lambda(-k)|_{\text{eff.}}(m-\gamma p) \times \eta_2(p)\eta_2(p')\gamma^0 J_{2\lambda}(k)|_{\text{eff.}}(-m-\gamma p')], \tag{5-4.147}$$

which is a rearranged version of

$$-\tfrac{1}{2}\left[\int d\omega_p\,\eta_1(-p)\gamma^0(m-\gamma p)\eta_2(p)\right]^2\int d\omega_k\,J_1{}^\lambda(-k)J_{2\lambda}(k). \tag{5-4.148}$$

It can again be written as

$$-(2e^2)^2\int dM^2\,d\omega_K\,A_1{}^\mu(-K)I_{\mu\nu}(K)A_2{}^\nu(K), \tag{5-4.149}$$

where now

$$I_{\mu\nu}(K) = \int d\omega_p\,d\omega_{p'}\,d\omega_k(2\pi)^3\,\delta(K-p-p'-k)\,\text{tr}_n\left\{\left[\gamma_\mu\frac{1}{\gamma(p+k)+m}\gamma^\lambda + \gamma^\lambda\frac{1}{-\gamma(p'+k)+m}\gamma_\mu\right](m-\gamma p)\left[\gamma_\lambda\frac{1}{\gamma(p+k)+m}\gamma_\nu + \gamma_\nu\frac{1}{-\gamma(p'+k)+m}\gamma_\lambda\right](-m-\gamma p')\right\}, \tag{5-4.150}$$

and $\mathrm{tr}_n$ indicates the trace, so normalized that

$$\mathrm{tr}_n\, 1 = 1. \tag{5-4.151}$$

The gauge invariance of the coupling (5-4.149), which implies the tensor structure

$$I_{\mu\nu}(K) = \left(g_{\mu\nu} + \frac{K_\mu K_\nu}{M^2}\right) I(M^2), \tag{5-4.152}$$

can be verified directly. What we must calculate is the scalar function

$$I(M^2) = \frac{1}{3}\int d\omega_p\, d\omega_{p'}\, d\omega_k\, (2\pi)^3\, \delta(K - p - p' - k)\, \mathrm{tr}_n\left\{\left[\gamma^\nu \frac{1}{\gamma(p+k)+m}\gamma^\lambda \right.\right.$$

$$\left. + \gamma^\lambda \frac{1}{-\gamma(p'+k)+m}\gamma^\nu\right](m - \gamma p)\left[\gamma_\lambda \frac{1}{\gamma(p+k)+m}\gamma_\nu\right.$$

$$\left.\left. + \gamma_\nu \frac{1}{-\gamma(p'+k)+m}\gamma_\lambda\right](-m - \gamma p')\right\}. \tag{5-4.153}$$

The matrix factors in brackets are reduced with the aid of the projection matrices in (5-4.153). Thus,

$$\gamma^\nu \frac{1}{\gamma(p+k)+m}\gamma^\lambda + \gamma^\lambda \frac{1}{-\gamma(p'+k)+m}\gamma^\nu$$

$$= \gamma^\nu \frac{m - \gamma(p+k)}{2pk}\gamma^\lambda + \gamma^\lambda \frac{m + \gamma(p'+k)}{2p'k}\gamma^\nu$$

$$\to \gamma^\nu \frac{2p^\lambda - \gamma k \gamma^\lambda}{2pk} + \frac{-2p'^\lambda + \gamma^\lambda \gamma k}{2p'k}\gamma^\nu \tag{5-4.154}$$

and

$$\gamma_\lambda \frac{1}{\gamma(p+k)+m}\gamma_\nu + \gamma_\nu \frac{1}{-\gamma(p'+k)+m}\gamma_\lambda$$

$$= \gamma_\lambda \frac{m - \gamma(p+k)}{2pk}\gamma_\nu + \gamma_\nu \frac{m + \gamma(p'+k)}{2p'k}\gamma_\lambda$$

$$\to \frac{2p_\lambda - \gamma_\lambda \gamma k}{2pk}\gamma_\nu + \gamma_\nu \frac{-2p'_\lambda + \gamma k \gamma_\lambda}{2p'k}. \tag{5-4.155}$$

The matrix product of (5-4.153) then becomes

$$\left[\left(\frac{p}{pk} - \frac{p'}{p'k}\right)^\lambda \gamma^\nu - \frac{\gamma^\nu \gamma k \gamma^\lambda}{2pk} + \frac{\gamma^\lambda \gamma k \gamma^\nu}{2p'k}\right](m - \gamma p)\left[\left(\frac{p}{pk} - \frac{p'}{p'k}\right)_\lambda \gamma_\nu\right.$$

$$\left. - \frac{\gamma_\lambda \gamma k \gamma_\nu}{2pk} + \frac{\gamma_\nu \gamma k \gamma_\lambda}{2p'k}\right](-m - \gamma p'). \tag{5-4.156}$$

We specifically note the appearance of the term

$$\left(\frac{p}{pk}-\frac{p'}{p'k}\right)^2 \gamma^\nu(m-\gamma p)\gamma_\nu(-m-\gamma p'), \qquad (5\text{–}4.157)$$

which is the expected, infra-red sensitive, radiative modification of the two-particle exchange mechanism.

There are two types of terms in (5–4.156) that involve a pair of $\gamma k$ factors. One is

$$\begin{aligned}&\gamma^\lambda\gamma k\gamma^\nu(m-\gamma p)\gamma_\lambda\gamma k\gamma_\nu(-m-\gamma p')\\ &= \gamma^\lambda\gamma k[2\gamma_\lambda\gamma k\gamma p + (m+\gamma p)\gamma^\nu\gamma_\lambda\gamma k\gamma_\nu](-m-\gamma p') = 0,\end{aligned} \qquad (5\text{–}4.158)$$

which holds since

$$\gamma^\lambda\gamma k\gamma_\lambda\gamma k = 2(\gamma k)^2 = 0, \qquad (5\text{–}4.159)$$

while

$$\gamma^\nu\gamma_\lambda\gamma k\gamma_\nu = 4k_\lambda \qquad (5\text{–}4.160)$$

also produces the null structure $(\gamma k)^2$. The other,

$$\gamma^\lambda\gamma k\gamma^\nu(m-\gamma p)\gamma_\nu\gamma k\gamma_\lambda(-m-\gamma p') \to \gamma k\gamma^\nu(m-\gamma p)\gamma_\nu\gamma k\gamma_\lambda(-m-\gamma p')\gamma^\lambda, \qquad (5\text{–}4.161)$$

since these are equivalent with respect to the trace, and the latter is further reduced to

$$\gamma k(-4m-2\gamma p)\gamma k(4m-2\gamma p') = 4kp\gamma k(4m-2\gamma p') \to 8kpkp'; \qquad (5\text{–}4.162)$$

the last step records the result of the trace operation.

An example of a term in (5–4.156) with one $\gamma k$ factor is

$$(m-\gamma p)\left(\frac{\gamma p}{pk}-\frac{\gamma p'}{p'k}\right)\gamma k\gamma_\nu(-m-\gamma p')\gamma^\nu, \qquad (5\text{–}4.163)$$

which has already exploited the cyclic property of the trace. If we write

$$\gamma_\nu(-m-\gamma p')\gamma^\nu = 2m + 2(m-\gamma p'), \qquad (5\text{–}4.164)$$

the expression in (5–4.163) decomposes into

$$2m(m-\gamma p)\left(\frac{\gamma p}{pk}-\frac{\gamma p'}{p'k}\right)\gamma k, \qquad (5\text{–}4.165)$$

and

$$2(m-\gamma p)\left(\frac{\gamma p}{pk}-\frac{\gamma p'}{p'k}\right)\gamma k(m-\gamma p')$$
$$=-2m\left(\frac{1}{pk}+\frac{1}{p'k}\right)(m-\gamma p)\gamma k(m-\gamma p')+4(m-\gamma p)(m-\gamma p'), \quad (5\text{–}4.166)$$

where the projection matrices have been used to simplify the structure. Now, the trace of the product of an odd number of $\gamma$-matrices is zero. The proof is an immediate generalization of that for one $\gamma$-matrix, based on anticommutativity with $\gamma_5$, which is given in Eq. (2–6.79). Hence the trace of (5–4.165) reduces to

$$2m^2\,\mathrm{tr}_n\left(\frac{\gamma p}{pk}-\frac{\gamma p'}{p'k}\right)\gamma k=0, \quad (5\text{–}4.167)$$

since

$$\mathrm{tr}_n\,\gamma A\gamma B=-AB, \quad (5\text{–}4.168)$$

while, in (5–4.166), we encounter

$$\mathrm{tr}_n(m-\gamma p)\gamma k(m-\gamma p')=-m\,\mathrm{tr}_n(\gamma p\gamma k+\gamma k\gamma p')$$
$$=m(pk+p'k), \quad (5\text{–}4.169)$$

and

$$\mathrm{tr}_n(m-\gamma p)(m-\gamma p')=m^2-pp'=\tfrac{1}{2}M'^2. \quad (5\text{–}4.170)$$

The immediate expression for the trace of the matrix in (5–4.153) is

$$\mathrm{tr}_n\{\ \}=\left(\frac{p}{pk}-\frac{p'}{p'k}\right)^2(M'^2+2m^2)+2\left(\frac{p'k}{pk}+\frac{pk}{p'k}\right)$$
$$+2m^2\left(\frac{1}{pk}+\frac{1}{p'k}\right)^2(pk+p'k)-2M'^2\left(\frac{1}{pk}+\frac{1}{p'k}\right). \quad (5\text{–}4.171)$$

This can be rearranged as

$$\mathrm{tr}_n\{\ \}=\left[-\frac{m^2}{(pk)^2}-\frac{m^2}{(p'k)^2}+\frac{M^2-2m^2}{pkp'k}\right](M'^2+2m^2)-4$$
$$-2m^2k(p-p')\left(\frac{1}{(pk)^2}-\frac{1}{(p'k)^2}\right)$$
$$-6m^2(M^2-M'^2)\frac{1}{pkp'k}+\tfrac{1}{2}(M^2-M'^2)^2\frac{1}{pkp'k}, \quad (5\text{–}4.172)$$

where one will recognize much of the spin 0 structure displayed in Eq. (5–4.51).

Indeed, no new integrals are encountered in evaluating the expectation value of this function, as required for the analogue of Eq. (5-4.66):

$$M^2a^{(3)}(M^2) = \frac{\alpha^2}{3\pi^2}\int\frac{dM'^2}{M^2}'\left(1-\frac{M'^2}{M^2}\right)'\left(1-\frac{4m^2}{M'^2}\right)^{1/2}\frac{1}{4}\langle \mathrm{tr}_n\{\ \}\rangle. \quad (5\text{-}4.173)$$

Following the spin 0 procedure, we first consider the domain $M - M' \gg \mu$, where

$$\begin{aligned}\frac{1}{4}\langle \mathrm{tr}_n\rangle = {} & \left(\frac{4m^2}{M^2-M'^2}\right)^2\frac{3-v'^2}{1-v'^2}\left[\frac{1+v^2}{1-v^2}\chi(v') - \frac{1}{1-v'^2}\right] \\ & + \frac{8m^2}{M^2-M'^2}\left(\chi(v') - \frac{1}{1-v'^2}\right) \\ & - \frac{24m^2}{M^2-M'^2}\chi(v') + 2\chi(v') - 1. \end{aligned} \quad (5\text{-}4.174)$$

This gives the following contribution to the coefficient of $\alpha^2/3\pi^2$ in (5-4.173):

$$\begin{aligned} 2(1-v^2)^2\int dv'\frac{v'^2}{(1-v'^2)^2}\bigg\{&(1-v^2)\frac{3-v'^2}{v^2-v'^2}\left[\frac{1+v^2}{1-v^2}\chi(v') - \frac{1}{1-v'^2}\right] - 4\chi(v') \\ & + 2\chi(v')\left(\frac{1}{1-v^2}-\frac{1}{1-v'^2}\right) - \frac{1}{1-v^2} - \frac{1}{1-v'^2}\bigg\}, \end{aligned} \quad (5\text{-}4.175)$$

where the integration domain is that described in the context of Eq. (5-4.71). Unlike the latter equation, the infra-red insensitive terms of (5-4.175) do not disappear on integration, and have been left intact.

The behavior in the region $M - M' \sim \mu$ is the same as with spin 0, except for the factor that expresses the different form of $a^{(2)}(M^2)$, $M^2 - 4m^2 \to M^2 + 2m^2$, where the additional factor of 4 is used to replace $\alpha^2/12\pi^2$ by $\alpha^2/3\pi^2$. Thus, with the multiplicative substitution in Eq. (5-4.80) of $v^2 \to \frac{1}{2}(3 - v^2)$, we get the following addition to (5-4.175),

$$(3-v^2)\left\{v[(1+v^2)\chi(v)-1]\log\left(\frac{2\delta M}{\mu}\right) + v\chi(v) - (1+v^2)\int_0^v dv'\frac{\chi(v')}{1-v'^2}\right\}. \quad (5\text{-}4.176)$$

It will cancel the parameter $\delta M$.

The form factor effect is a little more elaborate with spin $\frac{1}{2}$ particles since the additional magnetic moment coupling comes into play,

$$\gamma A \to F_1\gamma A + F_2\frac{\alpha}{2\pi}\frac{1}{2m}\sigma F. \quad (5\text{-}4.177)$$

The consequence for the trace calculation of Eqs. (4-3.20, 21) is indicated in

$$\mathrm{tr}_n[\gamma^\mu(m-\gamma p)\gamma_\mu(-m-\gamma p')] = M^2 + 2m^2$$

$$\to \mathrm{tr}_n\left[\left(F_1\gamma^\mu + \frac{\alpha}{4\pi m}F_2\gamma k\gamma^\mu\right)(m-\gamma p)\left(F_1\gamma_\mu + \frac{\alpha}{4\pi m}F_2\gamma_\mu\gamma k\right)(-m-\gamma p')\right], \tag{5–4.178}$$

which has exploited the conservation property of the structure to omit terms containing $k^\mu$. We again use the algebraic basis for this property, the projection matrices, in reducing the magnetic moment coupling. That is described by

$$\gamma k\gamma^\mu = (\gamma p + \gamma p')\gamma^\mu \to 2m\gamma^\mu - 2p^\mu, \tag{5–4.179}$$

and, similarly,

$$\gamma_\mu\gamma k = \gamma_\mu(\gamma p + \gamma p') \to 2m\gamma_\mu - 2p_\mu, \tag{5–4.180}$$

where the resulting $\gamma p$ combination can then be replaced by $-m$. This gives, for (5–4.178),

$$\left(F_1 + \frac{\alpha}{2\pi}F_2\right)^2 \mathrm{tr}_n[\gamma^\mu(m-\gamma p)\gamma_\mu(-m-\gamma p')]$$

$$+ \frac{\alpha}{\pi}F_1F_2\,\mathrm{tr}_n[(m-\gamma p)(-m-\gamma p')]$$

$$\cong \left(F_1^2 + \frac{\alpha}{\pi}F_2\right)(M^2+2m^2) + \frac{\alpha}{2\pi}F_2(M^2-4m^2), \tag{5–4.181}$$

or,

$$F_1^2(M^2+2m^2) + \frac{3\alpha}{2\pi}F_2M^2, \tag{5–4.182}$$

where only effects of order $\alpha$ have been retained. The form factors that appear here are

$$F_1(v) = 1 - \frac{\alpha}{2\pi}f_1(v), \tag{5–4.183}$$

with [cf. Eqs. (4–4.68, 77)]

$$f_1(v) = P\int_0^1 dv'\,\frac{(1+v'^2)\log[(4m^2/\mu^2)(v'^2/(1-v'^2))] - 1 - 2v'^2}{v^2-v'^2}$$

$$= f(v) + P\int_0^1 dv'\,\frac{1}{v^2-v'^2}, \tag{5–4.184}$$

and [Eq. (4–4.75)]

$$F_2(v) = -(1-v^2)P\int_0^1 dv'\,\frac{1}{v^2-v'^2} = -(1-v^2)\chi(v). \qquad (5\text{–}4.185)$$

The consequent change in

$$M^2a^{(2)}(M^2) = \frac{\alpha}{3\pi}v\tfrac{1}{2}(3-v^2) \qquad (5\text{–}4.186)$$

is given by

$$M^2\delta a^{(2)}(M^2) = -\frac{\alpha^2}{3\pi^2}v\tfrac{1}{2}(3-v^2)(f(v)+\chi(v)) - \frac{\alpha^2}{3\pi^2}v\tfrac{3}{2}(1-v^2)\chi(v). \qquad (5\text{–}4.187)$$

As in the spin 0 discussion, we shall first evaluate the integral, $\int_0^1 dv^2\, M^2a(M^2)$, which measures the vacuum polarization displacement of atomic energy levels. The $v^2$ integral of (5–4.175), produced by appropriate modification of (5–4.95), is

$$2\int_0^1 dv\, v^2(3-v^2)\left\{[(1+v^2)\chi(v)-1]\left[\log\left(\frac{m}{\delta M}\frac{1}{(1-v^2)^{1/2}}\right)\right.\right.$$
$$\left.\left. -\frac{3}{2}-\frac{1}{3}\right]+\frac{2}{3}\chi(v)\right\}-\frac{2}{3}, \qquad (5\text{–}4.188)$$

where the added constant, $-\frac{2}{3}$, gives the integrated value of the non-singular terms in (5–4.175):

$$\int_0^1 dv^2\, 2(1-v^2)^2\int_0^v dv'\,\frac{v'^2}{(1-v'^2)^2}\left\{-4\chi(v')+2\chi(v')\left(\frac{1}{1-v^2}-\frac{1}{1-v'^2}\right)\right.$$
$$\left.-\frac{1}{1-v^2}-\frac{1}{1-v'^2}\right\}$$
$$= 2\int_0^1 dv'\, v'^2\left\{-\chi(v')+\frac{4}{3}v'^2\chi(v')-\frac{5}{6}\right\} = -\frac{2}{3}. \qquad (5\text{–}4.189)$$

For the integral of (5–4.176), we observe that

$$\int_0^1 dv^2\,(3-v^2)(1+v^2)\int_0^v dv'\,\frac{\chi(v')}{1-v'^2} = \int_0^1 dv\left[\frac{11}{3}+\frac{2}{3}v^2-\frac{1}{3}v^4\right]\chi(v). \qquad (5\text{–}4.190)$$

The sum of the integrals of (5–4.175) and (5–4.176), from which $\delta M$ cancels, is then

$$2\int_0^1 dv\, v^2(3-v^2)[(1+v^2)\chi(v)-1]\log\left(\frac{2m}{\mu}\frac{1}{(1-v^2)^{1/2}}\right)-\frac{11}{3}\int_0^1 dv\,\chi(v)-\frac{29}{27}, \qquad (5\text{–}4.191)$$

where we have introduced the numerical values of all the integrals of type (5–4.102),

$\int_0^1 dv\, v^{2n}\chi(v)$, $n \geqslant 1$. As for the integral of (5–4.187), its contribution to the coefficient of $\alpha^2/3\pi^2$ is

$$-2\left(\log\frac{2m}{\mu}-1\right)\int_0^1 dv\, v^2(3-v^2)[(1+v^2)\chi(v)-1]-6\int_0^1 dv\, v^2\chi(v)$$
$$+4\int_0^1 dv\, v^4\chi(v)-\int_0^1 dv(1+v^2)\left[\frac{8}{3}-v^2-v^2(3-v^2)\chi(v)\right]\log\frac{v^2}{1-v^2}, \tag{5–4.192}$$

which uses the principal value integral

$$P\int_0^1 dv\,\frac{v^2(3-v^2)}{v^2-v'^2}=\frac{8}{3}-v'^2-v'^2(3-v'^2)\chi(v'). \tag{5–4.193}$$

The sum of (5–4.191) and (5–4.192), from which the fictitious photon mass finally cancels, is

$$\int_0^1 dv\, v^2(3-v^2)[(1+v^2)\chi(v)-1]\log\frac{1}{1-v^2}-\int_0^1 dv(1+v^2)\left[\frac{8}{3}-v^2\right.$$
$$\left.-v^2(3-v^2)\chi(v)\right]\log\frac{v^2}{1-v^2}-\frac{11}{3}\int_0^1 dv\,\chi(v)-\frac{14}{27}. \tag{5–4.194}$$

Some significant combinations for this evaluation are

$$\int_0^1 dv\, v^2(1+v^2)(3-v^2)\chi(v)\log v^2-\frac{11}{3}\int_0^1 dv\,\chi(v)=-\frac{44}{3}\left(\frac{1}{3}+\frac{1}{25}\right), \tag{5–4.195}$$

and

$$\int_0^1 dv\left(1-4v^2+\frac{5}{3}v^4\right)\log\frac{1}{1-v^2}=-\frac{8}{15}. \tag{5–4.196}$$

The result is expressed by

$$\int dv^2\, M^2a(M^2)=\frac{4\alpha}{15\pi}+\frac{82}{27}\frac{\alpha^2}{3\pi^2}=\frac{4\alpha}{15\pi}\left[1+\left(1+\frac{1}{81}\right)\frac{15\alpha}{4\pi}\right]. \tag{5–4.197}$$

This fractional increase is somewhat smaller than in the spin 0 situation, but it is still roughly one percent. The effect on the added constants of the energy displacement calculation is given by

$$-\frac{1}{5}\to-\frac{1}{5}-\left(1+\frac{1}{81}\right)\frac{3\alpha}{4\pi}, \tag{5–4.198}$$

where the unit [Eq. (4–11.114)] is 135.6 MHz. This represents a decrease in the 2*s*-level splitting of 0.24 MHz. It alters the last estimate, of Eq. (5–3.168), to

$$\text{H:}\quad E_{2s_{1/2}} - E_{2p_{1/2}} = 1057.93\ \text{MHz}, \tag{5-4.199}$$

which is strikingly close to the nominal experimental value of 1057.90 ± 0.10 MHz. The usual caveat about still unconsidered effects continues to apply, however.

The integrations required to exhibit $a(M^2)$ are very similar to those of the spin 0 situation. Such a relationship also appears in the results, for the substitution $v^2 \to \frac{1}{2}(3 - v^2)$, performed in all the terms of (5-4.132) that have such a factor, yields the precise spin $\frac{1}{2}$ counterparts, as displayed below:

$$\begin{aligned} M^2a(M^2) = \frac{\alpha}{3\pi} v\tfrac{1}{2}(3 - v^2) + \frac{\alpha^2}{3\pi^2}\Bigg\{&\frac{1}{2}(3 - v^2)(1 + v^2)\left[\frac{\pi^2}{6} + \log\frac{1+v}{2}\log\frac{1+v}{1-v}\right. \\ &\left. + 2l\left(\frac{1-v}{1+v}\right) + 2l\left(\frac{1+v}{2}\right) - 2l\left(\frac{1-v}{2}\right) - 4l(v) + l(v^2)\right] \\ &+ \left[\frac{11}{16}(3 - v^2)(1 + v^2) + \tfrac{1}{4}v^4 - \tfrac{3}{2}v(3 - v^2)\right]\log\frac{1+v}{1-v} \\ &+ 6v\frac{3 - v^2}{2}\log\frac{1+v}{2} - 4v\frac{3 - v^2}{2}\log v + \tfrac{3}{8}v(5 - 3v^2)\Bigg\}. \end{aligned} \tag{5-4.200}$$

The limiting behaviors here are

$$M^2 \gg (2m)^2:\quad M^2a(M^2) = \frac{\alpha}{3\pi} + \frac{\alpha^2}{4\pi^2} = \frac{\alpha}{3\pi}\left(1 + \frac{3\alpha}{4\pi}\right), \tag{5-4.201}$$

where the $\alpha^2$ contribution again comes entirely from the last term in the brace, and

$$M^2 \sim (2m)^2:\quad M^2a(M^2) = \frac{\alpha}{2\pi}v + \frac{\alpha^2}{4} = \frac{\alpha}{2\pi}v\left(1 + \frac{\pi\alpha}{2v}\right), \tag{5-4.202}$$

in which the $\alpha^2$ term continues to spring from the first bracket of the brace, with its origin in the form factor. Indeed, as was to be expected, the multiplicative factor of (5-4.202) is the same as with spin 0 [Eq. (5-4.134)]. A simple interpolation formula, which is weighted somewhat differently than for spin 0, is

$$M^2a(M^2) \sim \frac{\alpha}{3\pi}v\frac{3 - v^2}{2}\left[1 + \frac{\pi\alpha}{2v} - \frac{3+v}{4}\left(\frac{\pi}{2} - \frac{3}{4\pi}\right)\alpha\right]. \tag{5-4.203}$$

The reason for this shift in weight appears on comparing the two braces of Eqs. (5-4.132) and (5-4.200) in the following way:

$$\begin{aligned} \{\ \}_{\text{spin}\,\frac{1}{2}} - \frac{3 - v^2}{2v^2}\{\ \}_{\text{spin}\,0} = &\left[\frac{11}{16}(3 - v^2)(1 + v^2) + \tfrac{1}{4}v^4 - \frac{3 - v^2}{2v^2}\left(5\left(\frac{1 + v^2}{2}\right)^2 - 2\right)\right] \\ &\times \log\frac{1+v}{1-v} + \frac{3}{8}v(5 - 3v^2) - \frac{3}{2}\frac{3 - v^2}{2v}(1 + v^2) \end{aligned}$$

$$\cong - 3v, \quad v \ll 1. \tag{5-4.204}$$

When the interpolation formulas are used, with the weight factor $\frac{3}{4}$ symbolized by $\lambda$ for the moment, the above combination becomes

$$- \tfrac{3}{2}v[\lambda(\tfrac{1}{2}\pi^2 - \tfrac{3}{4}) - \tfrac{1}{2}(\tfrac{1}{2}\pi^2 - 3)], \quad v \ll 1. \tag{5-4.205}$$

The identification of the two expressions, for $v \ll 1$, then gives

$$\lambda = \frac{1}{2}\frac{\pi^2 + 2}{\pi^2 - \tfrac{3}{2}} = 0.71, \tag{5-4.206}$$

which, for simplicity, has been replaced with the nearby fraction $\frac{3}{4}$. When the interpolation formula (5-4.203) is used in the calculation of (5-4.197), the co-efficient of $\alpha^2/\pi^2$ is found to be

$$\frac{\pi^2}{4}\left(\frac{9}{40} + \frac{1}{9}\right) + \frac{1}{4}\left(\frac{3}{5} + \frac{7}{48}\right) = 1.016, \tag{5-4.207}$$

as compared with the exact answer,

$$\frac{82}{81} = 1.012. \tag{5-4.208}$$

Harold has a question.

H.: Perhaps I am overlooking a point, but shouldn't there be some mention of the annihilation scattering mechanism which accompanies the Coulomb scattering process that you have considered, in computing the vacuum polarization energy shift?

S.: Let me restate the question and, thereby, jog your memory. The modified photon propagation function has been exhibited in two forms. One [cf. Eq. (4-3.81)] is

$$\bar{D}_+(k) = \frac{1}{k^2}\frac{1}{1 - k^2 \displaystyle\int dM^2 \frac{a(M^2)}{k^2 + M^2}}, \tag{5-4.209}$$

and the other [Eq. (4-3.83)] is given by

$$\bar{D}_+(k) = \frac{1}{k^2} + \int dM^2 \frac{A(M^2)}{k^2 + M^2}, \tag{5-4.210}$$

where the connection between them [Eq. (4-3.85)] is repeated as

$$A(M^2) = \frac{a(M^2)}{\left[1 - M^2 P \displaystyle\int dM'^2 \frac{a(M'^2)}{M^2 - M'^2}\right]^2 + [\pi M^2 a(M^2)]^2}. \tag{5-4.211}$$

The weight function $a(M^2)$ characterizes an irreducible interaction process, the indefinite repetition of which is described by the denominator structure of (5-4.209). To the accuracy with which we have worked in this section, it suffices to expand the denominator factor:

$$\bar{D}_+(k) \simeq \frac{1}{k^2} + \int dM^2 \frac{a(M^2)}{k^2+M^2} + k^2\left[\int dM^2 \frac{a(M^2)}{k^2+M^2}\right]^2. \qquad (5\text{-}4.212)$$

It is the last term here that represents the annihilation interaction, the repetition of the two-particle exchange process. As we see, it does not contribute for $k = 0$, which is the approximate situation in the energy shift calculation. Now, one might ask how the same conclusion emerges on using the form (5-4.210), where the required quantity is the integral

$$\int dM^2 \frac{A(M^2)}{M^2}, \qquad (5\text{-}4.213)$$

since $A(M^2)$, as given by (5-4.211), certainly incorporates the repetition of the basic interaction process. Let us just note that, to the required order,

$$A(M^2) \simeq a(M^2) + 2M^2 a(M^2) P \int dM'^2 \frac{a(M'^2)}{M^2 - M'^2}, \qquad (5\text{-}4.214)$$

and, indeed,

$$\int dM^2\, a(M^2) P \int dM'^2 \frac{a(M'^2)}{M^2-M'^2} = P\int dM^2\, dM'^2 \frac{a(M^2)a(M'^2)}{M^2-M'^2} = 0. \quad (5\text{-}4.215)$$

Incidentally, I should draw attention to the relation (5-4.214), written as

$$A(M^2) \simeq \left[1 + M^2 P \int dM'^2 \frac{a(M'^2)}{M^2-M'^2}\right]^2 a(M^2), \qquad (5\text{-}4.216)$$

since it is analogous to the use already made of form factors, in improving the two particle exchange contribution. The form factor occurring here is the one that multiplies $D_+(k)$ to give $\bar{D}_+(k)$, evaluated at $k^2 = -M^2$,

$$F = \frac{1}{1 + M^2 \int dM'^2 \dfrac{a(M'^2)}{-M^2+M'^2-i\varepsilon}} = 1 - M^2 \int dM'^2 \frac{A(M'^2)}{-M^2+M'^2-i\varepsilon}. \qquad (5\text{-}4.217)$$

The relation (5-4.216) is an approximate one which, according to Eq. (5-4.211), is precisely stated as

$$A(M^2) = |F|^2 a(M^2). \qquad (5\text{-}4.218)$$

This prescription is physically sensible since, as a probability measure, the weight function $A(M^2)$ can be constructed from the absolute squares of emission probability amplitudes.

## 5-5 POSITRONIUM. MUONIUM

Electrodynamics, in its narrow sense, is concerned with the properties of those few particles whose dominant interaction mechanisms are electromagnetic in character. These are: the photon, the electron (positron), and the muon (positive-negative). There are also two kinds of unstable composite particles that have become accessible experimentally: positronium ($e^+e^-$) and muonium ($\mu^+e^-$). This section is mainly focused on positronium. It is the purest of electrodynamic systems. These atoms have fine and hyperfine structures that reflect completely known electromagnetic interactions, and their instability only involves decay into photons. In contrast, muonium invokes the weak interactions, which introduces the neutrino: $\mu^+e^- \to e^+ + e^- + 2\nu$.

The positronium structures are essentially non-relativistic, with a gross energy spectrum given by the Bohr formula that is appropriate to the reduced mass of $\frac{1}{2}m$. These binding energies are

$$|E_n| = \frac{1}{2n^2}\,\mathrm{Ry} = \frac{6.8029}{n^2}\,\mathrm{ev}. \tag{5-5.1}$$

The states of given principal quantum number $n = 1, 2, 3, \ldots$ can be further labeled by the quantum number $L = 0, 1, 2, \ldots$ of relative orbital angular momentum, the spin quantum number $S = 0, 1$, and the total angular momentum quantum number $J = 0, 1, 2, \ldots$. A particular state is designated as $n\,{}^{2S+1}L_J$. Relativistic effects and electromagnetic interactions other than the Coulomb attraction induce a fine structure splitting and a hyperfine structure splitting. Unlike hydrogen, with its large mass ratio, the fine and hyperfine structures in positronium are of the same order of magnitude. Particularly interesting is the hyperfine structure of the ground state, the splitting between the $1\,{}^3S_1$ and $1\,{}^1S_0$ levels. Positronium atoms formed in excited states will radiatively decay down to one of the hyperfine levels of the ground state. These atoms eventually annihilate completely into photons. We begin with a discussion of the annihilation mechanism.

The nature of the photon decay of positronium is governed by a selection rule associated with the concept of charge reflection. In general, charge reflection ($Q \to -Q$) converts a given state into a different one. But, for electrically neutral systems, another state of the same kind is produced and one can introduce the eigenvectors of the charge reflection operation. With two particles of opposite charge, as in positronium, there is a symmetrical and an antisymmetrical combination of the two charge assignments, corresponding to

$$r_q = \pm 1. \tag{5–5.2}$$

Now, the effect of interchanging all attributes of the two particles is controlled by the statistics of the particles, which, for F.D. particles, demands a net sign change. When the spatial coordinates are interchanged in a state of orbital quantum number $L$, the spherical harmonic governing the angle dependence responds with the factor $(-1)^L$ [cf. Eq. (2–7.21)]. As for the spin functions, triplet and singlet states are, respectively, symmetrical and antisymmetrical, as symbolized by the factor $-(-1)^S$. Thus, the full expression of F.D. statistics for the electron-positron system is contained in

$$-1 = r_q[-(-1)^S](-1)^L, \tag{5–5.3}$$

or

$$r_q = (-1)^{L+S}. \tag{5–5.4}$$

Accordingly, the $^1S_0$ state is charge symmetric ($r_q = +1$), and the $^3S_1$ state is charge antisymmetric ($r_q = -1$).

The state of a system of $n$ photons is represented by the product of the sources, $n$ in number, that emit or absorb these particles. Since every photon source, as an electric current, reverses sign under charge reflection, the charge parity of an $n$-photon state is

$$r_q = (-1)^n. \tag{5–5.5}$$

Hence, if charge parity is to be maintained in time, the $1\,^1S_0$ state, with $r_q = +1$, can only decay into an even number of photons, most probably $n = 2$, while the $1\,^3S_1$ state decay is restricted to an odd number of photons, most probably $n = 3$, since a single real photon is excluded. This inhibition in the decay mechanism of the $^3S_1$ state will result in a considerably slower rate of decay, compared to that of $1\,^1S_0$ positronium.

There is another reflection aspect of these states that deserves mention. It refers to space parity. The space reflection matrix is [Eq. (2–6.39)]

$$r_s = i\gamma^0, \tag{5–5.6}$$

which implies that the intrinsic parity, characterizing a particle at rest with $\gamma^{0\prime} = +1$, is $i$. This value of the intrinsic parity (which could equally well be $-i$) is independent of the electric charge value. Any arbitrariness in definition disappears for the two-particle positronium states, where the intrinsic parity becomes $(\pm i)^2 = -1$. That is superimposed on the orbital parity which, for a state of angular momentum quantum number $L$, is $(-1)^L$. Accordingly, the complete space parity is

$$P = -(-1)^L. \tag{5-5.7}$$

For consistency with conventional notation, we shall then designate the charge parity as $C$:

$$C = (-1)^{L+S}, \tag{5-5.8}$$

and note that

$$CP = -(-1)^S. \tag{5-5.9}$$

To the extent that $C$ and $P$, or at least the product $CP$, are exact quantum numbers, the distinction between singlet and triplet spin states is precisely maintained. The singlet and triplet classes of positronium are sometimes referred to as para and ortho positronium, respectively.

The $S$-levels that constitute the ground state of the gross structure have $P = -1$, which asserts the intrinsic parity of the two-particle system. Hence, the ${}^1S_0$ particle, with zero total angular momentum and odd parity, would be described by a pseudoscalar field ($\phi$), while the ${}^3S_1$ particle, a system with unit angular momentum and odd parity, is characterized by a vector field ($\phi_\mu$). A phenomenological description of the two-photon decay of ${}^1S_0$ positronium is provided by the gauge invariant coupling

$$\phi(-\tfrac{1}{4})^*F^{\mu\nu}F_{\mu\nu}, \tag{5-5.10}$$

and, indeed, this pseudoscalar type of coupling has already been exhibited in Eqs. (3-13.75, 76). As noted in that context, it implies that the two photons are orthogonally polarized. There are two possible gauge invariant combinations for the unit spin system, namely,

$$\tfrac{1}{2}(\partial_\mu\phi_\nu - \partial_\nu\phi_\mu)F^{\mu\nu}(-\tfrac{1}{4})F^{\kappa\lambda}F_{\kappa\lambda} \tag{5-5.11}$$

and

$$\tfrac{1}{2}(\partial_\mu\phi_\nu - \partial_\nu\phi_\mu)^*F^{\mu\nu}(-\tfrac{1}{4})^*F^{\kappa\lambda}F_{\kappa\lambda}, \tag{5-5.12}$$

corresponding to the two ways in which a four-dimensional rotationally invariant combination for a pair of photons can be formed. We shall later derive the precise combination that is applicable to ortho positronium, and also point out an essential difference in the nature of the two-photon and three-photon couplings.

The decay rate of para positronium can be quickly obtained from the annihilation cross section of free particles given in Eq. (3-13.74). In order to deal with annihilation in singlet states, the factor of $\frac{1}{4}$ that was introduced in averaging over all spin orientations must be removed. Then the rate of annihilation is produced by multiplying $4\sigma$ by the relative particle flux $v|\psi(0)|^2$, where $\psi$ is the non-relativistic wave function for relative motion. It is evaluated at the origin to represent the conditions of the relativistic annihilation process. Accordingly, the decay rate is

$$\gamma_{n^1S} = 4\pi \frac{\alpha^2}{m^2} |\psi_{nS}(0)|^2 = \frac{1}{2n^3} \alpha^5 m, \tag{5–5.13}$$

since

$$|\psi_{nS}(0)|^2 = \frac{1}{\pi(na)^3}, \tag{5–5.14}$$

in which the positronium Bohr radius is given by

$$\frac{1}{a} = \frac{1}{2a_0} = \frac{1}{2}\alpha m. \tag{5–5.15}$$

The implied lifetime for the ground level of para positronium is

$$\tau_{\text{para}} = \frac{2}{\alpha^5}\frac{1}{m} = 1.245 \times 10^{-10} \text{ sec}, \tag{5–5.16}$$

where it has been helpful to note that

$$\tfrac{1}{2}\alpha^5 m = \alpha^3 \text{ Ry} = 3\pi \times 135.6(2\pi \times 10^6) \text{ sec}^{-1}, \tag{5–5.17}$$

according to the energy unit of Eq. (4–11.113).

Now, let us repeat this derivation in the spirit of the phenomenological coupling (5–5.10), as derived from (3–13.75). The field product that appears here, a specialization of $\psi(x)\psi(x')$, is to be replaced by the two-particle field of the interacting system, $\psi(xx')$. The field of a given para positronium atom possesses a rather simple, non-relativistic character. For the two spinor indices we have, effectively, $\gamma^{0\prime} = +1$; the charge labels in this $C = 1$ state are combined in the symmetrical, normalized function

$$2^{-1/2}\,\delta_{q,-q'}; \tag{5–5.18}$$

and the normalized spin function of the antisymmetrical singlet state is

$$2^{-1/2}\sigma\,\delta_{\sigma,-\sigma'}. \tag{5–5.19}$$

There is also a normalized (equal-time) wave function for relative motion, $\psi(\mathbf{r})$, and a wave function for center of mass motion [cf. Eqs. (5–1.113, 116)] which in the rest frame is

$$(4m\,d\omega_P)^{1/2}\exp(iPx) = \left(\frac{(d\mathbf{P})}{(2\pi)^3}\right)^{1/2} \exp(-i2mx^0), \tag{5–5.20}$$

since the mass of positronium is very close to $2m$. We must finally remark that the normalization condition for two identical particles contains the factor $\frac{1}{2}$ to avoid repetitious counting. Thus, the field associated with a specific atom is indicated by

$$2^{1/2}[2^{-1/2}\,\delta_{q,-q'}][2^{-1/2}\sigma\,\delta_{\sigma,-\sigma'}]\psi(\mathbf{r})(4m\,d\omega_P)^{1/2}\exp(iPx), \qquad (5\text{–}5.21)$$

where the $\gamma^{0\prime} = +1$ restriction is implicit. Now, the antisymmetrical matrix $\gamma^0\gamma_5$ anticommutes with $\gamma^0$, but commutes with the charge and spin matrices. Accordingly, it connects equal values of $\gamma^0$ and opposite values of charge and spin quantum numbers, since all the associated matrices are antisymmetrical. [This is illustrated by $\psi\gamma^0\gamma_5 q\psi = \psi q\gamma^0\gamma_5\psi = (-q\psi)\gamma^0\gamma_5\psi$.] In the $\gamma^{0\prime} = +1$ subspace, the antisymmetrical $\gamma^0\gamma_5$ matrix effectively reduces to $\sigma\,\delta_{\sigma,-\sigma'}$, apart from a phase factor, as exhibited in Eq. (3–13.72), where helicity labels are used. This results in the equivalence

$$\tfrac{1}{2}\psi(x)\gamma^0\gamma_5\psi(x) \to 2^{-1/2}[2^{1/2}][2^{1/2}]\psi(0)(4m)^{1/2}\phi(x) = 2^{3/2}m^{1/2}\psi(0)\phi(x), \qquad (5\text{–}5.22)$$

where the phenomenological para positronium field $\phi(x)$ has been introduced to characterize the center of mass motion. The interaction term of Eq. (3–13.75) is thereby replaced by

$$W_{\text{para}} = \frac{4\pi\alpha}{m^2}\left(\frac{2}{m}\right)^{1/2}\psi(0)\int(dx)\phi(x)\mathbf{E}(x)\cdot\mathbf{H}(x). \qquad (5\text{–}5.23)$$

Its prediction for the decay rate is

$$\gamma_{\text{para}} = (4\pi\alpha)^2\frac{2}{m^2}|\psi(0)|^2\int d\omega_k\,d\omega_{k'}\,(2\pi)^4\,\delta(k+k'-P) = \frac{4\pi\alpha^2}{m^2}|\psi(0)|^2, \qquad (5\text{–}5.24)$$

which, naturally, coincides with (5–5.13).

The vacuum amplitude that describes three-photon decay is [cf. Eq. (3–12.24)]

$$i\tfrac{1}{2}\int(dx)(dx')(dx'')\psi(x)\gamma^0 eq\gamma A(x)G_+(x-x')eq\gamma A(x')G_+(x'-x'')eq\gamma A(x'')\psi(x''). \qquad (5\text{–}5.25)$$

We exhibit the coefficient of $iJ^*_{k\lambda}iJ^*_{k'\lambda'}iJ^*_{k''\lambda''}$, with the particle field taken as $\psi\exp[ip_0x]$, $p_0^2 + m^2 = 0$:

$$ie^2\left[\int(dx)\exp[i(2p_0-k-k'-k'')x]\right](d\omega_k\,d\omega_{k'}\,d\omega_{k''})^{1/2}\tfrac{1}{2}\psi\gamma^0 eq$$
$$\times\left\{\gamma e\frac{m-\gamma(k-p_0)}{-2kp_0}\gamma e'\frac{m-\gamma(p_0-k'')}{-2k''p_0}\gamma e''+\cdots\right\}\psi, \qquad (5\text{–}5.26)$$

which displays only one of the six ways of assigning the three photons. In the approximation of free-particle motion we have

$$(\gamma p_0 + m)\psi = 0, \qquad \psi\gamma^0(\gamma p_0 - m) = 0, \tag{5-5.27}$$

and we shall also adopt the gauge

$$p_0 e = 0, \tag{5-5.28}$$

for all photons. This leads to the substitutions

$$\begin{aligned}\psi\gamma^0\gamma e[m - \gamma(k - p_0)] &= \psi\gamma^0(m - \gamma p_0)\gamma e - \psi\gamma^0\gamma e\gamma k\\ &= \psi\gamma^0 i\sigma^{\mu\nu}e_\mu k_\nu\end{aligned} \tag{5-5.29}$$

and

$$\begin{aligned}[m - \gamma(p_0 - k'')]\gamma e''\psi &= \gamma k''\gamma e''\psi + \gamma e''(m + \gamma p_0)\psi\\ &= i\sigma^{\mu\nu}e''_\mu k''_\nu\psi,\end{aligned} \tag{5-5.30}$$

where the magnetic moment interaction with the electromagnetic field has become explicit. The latter will be expressed in three-dimensional form,

$$\sigma^{\mu\nu}e_\mu k_\nu = \boldsymbol{\sigma}\cdot\mathbf{e}\times\mathbf{k} - \gamma_5\boldsymbol{\sigma}\cdot\mathbf{e}k^0, \tag{5-5.31}$$

which uses the relation

$$\sigma^{0l} = -\gamma_5\sigma^l, \tag{5-5.32}$$

or, alternatively,

$$\boldsymbol{\gamma} = i\gamma^0\gamma_5\boldsymbol{\sigma}, \tag{5-5.33}$$

and applies the gauge condition (5-5.28) in the rest frame of $p_0$.

The braced expression of Eq. (5-5.26) can now be written out as

$$\begin{aligned}-\frac{i\gamma^0\gamma_5}{4kp_0k''p_0}&[(\boldsymbol{\sigma}\cdot\mathbf{e}\times\mathbf{k} + \gamma_5\boldsymbol{\sigma}\cdot\mathbf{e}k^0)\boldsymbol{\sigma}\cdot\mathbf{e}'(\boldsymbol{\sigma}\cdot\mathbf{e}''\times\mathbf{k}'' - \gamma_5\boldsymbol{\sigma}\cdot\mathbf{e}''k^{0''})\\ &+ (\boldsymbol{\sigma}\cdot\mathbf{e}''\times\mathbf{k}'' + \gamma_5\boldsymbol{\sigma}\cdot\mathbf{e}''k^{0''})\boldsymbol{\sigma}\cdot\mathbf{e}'(\boldsymbol{\sigma}\cdot\mathbf{e}\times\mathbf{k} - \gamma_5\boldsymbol{\sigma}\cdot\mathbf{e}k^0)] + \text{cycl. perm.},\end{aligned} \tag{5-5.34}$$

where we make explicit the result of interchanging $k$, $e$ and $k''$, $e''$. The various spin products can be reduced with the aid of the relations

$$\boldsymbol{\sigma}\cdot\mathbf{A}\boldsymbol{\sigma}\cdot\mathbf{B}\boldsymbol{\sigma}\cdot\mathbf{C} + \boldsymbol{\sigma}\cdot\mathbf{C}\boldsymbol{\sigma}\cdot\mathbf{B}\boldsymbol{\sigma}\cdot\mathbf{A} = 2\mathbf{A}\cdot\mathbf{B}\boldsymbol{\sigma}\cdot\mathbf{C} + 2\mathbf{B}\cdot\mathbf{C}\boldsymbol{\sigma}\cdot\mathbf{A} - 2\mathbf{A}\cdot\mathbf{C}\boldsymbol{\sigma}\cdot\mathbf{B} \tag{5-5.35}$$

and

$$\boldsymbol{\sigma}\cdot\mathbf{A}\boldsymbol{\sigma}\cdot\mathbf{B}\boldsymbol{\sigma}\cdot\mathbf{C} - \boldsymbol{\sigma}\cdot\mathbf{C}\boldsymbol{\sigma}\cdot\mathbf{B}\boldsymbol{\sigma}\cdot\mathbf{A} = 2i\mathbf{A}\times\mathbf{B}\cdot\mathbf{C}. \qquad (5\text{–}5.36)$$

The latter combination appears multiplied by $\gamma_5$. The resulting Dirac field structure, $\psi\gamma^0 eq\gamma^0\psi$, does not contribute under the assumed conditions, however. That follows by combining the equations of (5–5.27) into

$$0 = \psi\gamma^0 eq(\gamma p_0 + m)\psi + \psi\gamma^0(\gamma p_0 - m)eq\psi = -2m\psi\gamma^0 eq\gamma^0\psi, \qquad (5\text{–}5.37)$$

in which the last version is the $p_0$ rest frame evaluation. The form that emerges for (5–5.34) is

$$-\frac{1}{2m^2}\frac{1}{k^0k^{0\prime}k^{0\prime\prime}}[\mathbf{e}\times\mathbf{k}\cdot\mathbf{e}'k^{0\prime}\boldsymbol{\gamma}\cdot\mathbf{e}''\times\mathbf{k}'' + \mathbf{e}''\times\mathbf{k}''\cdot\mathbf{e}'k^{0\prime}\boldsymbol{\gamma}\cdot\mathbf{e}\times\mathbf{k}$$
$$- \mathbf{e}\times\mathbf{k}\cdot\mathbf{e}''\times\mathbf{k}''\boldsymbol{\gamma}\cdot\mathbf{e}'k^{0\prime} + k^0k^{0\prime}k^{0\prime\prime}(\mathbf{e}\cdot\mathbf{e}'\boldsymbol{\gamma}\cdot\mathbf{e}'' + \mathbf{e}'\cdot\mathbf{e}''\boldsymbol{\gamma}\cdot\mathbf{e}$$
$$- \mathbf{e}\cdot\mathbf{e}''\boldsymbol{\gamma}\cdot\mathbf{e}')] + \text{cycl. perm.} \qquad (5\text{–}5.38)$$

The effect of adding the three cyclic permutations is particularly simple for the second set of terms in (5–5.38), where it produces the symmetrical combination

$$k^0k^{0\prime}k^{0\prime\prime}(\mathbf{e}\cdot\mathbf{e}'\boldsymbol{\gamma}\cdot\mathbf{e}'' + \mathbf{e}'\cdot\mathbf{e}''\boldsymbol{\gamma}\cdot\mathbf{e} + \mathbf{e}''\cdot\mathbf{e}\boldsymbol{\gamma}\cdot\mathbf{e}'). \qquad (5\text{–}5.39)$$

In the elements $\mathbf{e}\times\mathbf{k}$ and $\mathbf{e}k^0$ we recognize the vectorial aspects of the magnetic and electric fields that are associated with a given photon. Indeed, in (5–5.38) we have a combination of individual photon fields that is produced by the total field structure

$$\tfrac{1}{2}(\mathbf{E}^2 - \mathbf{H}^2)\boldsymbol{\gamma}\cdot\mathbf{E} + \mathbf{E}\cdot\mathbf{H}\boldsymbol{\gamma}\cdot\mathbf{H}. \qquad (5\text{–}5.40)$$

With the proportional identification of $\frac{1}{2}\psi\gamma^0 q\gamma_k\psi$ with $\partial_0\phi_k - \partial_k\phi_0$, we encounter a particular linear combination of the two structures given in Eqs. (5–5.11, 12). But what is not anticipated in those phenomenological forms is the additional factor of $(kp_0k'p_0k''p_0)^{-1}$ that is exhibited in (5–5.38). This constitutes a form factor, coupling the photon fields non-locally to the ortho positronium field. In retrospect, we recognize that the two-photon probability amplitude, expressed in terms of field strengths, also has the factor $(kp_0k'p_0)^{-1}$, but the latter is completely fixed by the kinematics. We are being reminded that the intuition usually associated with phenomenological couplings refers to a system with inverse dimensions that are large compared with the momenta of the excitations that it emits or absorbs. Here we have the opposite extreme since [cf. Eq. (5–5.15)] $1/a$ is much smaller than $m$, the characteristic unit of the annihilation process. Accordingly, there is non-locality on the latter scale.

The importance of the form factor is emphasized by considering the limit as one photon energy approaches zero, thereby simulating a physical process in which the presence of a homogeneous magnetic field induces an ortho positronium atom to decay by two-photon emission. The singular limit of zero photon energy indicates a secular growth with time of the probability amplitude, which is actually limited by the finite splitting between the ortho and para ground levels, and constitutes the basis for measuring this quantity. If this homogeneous field is designated by $\mathbf{H}_0$, the field structure inferred from (5-5.40) is

$$\mathbf{E}\cdot\mathbf{H}\boldsymbol{\gamma}\cdot\mathbf{H}_0 - \mathbf{H}_0\cdot\mathbf{H}\boldsymbol{\gamma}\cdot\mathbf{E} + \mathbf{H}_0\cdot\mathbf{E}\boldsymbol{\gamma}\cdot\mathbf{H}, \tag{5-5.41}$$

or

$$\mathbf{E}\cdot\mathbf{H}\boldsymbol{\gamma}\cdot\mathbf{H}_0 - (\mathbf{E}\times\mathbf{H})\cdot(\boldsymbol{\gamma}\times\mathbf{H}_0). \tag{5-5.42}$$

The last term does not contribute in two-body decay, where the photons have equal and opposite momenta, since

$$\mathbf{e}\times(\mathbf{e}'\times\mathbf{k}') + \mathbf{e}'\times(\mathbf{e}\times\mathbf{k}) = -(\mathbf{k}+\mathbf{k}')\mathbf{e}\cdot\mathbf{e}' = 0. \tag{5-5.43}$$

What remains is just the pseudo-scalar coupling that characterizes the two-photon decay of para positronium. This stimulated decay takes place in ortho positronium atoms that are polarized parallel to the magnetic field, or, equivalently, have zero magnetic quantum number relative to the magnetic field direction. We have arrived at a description of the magnetic field induced mixing of para and ortho positronium, which is customarily handled by atomic perturbation theory. Some details of the latter, including the removal of the limitation to weak magnetic fields, will be given later.

To proceed with the calculation of the three-photon decay rate, we consider the particular state of the $^3S_1$, $C = -1$, system that has zero magnetic quantum number, as above, relative to an arbitrary direction with unit vector $\boldsymbol{\nu}$. The associated field, analogous to (5-5.21), is

$$2^{1/2}[2^{-1/2}q\,\delta_{q,-q'}][2^{-1/2}\,\delta_{\sigma,-\sigma'}]\psi(\mathbf{r})(4m\,d\omega_P)^{1/2}\exp(iPx), \tag{5-5.44}$$

and

$$\tfrac{1}{2}\psi(x)\gamma^0 q\boldsymbol{\nu}\cdot\boldsymbol{\gamma}\psi(x) \to 2^{3/2}m^{1/2}\psi(0)\boldsymbol{\nu}\cdot\boldsymbol{\phi}(x). \tag{5-5.45}$$

The decay rate is then inferred as

$$\gamma_{\text{ortho}} = (4\pi\alpha)^3\frac{1}{2m^4}|\psi(0)|^2\frac{1}{6}\int d\omega_k\,d\omega_{k'}\,d\omega_{k''}\,(2\pi)^4\,\delta(k+k'+k''-P)\sum|M|^2, \tag{5-5.46}$$

where, introducing unit vectors $\mathbf{n}$ along the photon propagation directions, we have

$$M = (\mathbf{e}\times\mathbf{n}\cdot\mathbf{e}' + \mathbf{e}'\times\mathbf{n}'\cdot\mathbf{e})\boldsymbol{\nu}\cdot\mathbf{e}''\times\mathbf{n}'' + (\mathbf{e}\cdot\mathbf{e}' - \mathbf{e}\times\mathbf{n}\cdot\mathbf{e}'\times\mathbf{n}')\boldsymbol{\nu}\cdot\mathbf{e}'' + \text{cycl. perm.}, \tag{5-5.47}$$

and the summation in (5-5.46) is extended over all possible polarizations. The factor of $\frac{1}{8}$ then removes the repetitious counting of the photons. Since the final result is independent of the vector $\boldsymbol{\nu}$, it is convenient to average over that direction, first. The polarization summations are performed using the dyadic relation

$$\sum_{\mathbf{e}} \mathbf{e}\mathbf{e} = \sum_{\mathbf{e}} \mathbf{e}\times\mathbf{n}\,\mathbf{e}\times\mathbf{n} = \mathbf{1} - \mathbf{n}\mathbf{n}, \tag{5-5.48}$$

which expresses the completeness of the two $\mathbf{e}$ vectors and $\mathbf{n}$. These summations are illustrated by

$$\begin{aligned}
&\sum_{\mathbf{e}\mathbf{e}'\mathbf{e}''} (\mathbf{e}\times\mathbf{n}\cdot\mathbf{e}' + \mathbf{e}'\times\mathbf{n}'\cdot\mathbf{e})^2(\mathbf{e}''\times\mathbf{n}'')^2 \\
&= 2\sum_{\mathbf{e}\mathbf{e}'} [(\mathbf{e}\times\mathbf{n}\cdot\mathbf{e}')^2 + (\mathbf{e}'\times\mathbf{n}'\cdot\mathbf{e})^2 + 2\mathbf{e}\times\mathbf{n}\cdot\mathbf{e}'\,\mathbf{e}'\times\mathbf{n}'\cdot\mathbf{e}] \\
&= 2\sum_{\mathbf{e}} [1 - (\mathbf{n}'\cdot\mathbf{e}\times\mathbf{n})^2 + 1 - (\mathbf{n}'\cdot\mathbf{e})^2 - 2\mathbf{n}\cdot\mathbf{n}'] = 4(1-\mathbf{n}\cdot\mathbf{n}')^2,
\end{aligned} \tag{5-5.49}$$

and

$$\begin{aligned}
&\sum_{\mathbf{e}\mathbf{e}'\mathbf{e}''} (\mathbf{e}\cdot\mathbf{e}' - (\mathbf{e}\times\mathbf{n})\cdot(\mathbf{e}'\times\mathbf{n}'))^2(\mathbf{e}'')^2 \\
&= 2\sum_{\mathbf{e}\mathbf{e}'} [(1-\mathbf{n}\cdot\mathbf{n}')^2(\mathbf{e}\cdot\mathbf{e}')^2 + (\mathbf{e}\cdot\mathbf{n}')^2(\mathbf{e}'\cdot\mathbf{n})^2 + 2(1-\mathbf{n}\cdot\mathbf{n}')\mathbf{e}\cdot\mathbf{e}'\,\mathbf{e}\cdot\mathbf{n}'\,\mathbf{e}'\cdot\mathbf{n}] \\
&= 2\sum_{\mathbf{e}} [(1-\mathbf{n}\cdot\mathbf{n}')^2(1-(\mathbf{n}'\cdot\mathbf{e})^2) + (1-(\mathbf{n}\cdot\mathbf{n}')^2)(\mathbf{n}'\cdot\mathbf{e})^2 - 2(1-\mathbf{n}\cdot\mathbf{n}')\mathbf{n}\cdot\mathbf{n}'(\mathbf{n}'\cdot\mathbf{e})^2] \\
&= 4(1-\mathbf{n}\cdot\mathbf{n}')^2.
\end{aligned} \tag{5-5.50}$$

It turns out that all the other types of terms combine to cancel, thus giving

$$\sum |M|^2 = 8(1-\mathbf{n}\cdot\mathbf{n}')^2, \tag{5-5.51}$$

which has been written in the unsymmetrical form permitted by the equivalence of all the photons with respect to the integration. This stage of the calculation is expressed by

$$\gamma_{\text{ortho}} = (4\pi\alpha)^3 \frac{4}{m^4} |\psi(0)|^2 \frac{1}{6} \int d\omega_k\, d\omega_{k'}\, d\omega_{k''}\, (2\pi)^4\, \delta(k+k'+k''-P)(1-\mathbf{n}\cdot\mathbf{n}')^2. \tag{5-5.52}$$

One gives invariant form to $1-\mathbf{n}\cdot\mathbf{n}'$ as

$$1-\mathbf{n}\cdot\mathbf{n}' = -4m^2\frac{kk'}{kPk'P} = -4m^2\frac{kk'}{(kk'+kk'')(kk'+k'k'')}. \tag{5-5.53}$$

We shall group two of the photons into a system of mass $M$, as indicated by

$$\int d\omega_k\, d\omega_{k'}\, d\omega_{k''}\, (2\pi)^4\, \delta(k + k' + k'' - P)$$

$$= \int d\omega_k\, d\omega_{k'}\, (2\pi)^3\, \delta(k + k' - K)\, dM^2\, d\omega_K\, d\omega_{k''}\, (2\pi)^4\, \delta(K + k'' - P), \quad (5\text{-}5.54)$$

where

$$M^2 = -(k + k')^2 = -2kk' \quad (5\text{-}5.55)$$

and

$$(2m)^2 = -(K + k'')^2 = M^2 - 2Kk''. \quad (5\text{-}5.56)$$

In the rest frame of $K$, we have

$$\int d\omega_k\, d\omega_{k'}\, (2\pi)^3\, \delta(k + k' - K)(1 - \mathbf{n}\cdot\mathbf{n}')^2$$

$$= \frac{1}{(4\pi)^2}\int_{-1}^{1}\frac{1}{2}\,dz\left[\frac{4m^2\frac{1}{2}M^2}{\left(\frac{1}{2}M^2 + \frac{4m^2 - M^2}{4}(1 - z)\right)\left(\frac{1}{2}M^2 + \frac{4m^2 - M^2}{4}(1 + z)\right)}\right]^2, \quad (5\text{-}5.57)$$

where $z$ is the cosine of the angle between $\mathbf{k} = -\mathbf{k}'$ and $\mathbf{k}''$. Performing the $z$ integration gives

$$\frac{1}{\pi^2}\frac{(4m^2)^2M^4}{(4m^2 + M^2)^2}\left[\frac{\log 4m^2/M^2}{(4m^2 + M^2)(4m^2 - M^2)} + \frac{1}{2}\frac{1}{4m^2M^2}\right]. \quad (5\text{-}5.58)$$

Then, since

$$\int d\omega_K\, d\omega_{k''}\, (2\pi)^3\, \delta(K + k'' - P) = \frac{1}{(4\pi)^2}\left(1 - \frac{M^2}{4m^2}\right), \quad (5\text{-}5.59)$$

the remaining integration is proportional to

$$\int_0^{(2m)^2} dM^2\left(1 - \frac{M^2}{4m^2}\right)\frac{4m^2M^4}{(4m^2 + M^2)^2}\left[\frac{\log 4m^2/M^2}{(4m^2 + M^2)(4m^2 - M^2)} + \frac{1}{2}\frac{1}{4m^2M^2}\right]$$

$$= \int_0^1 du\left[\frac{u^2}{(1 + u)^3}\log\frac{1}{u} + \frac{1}{2}\frac{u(1 - u)}{(1 + u)^2}\right], \quad (5\text{-}5.60)$$

in which

$$u = M^2/4m^2. \quad (5\text{-}5.61)$$

Successive partial integrations reduce the latter integral to

$$\int_0^1 du \frac{1}{1+u} \log \frac{1}{u} - \frac{3}{4} = \frac{1}{12}(\pi^2 - 9), \tag{5–5.62}$$

which evaluation makes use of yet another partial integration to get

$$\int_0^1 du \frac{1}{1+u} \log \frac{1}{u} = \int_0^1 du \frac{1}{u} \log(1+u) = \frac{\pi^2}{12}, \tag{5–5.63}$$

as an application of Eq. (5–4.115). Putting things together we find the decay rate to be

$$\gamma_{n^3S} = \frac{16}{9}(\pi^2 - 9)\frac{\alpha^3}{m^2}|\psi(0)|^2 = \frac{2}{9\pi}(\pi^2 - 9)\frac{1}{n^3}\alpha^6 m, \tag{5–5.64}$$

and the lifetime for the ground level of ortho positronium is

$$\tau_{\text{ortho}} = \frac{9\pi}{4}\frac{1}{\pi^2 - 9}\frac{1}{\alpha}\tau_{\text{para}} = 1.387 \times 10^{-7}\,\text{sec}. \tag{5–5.65}$$

The energy spectrum of positronium is first approached by applying the results of Section 5–2, specifically the rest frame energy operator of Eq. (5–2.134), where we now have

$$m_1 = m_2 = m, \qquad \mu = \tfrac{1}{2}m, \qquad e_1 = -e_2 = e. \tag{5–5.66}$$

This gives

$$\begin{aligned} H_0 = 2m + \frac{\mathbf{p}^2}{m} - \frac{1}{4m^3}(\mathbf{p}^2)^2 - \alpha\Bigg\{&\frac{1}{r} + \frac{1}{2m^2}\mathbf{p}\cdot\left(\frac{1}{r} + \frac{\mathbf{rr}}{r^3}\right)\cdot\mathbf{p} - \frac{\pi}{m^2}\delta(\mathbf{r}) \\ &- \frac{3}{4}\frac{1}{m^2}\frac{(\boldsymbol{\sigma}_1 + \boldsymbol{\sigma}_2)\cdot\mathbf{r}\times\mathbf{p}}{r^3} - \frac{1}{4m^2}\left(3\frac{\boldsymbol{\sigma}_1\cdot\mathbf{r}\boldsymbol{\sigma}_2\cdot\mathbf{r}}{r^5} - \frac{\boldsymbol{\sigma}_1\cdot\boldsymbol{\sigma}_2}{r^3}\right) \\ &- \frac{2\pi}{3}\frac{1}{m^2}\boldsymbol{\sigma}_1\cdot\boldsymbol{\sigma}_2\,\delta(\mathbf{r})\Bigg\}. \end{aligned} \tag{5–5.67}$$

The simplest application is to the singlet levels of para positronium where, effectively,

$$\boldsymbol{\sigma}_1 + \boldsymbol{\sigma}_2 = 0, \tag{5–5.68}$$

thus reducing (5–5.67) to

$$H_{0\,\text{para}} = 2m + \frac{\mathbf{p}^2}{m} - \frac{1}{4m}\left(\frac{\mathbf{p}^2}{m}\right)^2 - \alpha\left\{\frac{1}{r} + \frac{1}{2m^2}\mathbf{p}\cdot\left(\frac{1}{r} + \frac{\mathbf{rr}}{r^3}\right)\cdot\mathbf{p} + \frac{\pi}{m^2}\delta(\mathbf{r})\right\}. \tag{5–5.69}$$

In order to find the first deviations from the gross structure, the spectrum non-relativistic energy operator

$$\frac{\mathbf{p}^2}{m} - \frac{\alpha}{r} = T + V, \tag{5–5.70}$$

we apply the result of Eq. (5–2.148), with $m$ replaced by $\frac{1}{2}m$ in accordance with its origin in the non-relativistic energy operator [Eq. (5–2.144)]. The consequence, expressed as

$$-\frac{2\alpha}{m}\left\langle \mathbf{p}\cdot\left(\frac{1}{r} + \frac{\mathbf{rr}}{r^3}\right)\cdot\mathbf{p}\right\rangle - \frac{4\pi\alpha}{m}\langle\delta(\mathbf{r})\rangle = \langle 2\{T, V\} + V^2\rangle, \tag{5–5.71}$$

enables us to present the expectation value of (5–5.69) in the form

$$E_{\text{rel.}} - 2m = E_{\text{non-rel.}} + \frac{1}{4m}\langle 2\{T, V\} + V^2 - T^2\rangle. \tag{5–5.72}$$

The elimination of the potential energy gives

$$\begin{aligned}\langle 2\{T, V\} + V^2 - T^2\rangle &= E^2_{\text{non-rel.}} + 2E_{\text{non-rel.}}\langle T\rangle - 4\langle T^2\rangle \\ &= -E^2_{\text{non-rel.}} - 4\langle T^2\rangle\end{aligned} \tag{5–5.73}$$

since, in the Coulomb field [Eq. (5–2.152)],

$$\langle T\rangle = -E_{\text{non-rel.}}. \tag{5–5.74}$$

We also know that, in the state of orbital quantum number $L$ [Eq. (5–2.155), with $l \to L$, $m \to \frac{1}{2}m$],

$$\langle T^2\rangle = \frac{m^2\alpha^4}{4n^3}\left(\frac{1}{L+\frac{1}{2}} - \frac{3}{4n}\right). \tag{5–5.75}$$

The outcome is an expression for the first terms in a power series expansion of the para positronium fine structure:

$$E_{\text{para}}(n, L) = 2m - \frac{m\alpha^2}{4n^2} - \frac{m\alpha^4}{4n^3}\left(\frac{1}{L+\frac{1}{2}} - \frac{11}{16n}\right). \tag{5–5.76}$$

Notice that the reduced mass formula of Eq. (5–2.163), with $M = m$, $\mu = \frac{1}{2}m$, $j = L$, reproduces this result, except for the numerical coefficient 11/16. [That too would be right if, consistent with its reference to $M \gg m$, the last term of (5–2.163) were replaced by $-\mu^2\alpha^4/8(M+m)n^4$.]

The spectrum of ortho positronium is considerably more elaborate. We must take into account the spin-orbit term of Eq. (5–5.67),

$$\frac{3}{4}\frac{\alpha}{m^2}\frac{(\boldsymbol{\sigma}_1 + \boldsymbol{\sigma}_2)\cdot\mathbf{r}\times\mathbf{p}}{r^3} = \frac{3}{2}\frac{\alpha}{m^2}\frac{\mathbf{S}\cdot\mathbf{L}}{r^3}, \tag{5–5.77}$$

the tensor spin-spin coupling also exhibited there:

$$\frac{\alpha}{4m^2}\left(3\frac{\boldsymbol{\sigma}_1\cdot\mathbf{r}\boldsymbol{\sigma}_2\cdot\mathbf{r}}{r^5}-\frac{\boldsymbol{\sigma}_1\cdot\boldsymbol{\sigma}_2}{r^3}\right)=\frac{\alpha}{2m^2}\left(3\frac{(\mathbf{S}\cdot\mathbf{r})^2}{r^5}-\frac{\mathbf{S}^2}{r^3}\right). \tag{5-5.78}$$

For a given total angular momentum quantum number $J$, which values generally are $J = L + 1, L, L - 1$, one has only to take the expectation value of (5–5.77) in the state $nLJ$, according to

$$\left\langle\frac{\mathbf{L}\cdot\mathbf{S}}{r^3}\right\rangle_{nLJ} = \tfrac{1}{2}[J(J+1)-L(L+1)-2]\left\langle\frac{1}{r^3}\right\rangle_{nL}. \tag{5-5.79}$$

The factors appearing here are

$$\tfrac{1}{2}[J(J+1)-L(L+1)-2]=\begin{cases} J=L+1 : L\\ J=L \quad\;\; : -1\\ J=L-1 : -L-1\end{cases}, \tag{5-5.80}$$

and [Eq. (4–11.107), with $a_0 \to 2a_0$]

$$\left\langle\frac{1}{r^3}\right\rangle_{nL}=\frac{\alpha^3m^3}{8n^3}\frac{1}{L(L+\frac{1}{2})(L+1)}. \tag{5-5.81}$$

For the particular example $J = L = 1, 2,\ldots$, the spin-orbit energy shift is

$$\left\langle\frac{3}{2}\frac{\alpha}{m^2}\frac{\mathbf{L}\cdot\mathbf{S}}{r^3}\right\rangle_{nJJ}=-\frac{3}{2}\frac{\alpha}{m^2}\left\langle\frac{1}{r^3}\right\rangle_{nJ}=-\frac{3}{16}\frac{\alpha^4m}{n^3}\frac{1}{J(J+\frac{1}{2})(J+1)}. \tag{5-5.82}$$

The tensor interaction (5–5.78) is more complicated than the spin-orbit coupling, for it can change the orbital angular momentum while maintaining the orbital parity $(-1)^L$, thus mixing the two types of states with $L = J \pm 1$. Since there is no mixing for $J = L$, the spin-angle factor

$$3(\mathbf{S}\cdot\mathbf{n})^2-\mathbf{S}^2, \qquad \mathbf{n}=\mathbf{r}/r, \tag{5-5.83}$$

must have an eigenvalue in that kind of state. There are just two eigenvalues for this combination, corresponding to the unit spin possibilities

$$(\mathbf{S}\cdot\mathbf{n})^{2\prime}=1,0, \tag{5-5.84}$$

which yields

$$(3(\mathbf{S}\cdot\mathbf{n})^2-\mathbf{S}^2)'=1,-2. \tag{5-5.85}$$

To learn which of these is the correct eigenvalue for $J = L$, it suffices to use qualitative arguments that are asymptotically accurate for large $L$. The eigenvalues of (5–5.84) distinguish two situations that can be described as $\mathbf{S}$ being parallel (antiparallel) to $\mathbf{n}$, or orthogonal to $\mathbf{n}$, respectively. Since the unit radial vector **n** is orthogonal to **L**, the first of the two situations detailed in Eqs. (5–5.84, 85) can be characterized as one in which **L** and **S** are orthogonal. On inspecting the eigenvalues of $\mathbf{L}\cdot\mathbf{S}$ exhibited in (5–5.80), we recognize that $(\mathbf{L}\cdot\mathbf{S})'/L$, for $L \gg 1$, is $1, 0, -1$, corresponding to $J = L+1, L, L-1$, respectively. Accordingly, the asymptotic situation $\mathbf{L}\cdot\mathbf{S} \sim 0$, in which (5–5.83) has the eigenvalue 1, occurs for $J = L$. Naturally, the unit eigenvalue that appears for $J = L$ can be derived in a more formal way, which is not very lengthy, but such an approach gives no understanding of why that particular eigenvalue appears. The energy shift now deduced from (5–5.78) is

$$\left\langle\frac{\alpha}{2m^2}\left(3\frac{(\mathbf{S}\cdot\mathbf{r})^2}{r^5}-\frac{\mathbf{S}^2}{r^3}\right)\right\rangle_{nJJ} = \frac{\alpha}{2m^2}\left\langle\frac{1}{r^3}\right\rangle_{nJ} = \frac{1}{16}\frac{\alpha^4 m}{n^3}\frac{1}{J(J+\frac{1}{2})(J+1)}. \tag{5–5.86}$$

On adding (5–5.82) and (5–5.86), we get the energy displacement of the non-$S$ ortho positronium ${}^3J_J$ states relative to the para positronium ${}^1J_J$ levels:

$$E_{\text{ortho}}(nJJ) - E_{\text{para}}(nJJ) = -\frac{1}{8}\frac{\alpha^4 m}{n^3}\frac{1}{J(J+\frac{1}{2})(J+1)}. \tag{5–5.87}$$

The first examples of mixed ortho positronium states are: ${}^3S_1 + {}^3D_1, {}^3P_2 + {}^3F_2$, where both orbital states must belong to the same gross structure level, if the mixing is to be appreciable. Hence, for $n = 1$, where only $L = 0$ occurs, and $n = 2$ with $L = 0, 1$, no such mixing can appear. Since available experimental data are limited to $n = 1$, we give no further details about the mixing of levels. The ortho-para splitting of the ground $S$-level comes entirely from the last term of (5–5.67). Recalling that

$$(\boldsymbol{\sigma}_1\cdot\boldsymbol{\sigma}_2)' = \begin{cases} S = 1: & 1 \\ S = 0: & -3 \end{cases}, \tag{5–5.88}$$

we get

$$E_{\text{ortho}} - E_{\text{para}} = \frac{8\pi}{3}\frac{\alpha}{m^2}|\psi(0)|^2 = \tfrac{1}{3}\alpha^4 m, \tag{5–5.89}$$

which is also produced by the hyperfine structure formula of Eq. (4–17.16) on placing $M_p = m$, $Z = 1$, $S = \frac{1}{2}$, $g_S = 2$. Even at the present level of accuracy, this is not the complete story, however.

Ortho positronium, with $C = -1$, decays into three photons because a single real photon is excluded kinematically. But a single virtual photon can be emitted and reabsorbed, which leads to an additional energy displacement relative to para positronium. The exchange of a virtual photon is described by the interaction

$$W = \frac{1}{2}\int (dx)(dx')j^\mu(x)\, D_+(x-x')\, j_\mu(x'), \tag{5–5.90}$$

where, in consequence of the covariantly stated correspondence of Eq. (5–5.45),

$$j^\mu(x) \to e\,2^{3/2}m^{1/2}\psi(0)\phi^\mu(x). \tag{5–5.91}$$

Since the annihilation mechanism involves the exchange of the mass $2m$, we have, effectively,

$$D_+(x-x') = \int \frac{(dk)}{(2\pi)^4}\frac{\exp[ik(x-x')]}{k^2} \to -\frac{1}{(2m)^2}\,\delta(x-x'), \tag{5–5.92}$$

and the annihilation coupling becomes

$$\begin{aligned} W_{\text{annih.}} &= -\frac{8\pi\alpha}{m}\,|\psi(0)|^2 \int (dx)\tfrac{1}{2}\phi^\mu(x)\phi_\mu(x) \\ &= \int (dx)[-\,\delta m^2_{\text{ortho}}\tfrac{1}{2}\phi^\mu(x)\phi_\mu(x)]; \end{aligned} \tag{5–5.93}$$

the last expression states the phenomenological interpretation of this term, as a mass displacement. Hence,

$$\delta m_{\text{ortho}} = (E_{\text{ortho}} - E_{\text{para}})_{\text{annih.}} = \frac{2\pi\alpha}{m^2}\,|\psi(0)|^2 = \tfrac{1}{4}\alpha^4 m, \tag{5–5.94}$$

and the complete statement that replaces (5–5.89) is

$$E_{\text{ortho}} - E_{\text{para}} = (\tfrac{1}{3}+\tfrac{1}{4})\alpha^4 m = \frac{7}{6}\alpha^2\,\text{Ry} = 2.0439\times 10^5\ \text{MHz}. \tag{5–5.95}$$

A recent experimental value is

$$E_{\text{ortho}} - E_{\text{para}} = (2.0340 \pm 0.0001)\times 10^5\ \text{MHz}. \tag{5–5.96}$$

The agreement to within $\frac{1}{2}$ percent is very gratifying, particularly in view of the anticipated presence of theoretical modifications of relative order $\alpha$.

But before considering the latter, let us fulfill the promise to discuss the effect of a magnetic field in mixing ortho and para positronium. In the $S$-states of relative motion, the coupling with a magnetic field comes entirely from the spin magnetic moments. This interaction energy, with a particular assignment of the charge labels, is

$$-\frac{e}{2m}\,(\boldsymbol{\sigma}_1 - \boldsymbol{\sigma}_2)\cdot\mathbf{H}. \tag{5–5.97}$$

The antisymmetry in the two spins implies a vanishing expectation value in the singlet and triplet states. Indeed, the sole effect of (5–5.97) is the expected one, of mixing the ${}^1S_0$, $C = 1$, state with the ${}^3S_1$, $C = -1$, state. Only the magnetic level $m = 0$ of the ${}^3S_1$ state is coupled with the ${}^1S_0$ state since the angular momentum about the $z$-axis, the magnetic field direction, is still conserved. The matrix element can be inferred by noting that

$$\langle {}^1S|[(\boldsymbol{\sigma}_1 - \boldsymbol{\sigma}_2)_z]^2|{}^1S\rangle = \langle {}^1S|(\boldsymbol{\sigma}_1 - \boldsymbol{\sigma}_2)_z|{}^3S\rangle\langle {}^3S|(\boldsymbol{\sigma}_1 - \boldsymbol{\sigma}_2)_z|{}^1S\rangle$$
$$= 4, \tag{5-5.98}$$

since $\boldsymbol{\sigma}_1 + \boldsymbol{\sigma}_2$ vanishes in the singlet state. Accordingly, with a permissible choice of phase, the submatrix of the energy operator for this pair of levels is

$$\begin{pmatrix} E_{\text{para}} & -eH/m \\ -eH/m & E_{\text{ortho}} \end{pmatrix}. \tag{5-5.99}$$

Its eigenvalues are

$$E = \tfrac{1}{2}(E_{\text{ortho}} + E_{\text{para}}) \pm [\tfrac{1}{4}(E_{\text{ortho}} - E_{\text{para}})^2 + (eH/m)^2]^{1/2}, \tag{5-5.100}$$

while the amplitudes of the two states, as determined by the eigenvector equations and the normalization condition, are given by

$$\psi_{\text{para}} = \frac{eH/m}{E_{\text{para}} - E}\psi_{\text{ortho}}, \qquad \left[1 + \left(\frac{eH/m}{E_{\text{para}} - E}\right)^2\right]|\psi_{\text{ortho}}|^2 = 1, \tag{5-5.101}$$

or

$$\psi_{\text{ortho}} = \frac{eH/m}{E_{\text{ortho}} - E}\psi_{\text{para}}, \qquad \left[1 + \left(\frac{eH/m}{E_{\text{ortho}} - E}\right)^2\right]|\psi_{\text{para}}|^2 = 1. \tag{5-5.102}$$

These amplitudes enable one to compute the decay rate of a mixed state:

$$\gamma = |\psi_{\text{para}}|^2\gamma_{\text{para}} + |\psi_{\text{ortho}}|^2\gamma_{\text{ortho}}. \tag{5-5.103}$$

For weak magnetic fields, as defined by

$$eH/m \ll \tfrac{1}{2}\Delta E, \tag{5-5.104}$$

where

$$\Delta E = E_{\text{ortho}} - E_{\text{para}}, \tag{5-5.105}$$

the energy eigenvalues are

$$E \simeq E_{\text{ortho}} + \frac{(eH/m)^2}{\Delta E}, \qquad E_{\text{para}} - \frac{(eH/m)^2}{\Delta E}. \tag{5-5.106}$$

Corresponding to these two alternatives, which describe perturbed ortho and para levels, respectively, we have

$$|\psi_{\text{para}}|^2 \simeq \left(\frac{eH/m}{\Delta E}\right)^2, \qquad |\psi_{\text{ortho}}|^2 \simeq 1 - \left(\frac{eH/m}{\Delta E}\right)^2 \tag{5-5.107}$$

and

$$|\psi_{\text{ortho}}|^2 \cong \left(\frac{eH/m}{\Delta E}\right)^2, \qquad |\psi_{\text{para}}|^2 \cong 1 - \left(\frac{eH/m}{\Delta E}\right)^2. \tag{5-5.108}$$

The associated decay rates are

$$\gamma \cong \gamma_{\text{ortho}} + \left(\frac{eH/m}{\Delta E}\right)^2 (\gamma_{\text{para}} - \gamma_{\text{ortho}}) \tag{5-5.109}$$

and

$$\gamma \cong \gamma_{\text{para}} - \left(\frac{eH/m}{\Delta E}\right)^2 (\gamma_{\text{para}} - \gamma_{\text{ortho}}), \tag{5-5.110}$$

the first of which describes the increased rate for ortho positronium decay owing to the induced process of two-photon emission. Here is the mechanism that was earlier discussed qualitatively as an application of the form factor associated with three-photon decay. The general situation of arbitrary magnetic field strength is described by the perturbation theory results just obtained. In particular, the strong field limit, where the inequality of (5-5.104) is reversed, is such that $\psi_{\text{para}} = \mp\, \psi_{\text{ortho}}$ and, thus, both decay rates tend to a common limit, the equally weighted average $\frac{1}{2}(\gamma_{\text{para}} + \gamma_{\text{ortho}})$.

As a first step toward evaluating the modification of order $\alpha$ in the ortho-para splitting energy, we consider the single photon exchange of the annihilation mechanism. The two elements that have been combined in the calculation are: the primitive interaction describing the interconversion of photon and electron-positron pair; the photon propagation function. For the latter we must now use the modified propagation function

$$\bar{D}(k) = \frac{1}{k^2} + \frac{1}{4m^2}\frac{\alpha}{\pi}\int_0^1 dv\, \frac{v^2(1-\frac{1}{3}v^2)}{1+(k^2/4m^2)(1-v^2)}, \tag{5-5.111}$$

while the primitive interaction is altered in accordance with the form factor generalization

$$\gamma A(k) \to F_1(k)\gamma A(k) + \frac{\alpha}{2\pi}\frac{1}{2m}F_2(k)\sigma F(k). \tag{5-5.112}$$

The propagation function calculation is immediate. The evaluation for $k^2 = -4m^2$ gives

$$\bar{D}(k) = -\frac{1}{4m^2}\left[1 - \frac{\alpha}{\pi}\int_0^1 dv\,(1-\tfrac{1}{3}v^2)\right], \tag{5-5.113}$$

which constitutes a decrease in the annihilation contribution to the ortho-para splitting by the factor

$$1 - \frac{8}{9}\frac{\alpha}{\pi}. \tag{5–5.114}$$

Turning to the additional magnetic moment coupling that appears in (5–5.112), let us note that

$$\sigma F(k) = \tfrac{1}{2}[-\gamma k, \gamma A(k)]. \tag{5–5.115}$$

In the rest frame creation process, for example,

$$-\gamma k = 2m\gamma^0, \tag{5–5.116}$$

and

$$\psi^*\gamma^0 = \psi^*, \qquad \gamma^0\psi^* = -\psi^*, \tag{5–5.117}$$

which means that the primitive coupling $\gamma A(k)$ is multiplied by the effective form factor

$$F_e(k) = F_1(k) + \frac{\alpha}{2\pi}F_2(k). \tag{5–5.118}$$

The components of this form factor, of which only the real part is significant, are given in Eqs. (5–4.183, 184, 185) as

$$F_1 = 1 - \frac{\alpha}{2\pi}f_1(v), \qquad f_1(v) = f(v) + \chi(v),$$

$$F_2 = -(1 - v^2)\chi(v), \tag{5–5.119}$$

which leads to

$$F_e = 1 - \frac{\alpha}{2\pi}[f(v) + (2 - v^2)\chi(v)]. \tag{5–5.120}$$

In the non-relativistic situation of interest $(v \ll 1)$, $\chi(v)$ reduces to unity, and [Eqs. (5–4.87, 89)]

$$f(v) = P\int_0^1 dv' \; \frac{(1 + v'^2)\log\dfrac{v'^2}{1 - v'^2}}{v^2 - v'^2} + 2\left(\log\frac{2m}{\mu} - 1\right)((1 + v^2)\chi(v) - 1)$$

$$\cong -\frac{\pi^2}{2v} + 2, \tag{5–5.121}$$

according to the small $v$ limit of (5–4.131), which uses the properties

$$\left.\frac{d}{dv}l(v)\right|_{v=0} = 1, \qquad \left.\frac{d}{dv}l(v)\right|_{v=1/2} = 2\log 2. \tag{5–5.122}$$

Notice that the reference to the photon mass, the sign of infra-red sensitivity, has disappeared at this level of accuracy. The implication for the modification of the single photon exchange process is given by the factor

$$[F_e]^2 \cong 1 - \frac{\alpha}{\pi}\left(-\frac{\pi^2}{2v} + 4\right) \cong \left(1 + \frac{\pi\alpha}{2v}\right)\left(1 - \frac{4\alpha}{\pi}\right). \tag{5–5.123}$$

We recognize in $1 + (\pi\alpha/2v)$ the function that gives the non-relativistic evaluation of $|\psi(0)|^2$ for free particles [cf. Eqs. (5–4.135, 138)]. The replacement of this evaluation by the one appropriate to the bound positronium atom is already incorporated in the initial calculation. Accordingly, the actual modification of that calculation is given by the other factor of (5–5.123),

$$1 - \frac{4\alpha}{\pi}. \tag{5–5.124}$$

The complete modification of the single-photon annihilation part of the ortho-para splitting is, therefore, provided by the factor

$$\left(1 - \frac{8}{9}\frac{\alpha}{\pi}\right)\left(1 - \frac{4\alpha}{\pi}\right) \cong 1 - \frac{44}{9}\frac{\alpha}{\pi}. \tag{5–5.125}$$

With the level of description altered by the additional factor of $\alpha$, one must now also consider two-photon processes, such as the annihilation mechanism of para positronium. The effective two-photon source there is

$$\begin{aligned} iJ^\mu(x)J^\nu(x')|_{\text{eff.}} = \tfrac{1}{2}e^2[\psi(x)\gamma^0\gamma^\mu G_+(x - x')\gamma^\nu\psi(x') \\ + \psi(x')\gamma^0\gamma^\nu G_+(x' - x)\gamma^\mu\psi(x)], \end{aligned} \tag{5–5.126}$$

and the vacuum amplitude describing a two-photon exchange is obtained from

$$\tfrac{1}{2}\int (dx)\cdots(dx''')iJ_1{}^\mu(x)J_1{}^\nu(x')|_{\text{eff.}}\, D_+(x - x'')\, D_+(x' - x''')iJ_{2\mu}(x'')J_{2\nu}(x''')|_{\text{eff.}}. \tag{5–5.127}$$

We have the option of performing a causal or a non-causal evaluation of this coupling. The latter facilitates simplifications based on the special nature of the $^1S$ state, and we introduce it by replacing (5–5.127) with the momentum version

$$\frac{1}{2}\int \frac{(dk)}{(2\pi)^4}\frac{(dk')}{(2\pi)^4}\frac{1}{k^2 - i\varepsilon}\frac{1}{k'^2 - i\varepsilon}iJ_1{}^\mu(-k)J_1{}^\nu(-k')|_{\text{eff.}}\, iJ_{2\mu}(k)J_{2\nu}(k')|_{\text{eff.}}. \tag{5–5.128}$$

The space-time structure of the particle fields is sufficiently indicated by

$$\psi_1(x) = \exp(-i\tfrac{1}{2}p_1x)\,\psi^*, \qquad \psi_2(x) = \exp(i\tfrac{1}{2}p_2x)\,\psi \tag{5-5.129}$$

where, in the positronium rest frame,

$$p_1{}^0 = p_2{}^0 \simeq 2m. \tag{5-5.130}$$

Accordingly,

$$iJ_1{}^\mu(-k)J_1{}^\nu(-k')|_{\text{eff.}} = (2\pi)^4\,\delta(k+k'-p_1)2\pi\alpha\psi^*\gamma^0\left[\gamma^\mu\frac{1}{-\gamma(k-\frac{1}{2}p_1)+m}\gamma^\nu \right.$$
$$\left. + \gamma^\nu\frac{1}{-\gamma(k'-\frac{1}{2}p_1)+m}\gamma^\mu\right]\psi^* \tag{5-5.131}$$

and

$$iJ_2{}^\mu(k)J_2{}^\nu(k')|_{\text{eff.}} = (2\pi)^4\,\delta(k+k'-p_2)2\pi\alpha\psi\gamma^0\left[\gamma^\mu\frac{1}{\gamma(k-\frac{1}{2}p_2)+m}\gamma^\nu \right.$$
$$\left. + \gamma^\nu\frac{1}{\gamma(k'-\frac{1}{2}p_2)+m}\gamma^\mu\right]\psi. \tag{5-5.132}$$

In view of the momentum restrictions expressed by the delta functions, only one four-dimensional momentum integral occurs. We shall express this through the change of variables ($p_1 = p_2 = p$)

$$k \to \tfrac{1}{2}p + k, \qquad k' \to \tfrac{1}{2}p - k, \tag{5-5.133}$$

and thereby present (5-5.128) as [$-i\varepsilon$ is understood]

$$(2\pi)^4\,\delta(p_1-p_2)(2\pi\alpha)^2\frac{1}{2}\int\frac{(dk)}{(2\pi)^4}\frac{1}{(k+\frac{1}{2}p)^2}\frac{1}{(k-\frac{1}{2}p)^2}\psi^*\gamma^0\left[\gamma^\mu\frac{m+\gamma k}{k^2+m^2}\gamma^\nu\right.$$
$$\left.+\gamma^\nu\frac{m-\gamma k}{k^2+m^2}\gamma^\mu\right]\psi^*\psi\gamma^0\left[\gamma_\mu\frac{m-\gamma k}{k^2+m^2}\gamma_\nu+\gamma_\nu\frac{m+\gamma k}{k^2+m^2}\gamma_\mu\right]\psi. \tag{5-5.134}$$

The nature of the $^1S$ state is such that only pseudoscalar and pseudovector combinations of the particle fields can be formed. The products of three $\gamma$-matrices in (5-5.134) meet this requirement, since

$$\gamma^\mu\gamma^\lambda\gamma^\nu = \cdots + \varepsilon^{\mu\lambda\nu\kappa}\gamma_\kappa\gamma_5, \tag{5-5.135}$$

where the dots indicate other terms involving a single $\gamma$-matrix. The introduction of this simplification in (5-5.134) reduces the latter to

$$(2\pi)^4\,\delta(p_1-p_2)(4\pi\alpha)^2\int\frac{(dk)}{(2\pi)^4}\frac{1}{(k+\frac{1}{2}p)^2}\frac{1}{(k-\frac{1}{2}p)^2}\frac{1}{(k^2+m^2)^2}(k^2g_{\mu\nu}-k_\mu k_\nu)$$
$$\times\,\psi^*\gamma^0\gamma^\mu\gamma_5\psi^*\psi\gamma^0\gamma^\nu\gamma_5\psi, \tag{5-5.136}$$

which uses the relation

$$-\tfrac{1}{2}\varepsilon^{\mu\lambda\nu\kappa}\varepsilon_{\mu\lambda'\nu\kappa'} = \delta^{\lambda}_{\lambda'}\,\delta^{\kappa}_{\kappa'} - \delta^{\kappa}_{\lambda'}\,\delta^{\lambda}_{\kappa'}. \qquad (5\text{–}5.137)$$

Let us note here that, in accordance with the spin 0 character of the system, the pseudovector combination is proportional to the gradient of the pseudoscalar, or

$$\psi\gamma^0\gamma^\nu\gamma_5\psi = -\frac{p^\nu}{2m}\,\psi\gamma^0\gamma_5\psi. \qquad (5\text{–}5.138)$$

This is verified in the rest frame of the vector $p$, where the property $\gamma^0\psi = \psi$, and the antisymmetry of $\gamma^0$, implies the vanishing of the field structure containing $\gamma_k\gamma_5$, which commutes with $\gamma^0$. The analogous relation

$$\psi^*\gamma^0\gamma^\mu\gamma_5\psi^* = \frac{p^\mu}{2m}\,\psi^*\gamma^0\gamma_5\psi^* \qquad (5\text{–}5.139)$$

makes similar reference to the property $\psi^*\gamma^0 = \psi^*$. A particular consequence is given by

$$\psi^*\gamma^0\gamma^\mu\gamma_5\psi^*\psi\gamma^0\gamma_\mu\gamma_5\psi = \psi^*\gamma^0\gamma_5\psi^*\psi\gamma^0\gamma_5\psi. \qquad (5\text{–}5.140)$$

To evaluate the momentum integral of (5–5.136), we use the representation

$$\frac{1}{(k+\frac{1}{2}p)^2}\,\frac{1}{(k-\frac{1}{2}p)^2}\,\frac{1}{(k^2+m^2)^2}$$

$$= \int_0^\infty ds_1\,ds_2\,ds_3\,s_3\exp[-is_1(k+\tfrac{1}{2}p)^2]\exp[-is_2(k-\tfrac{1}{2}p)^2]\exp[-is_3(k^2+m^2)]$$

$$= \int_0^\infty ds\,s^3\int_0^1 du\,u(1-u)\int_{-1}^1\frac{1}{2}dv\exp\left\{-is\left[\frac{1+v}{2}u(k+\tfrac{1}{2}p)^2\right.\right.$$

$$\left.\left.+\frac{1-v}{2}u(k-\tfrac{1}{2}p)^2+(1-u)(k^2+m^2)\right]\right\}. \qquad (5\text{–}5.141)$$

On introducing the property

$$p^2 = -4m^2, \qquad (5\text{–}5.142)$$

the coefficient of $-is$ in the exponent becomes

$$k^2 + uvkp + (1-2u)m^2 = \left(k+\frac{uv}{2}p\right)^2 + (1-2u+u^2v^2)m^2. \qquad (5\text{–}5.143)$$

The basic structure of the momentum integral is, then,

$$\int \frac{(dk)}{(2\pi)^4} k_\mu k_\nu \exp[-is(k+\tfrac{1}{2}uvp)^2] = \int \frac{(dk)}{(2\pi)^4} (k-\tfrac{1}{2}uvp)_\mu (k-\tfrac{1}{2}uvp)_\nu \exp(-isk^2)$$

$$= \left(-\frac{i}{2s} g_{\mu\nu} + \tfrac{1}{4}u^2v^2 p_\mu p_\nu\right) \frac{1}{(4\pi)^2} \frac{1}{is^2}, \qquad (5\text{-}5.144)$$

according to the momentum integrals of (4-8.57) and (4-14.75). More specifically, we have

$$\int \frac{(dk)}{(2\pi)^4} (k^2 g_{\mu\nu} - k_\mu k_\nu) \exp[-is(k+\tfrac{1}{2}uvp)^2]$$

$$= \left[-\frac{3i}{2s} g_{\mu\nu} + \tfrac{1}{4}u^2v^2(p^2 g_{\mu\nu} - p_\mu p_\nu)\right] \frac{1}{(4\pi)^2} \frac{1}{is^2}. \qquad (5\text{-}5.145)$$

The combination $p^2 g_{\mu\nu} - p_\mu p_\nu$ will not contribute because of the null curl form of the vector (5-5.138). This gives the effective evaluation

$$\int \frac{(dk)}{(2\pi)^4} \frac{1}{(k+\frac{1}{2}p)^2} \frac{1}{(k-\frac{1}{2}p)^2} \frac{1}{(k^2+m^2)^2} (k^2 g_{\mu\nu} - k_\mu k_\nu)$$

$$\to -\frac{1}{(4\pi)^2} \tfrac{3}{2} g_{\mu\nu} \int_0^\infty ds \int_0^1 du\, u(1-u) \int_0^1 dv \exp[-ism^2(1-2u+u^2v^2)]$$

$$= i \frac{1}{(4\pi)^2} \tfrac{3}{2} g_{\mu\nu} \frac{1}{m^2} I, \qquad (5\text{-}5.146)$$

where

$$I = \int_0^1 du\, u(1-u) \int_0^1 dv \frac{1}{1-2u+u^2v^2-i\varepsilon}. \qquad (5\text{-}5.147)$$

On employing (5-5.140), and introducing the pseudoscalar field correspondence of Eq. (5-5.22), the vacuum amplitude (5-5.136) becomes

$$i \left[\int (dx) \exp[i(-p_1+p_2)x]\, \phi_1 \phi_2\right] 48 \frac{\alpha^2}{m} |\psi(0)|^2 I, \qquad (5\text{-}5.148)$$

from which we infer the action contribution

$$48 \frac{\alpha^2}{m} |\psi(0)|^2 I \int (dx) \tfrac{1}{2}(\phi(x))^2. \qquad (5\text{-}5.149)$$

This has the structure of a mass term and identifies a (mass)$^2$ displacement, which is complex, corresponding to the instability of the particle. Thus,

$$\delta m^2_{\text{para}} - i m_{\text{para}} \gamma_{\text{para}} = -48 \frac{\alpha^2}{m} |\psi(0)|^2 I \qquad (5\text{-}5.150)$$

and ($m_{\text{para}} \simeq 2m$)

$$\gamma_{\text{para}} = 24\frac{\alpha^2}{m^2}|\psi(0)|^2 \operatorname{Im} I, \qquad \delta m_{\text{para}} = -12\frac{\alpha^2}{m^2}|\psi(0)|^2 \operatorname{Re} I. \tag{5-5.151}$$

For a consistency check with the earlier calculation that is recorded in Eq. (5-5.24), we note that

$$\begin{aligned}\operatorname{Im} I &= \pi\int_0^1 du\, u(1-u)\int_0^1 dv\, \delta(1-2u+u^2v^2) \\ &= \frac{\pi}{2}\int_{1/2}^1 du\,\frac{1-u}{(2u-1)^{1/2}} = \frac{\pi}{2}\int_0^1 dx\,\tfrac{1}{2}(1-x^2) = \frac{\pi}{6},\end{aligned} \tag{5-5.152}$$

where the final evaluation employs the substitution

$$2u - 1 = x^2. \tag{5-5.153}$$

The result inferred from (5-5.151) does indeed agree with (5-5.24). The evaluation of the real part of $I$,

$$\operatorname{Re} I = P\int_0^1 du\, u(1-u)\int_0^1 dv\,\frac{1}{1-2u+u^2v^2}, \tag{5-5.154}$$

proceeds by separating the two regions, $u < \frac{1}{2}$ and $u > \frac{1}{2}$. In the first of these we have

$$u\int_0^1 dv\,\frac{1}{1-2u+u^2v^2} = \frac{1}{(1-2u)^{1/2}}\tan^{-1}\frac{u}{(1-2u)^{1/2}}, \tag{5-5.155}$$

and the introduction of the transformation

$$1 - 2u = x^2 \tag{5-5.156}$$

brings this contribution to the form

$$\int_0^1 dx\,\tfrac{1}{2}(1+x^2)\tan^{-1}\frac{1-x^2}{2x} = \int_0^1 dx\,\frac{x+\frac{1}{3}x^3}{1+x^2} = \tfrac{1}{3}(\log 2 + \tfrac{1}{2}). \tag{5-5.157}$$

For the region $u > \frac{1}{2}$, we replace (5-5.155) with

$$uP\int_0^1 dv\,\frac{1}{1-2u+u^2v^2} = -\frac{1}{(2u-1)^{1/2}}\log\frac{1+(2u-1)^{1/2}}{1-(2u-1)^{1/2}}. \tag{5-5.158}$$

Using the transformation of (5-5.153), this contribution to $\operatorname{Re} I$ is found to be

$$-\int_0^1 dx\,\tfrac{1}{2}(1-x^2)\log\frac{1+x}{1-x} = -\tfrac{1}{3}(2\log 2 - \tfrac{1}{2}), \tag{5-5.159}$$

and

$$\operatorname{Re} I = \tfrac{1}{3}(1 - \log 2). \tag{5–5.160}$$

This gives, as the two-photon annihilation contribution,

$$\delta m_{\text{para}} = -4(1 - \log 2)\frac{\alpha^2}{m^2}|\psi(0)|^2 = -\frac{1 - \log 2}{\pi}\alpha^3 \,\text{Ry}. \tag{5–5.161}$$

In addition to the annihilation processes characteristic of positronium, there are conventional interaction mechanisms in which the two particles maintain their existence. We shall follow the classification of Section **5–3**, where the exchange of transverse photons was superimposed on an initial description employing the instantaneous Coulomb interaction. But, in view of the high energy nature of the process now under consideration, we prefer, from the beginning, to regard the particles as essentially free during the photon exchange acts. Besides the exchange of two transverse photons, we must consider those effects of the Coulomb interaction on single transverse photon exchange that are not summarized by the use of the wave function $\psi(0)$, in conjunction with a static spin interaction. There is, for example, the possibility of an additional instantaneous Coulomb interaction while the transverse photon is in flight. And, we must cease to ignore completely the momentum associated with the relative motion of the particles [this is the use of $\psi(0)$]. As we have often exploited in single-particle contexts, the desired short distance behavior is introduced by a first iteration of the Coulomb interaction on the wave function $\psi(0)$. The processes we have just enumerated constitute all possible ways in which a transverse photon can be combined with an instantaneous Coulomb interaction. In effect, then, we are interested in the totality of two-photon exchanges. [The inclusion of a repeated Coulomb interaction does no harm, since it contains no spin-spin interaction.] It is then slightly simpler not to use the decomposition into instantaneous and transverse interactions of Eqs. (**5–3.131, 132**), but to work with the covariant propagation function

$$D_+(k)_{\mu\nu} = g_{\mu\nu}\frac{1}{k^2 - i\varepsilon}. \tag{5–5.162}$$

With the introduction of causal labels for the particle fields, the effective source of (**5–5.126**) becomes

$$\begin{aligned} iJ^\mu(x)J^\nu(x')\big|_{\text{eff.}} = e^2[&\psi_1(x)\gamma^0\gamma^\mu G_+(x - x')\gamma^\nu\psi_2(x') \\ &+ \psi_1(x')\gamma^0\gamma^\nu G_+(x' - x)\gamma^\mu\psi_2(x)]. \end{aligned} \tag{5–5.163}$$

The two particles will be distinguished by subscripts $a$ and $b$, so that the vacuum amplitude for two-photon exchange is indicated by

$$\frac{1}{2}\int\frac{(dk)}{(2\pi)^4}\frac{(dk')}{(2\pi)^4}\frac{1}{k^2-i\varepsilon}\frac{1}{k'^2-i\varepsilon}iJ^\mu(-k)J^\nu(-k')|_a iJ_\mu(k)J_\nu(k')|_b. \quad (5\text{–}5.164)$$

It suffices again to use the simple form of the particle fields given in Eq. (5–5.129). As a consequence, we have $k' = -k$, and the vacuum amplitude reduces to

$$(2\pi)^4\,\delta(p_1-p_2)(4\pi\alpha)^2\frac{1}{2}\int\frac{(dk)}{(2\pi)^4}\left(\frac{1}{k^2}\right)^2\psi^*\gamma^0\left(\gamma^\mu\frac{1}{\gamma(\frac{1}{2}p-k)+m}\gamma^\nu\right.$$
$$\left.+\gamma^\nu\frac{1}{\gamma(\frac{1}{2}p+k)+m}\gamma^\mu\right)\psi\Big|_a\psi^*\gamma^0\left(\gamma_\mu\frac{1}{\gamma(\frac{1}{2}p+k)+m}\gamma_\nu\right.$$
$$\left.+\gamma_\nu\frac{1}{\gamma(\frac{1}{2}p-k)+m}\gamma_\mu\right)\psi\Big|_b. \quad (5\text{–}5.165)$$

When space and time components are exhibited separately, and the term with only time components discarded, this expression becomes

$$(2\pi)^4\,\delta(p_1-p_2)(4\pi\alpha)^2\frac{1}{2}\int\frac{(dk)}{(2\pi)^4}\left(\frac{1}{k^2}\right)^2\psi^*\gamma^0\left(\gamma_k\frac{1}{\gamma(\frac{1}{2}p-k)+m}\gamma_l\right.$$
$$\left.+\gamma_l\frac{1}{\gamma(\frac{1}{2}p+k)+m}\gamma_k\right)\psi\Big|_a\psi^*\gamma^0\left(\gamma_k\frac{1}{\gamma(\frac{1}{2}p+k)+m}\gamma_l\right.$$
$$\left.+\gamma_l\frac{1}{\gamma(\frac{1}{2}p-k)+m}\gamma_k\right)\psi\Big|_b-(2\pi)^4\,\delta(p_1-p_2)(4\pi\alpha)^2\int\frac{(dk)}{(2\pi)^4}\left(\frac{1}{k^2}\right)^2$$
$$\times\psi^*\gamma^0\left(\gamma_k\frac{1}{\gamma(\frac{1}{2}p-k)+m}\gamma^0+\gamma^0\frac{1}{\gamma(\frac{1}{2}p+k)+m}\gamma_k\right)\psi\Big|_a$$
$$\times\psi^*\gamma^0\left(\gamma_k\frac{1}{\gamma(\frac{1}{2}p+k)+m}\gamma^0+\gamma^0\frac{1}{\gamma(\frac{1}{2}p-k)+m}\gamma_k\right)\psi\Big|_b. \quad (5\text{–}5.166)$$

In view of the $\gamma^0$ eigenvector properties of $\psi$ and $\psi^*$ in the rest frame of $p$, only an even number of $\gamma$ matrices can survive in the individual terms of (5–5.166). The resulting simplifications are illustrated by

$$\gamma_k\frac{1}{\gamma(\frac{1}{2}p-k)+m}\gamma_l=\gamma_k\frac{m-\gamma(\frac{1}{2}p-k)}{k^2-pk}\gamma_l\to\frac{k^0}{k^2-pk}(-i\sigma_{kl}) \quad (5\text{–}5.167)$$

and

$$\gamma_k\frac{1}{\gamma(\frac{1}{2}p-k)+m}\gamma^0\to\frac{1}{k^2-pk}(-i\sigma_{kl}k_l), \quad (5\text{–}5.168)$$

in which only the desired spin structure has been retained. The space vector of the latter result can then be rotationally averaged,

$$k_lk_m\to\tfrac{1}{3}\,\delta_{lm}\,\mathbf{k}^2. \quad (5\text{–}5.169)$$

What emerges for the two terms of the vacuum amplitude (5-5.166) is

$$(2\pi)^4 \delta(p_1 - p_2)(\psi_a\psi_b)^* \boldsymbol{\sigma}_a \cdot \boldsymbol{\sigma}_b \psi_a\psi_b \frac{(4\pi\alpha)^2}{m^2}(I_1 + I_2), \tag{5-5.170}$$

where, written covariantly,

$$\begin{aligned} I_1 &= \int \frac{(dk)}{(2\pi)^4}\left(\frac{1}{k^2}\right)^2 \frac{(pk)^2}{4}\left(\frac{1}{k^2 - pk} + \frac{1}{k^2 + pk}\right)^2 \\ &= \int \frac{(dk)}{(2\pi)^4} \frac{(pk)^2}{(k^2 - pk)^2(k^2 + pk)^2} \end{aligned} \tag{5-5.171}$$

and

$$\begin{aligned} I_2 &= -\tfrac{2}{3}m^2 \int \frac{(dk)}{(2\pi)^4}\left(\frac{1}{k^2}\right)^2\left(k^2 + \frac{(pk)^2}{4m^2}\right)\left(\frac{1}{k^2 - pk} + \frac{1}{k^2 + pk}\right)^2 \\ &= -\frac{2}{3}\int \frac{(dk)}{(2\pi)^4} \frac{(pk)^2 + 4m^2k^2}{(k^2 - pk)^2(k^2 + pk)^2}. \end{aligned} \tag{5-5.172}$$

The sum of the two terms is

$$I = I_1 + I_2 = \tfrac{1}{3}\int \frac{(dk)}{(2\pi)^4} \frac{(pk)^2 - 8m^2k^2}{(k^2 - pk)^2(k^2 + pk)^2}, \tag{5-5.173}$$

and the two contributions that appear under the integral sign can be identified with the exchange of two transverse photons and one transverse photon, respectively.

The integrals are evaluated with the aid of the representation

$$\begin{aligned} \frac{1}{(k^2 - pk)^2}\frac{1}{(k^2 + pk)^2} &= \int_0^\infty ds_1\, s_1 \int_0^\infty ds_2\, s_2 \exp\{-i[s_1(k^2 - pk) + s_2(k^2 + pk)]\} \\ &= \int_0^\infty ds\, s^3 \int_{-1}^1 \frac{1}{2} dv\, \frac{1 - v^2}{4} \exp\{-is(k^2 - pkv)\}, \end{aligned} \tag{5-5.174}$$

where

$$k^2 - pkv = (k - \tfrac{1}{2}pv)^2 + m^2v^2. \tag{5-5.175}$$

After the redefinition $k - \frac{1}{2}pv \to k$, the basic momentum integral encountered in $I$ is

$$\begin{aligned} &\frac{1}{3}\int \frac{(dk)}{(2\pi)^4}[(pk - 2m^2v)^2 - 8m^2(k + \tfrac{1}{2}pv)^2] \exp(-isk^2) \\ &= \frac{1}{3}\int \frac{(dk)}{(2\pi)^4}[(pk)^2 - 8m^2k^2 + 12m^4v^2] \exp(-isk^2) \end{aligned}$$

$$= \frac{1}{(4\pi)^2}\frac{4m^2}{is^2}\left(\frac{3i}{2s}+m^2v^2\right). \tag{5-5.176}$$

Accordingly, we have

$$\begin{aligned} I &= -\frac{i}{(4\pi)^2}4m^2\int_0^1 dv\,\frac{1-v^2}{4}\int_0^\infty ds\left(\frac{3i}{2}+m^2v^2s\right)\exp(-ism^2v^2) \\ &= -\frac{i}{(4\pi)^2}\int_0^1 dv\,\tfrac{1}{2}(1-v^2)\left[\frac{3}{v^2-i\varepsilon}-\frac{2v^2}{(v^2-i\varepsilon)^2}\right]. \end{aligned} \tag{5-5.177}$$

Now note that

$$\begin{aligned} \int_0^1 dv\,\tfrac{1}{2}(1-v^2)\left[\frac{1}{v^2-i\varepsilon}-\frac{2v^2}{(v^2-i\varepsilon)^2}\right] &= \int_0^1 dv\,\tfrac{1}{2}(1-v^2)\frac{d}{dv}\frac{v}{v^2-i\varepsilon} \\ &= \int_0^1 dv\,\frac{v^2}{v^2-i\varepsilon} = 1, \end{aligned} \tag{5-5.178}$$

which converts (5-5.177) into

$$I = -\frac{i}{(4\pi)^2}\left[\int_0^1 dv\,\frac{1-v^2}{v^2-i\varepsilon}+1\right] = -\frac{i}{(4\pi)^2}\int_0^1 dv\,\frac{1}{v^2-i\varepsilon}. \tag{5-5.179}$$

Harold seems upset.

H.: Surely something is wrong here? That last integral doesn't exist!

S.: Indeed. But I was about to recall the special calculational rule that accompanies the technique for introducing the short distance behavior of the wave function. It is stated in Eq. (4-15.45) and implies that the particle propagation function $(p^2+m^2-i\varepsilon)^{-1}$, evaluated for $p^0 = m$, is to be replaced by $(\mathbf{p}^2-\varepsilon^2)^{-1}$, $\varepsilon \to 0$, which enters integrals as a Cauchy principal value. That is what we are encountering here, with the quantity $m^2v^2$ playing the role of $\mathbf{p}^2$. As physical evidence for this identification, note that the first term on the right side of (5-5.173), which is contributed by double transverse photon exchange, does not produce this kind of integral:

$$\begin{aligned} \int\frac{(dk)}{(2\pi)^4}\frac{(pk)^2}{(k^2-pk)^2(k^2+pk)^2} &= -\frac{i}{(4\pi)^2}\int_0^1 dv\,\tfrac{1}{2}(1-v^2)\left(\frac{1}{v^2-i\varepsilon}-\frac{2v^2}{(v^2-i\varepsilon)^2}\right) \\ &= -\frac{i}{(4\pi)^2}. \end{aligned} \tag{5-5.180}$$

It is the second term of (5-5.173), which combines one transverse photon with the instantaneous Coulomb interaction, that is responsible for the singular integral of (5-5.179). It is therefore correct to invoke the principal value rule.

One could have incorporated it explicitly earlier in the calculation, but it seemed simpler to wait until the need became evident. The integral that should appear in (5-5.179) is, then,

$$P\int_0^1 dv\,\frac{1}{v^2-\varepsilon^2}\bigg|_{\varepsilon\to 0} = -\operatorname*{Lim}_{\varepsilon\to 0}\operatorname{Re}\frac{1}{2\varepsilon}\log\frac{v+\varepsilon}{v-\varepsilon}\bigg|_0^1 = -1, \qquad (5\text{-}5.181)$$

and

$$I = \frac{i}{(4\pi)^2}. \qquad (5\text{-}5.182)$$

The vacuum amplitude (5-5.170) has now been evaluated as

$$i\int (dx)\exp[i(-p_1+p_2)x]\,(\psi_a\psi_b)^*\boldsymbol{\sigma}_a\cdot\boldsymbol{\sigma}_b\psi_a\psi_b\,\frac{\alpha^2}{m^2}. \qquad (5\text{-}5.183)$$

We recognize in $\psi_a\psi_b\exp(ip_2x)$ the free-particle form of the two-particle field $\psi(xx)$ that is associated with an emitted positronium atom. The field $(\psi_a\psi_b)^*\times\exp(-ip_1x)$ refers similarly to a detected atom. What replaces them, in describing the bound system, factors into a normalized center of mass function, which is removed by the spatial integration of (5-5.183), and a wave function for relative motion that is evaluated at the origin: $\psi(0)$. The resulting coefficient of $-i\int dx^0$ in (5-5.183) is the desired energy shift,

$$\delta E = -\frac{\alpha^2}{m^2}|\psi(0)|^2\boldsymbol{\sigma}_a\cdot\boldsymbol{\sigma}_b, \qquad (5\text{-}5.184)$$

with

$$\boldsymbol{\sigma}_a\cdot\boldsymbol{\sigma}_b = \begin{cases}\text{ortho:} & 1\\ \text{para:} & -3\end{cases}. \qquad (5\text{-}5.185)$$

This contribution to the ortho-para splitting is, therefore,

$$-\frac{4\alpha^2}{m^2}|\psi(0)|^2 = -\frac{1}{\pi}\alpha^3\,\text{Ry}. \qquad (5\text{-}5.186)$$

There is one other effect to be considered at this level of description. It is the $\alpha/2\pi$ modification of the magnetic moment, which multiplies (5-5.89) by

$$\left(1+\frac{\alpha}{2\pi}\right)^2 \simeq 1+\frac{\alpha}{\pi}. \qquad (5\text{-}5.187)$$

The various contributions to the ortho-para splitting of order $\alpha^3$ Ry, as contained in Eqs. (5-5.125, 161, 186, 187) are made explicit in

$$\frac{1}{\pi}\alpha^3\,\text{Ry}\left[\frac{1}{2}\left(-\frac{44}{9}\right)+(1-\log 2)-1+\frac{2}{3}\right], \tag{5–5.188}$$

and the resulting modification of Eq. (5–5.95) is

$$E_{\text{ortho}}-E_{\text{para}}=\left[\frac{7}{6}-\left(\frac{16}{9}+\log 2\right)\frac{\alpha}{\pi}\right]\alpha^2\,\text{Ry}. \tag{5–5.189}$$

This represents a decrease that is close to ½ percent, reducing the numerical value of (5–5.95) to

$$E_{\text{ortho}}-E_{\text{para}}=2.0338\times 10^5\,\text{MHz}, \tag{5–5.190}$$

which greatly improves the comparison with the experimental value given in Eq. (5–5.96).

H.: I see that you have reproduced the old Karplus-Klein result [cf. *Selected Papers on Quantum Electrodynamics*, Dover, 1958]. But, with no reference to such irrelevancies as divergences, heavy photons and infra-red cutoffs, there is quite a conceptual improvement. Also, since elementary arguments replace the machinery of the two-particle equation, the calculation is greatly simplified. Presumably one could now go on to the next level of description in much the same way?

S.: I should think so, at least with regard to effects of relative order $\alpha^2 \log 1/\alpha$, but we are not yet prepared for an $\alpha^2$ computation. However, I would prefer now to discuss the closely related hyperfine splitting of muonium.

For most purposes, muonium behaves like hydrogen with a lighter nucleus $[m_\mu/m_e = 206.77,\ m_{\text{prot.}}/m_e = 1836.1]$. In particular, the hyperfine splitting of the ground state, which is the quantity accessible to measurement, should be describable in large part by the theory developed in Section 4–17 for an immobile nucleus. But there are dynamical modifications involving the mass ratio $m_e/m_\mu$, and we now proceed to evaluate them in the spirit of the preceding positronium discussion. There is no counterpart to the annihilation mechanism, of course, and our attention moves directly to the discussion of two-photon interaction processes as described, for positronium, in Eq. (5–5.165). The only changes that must be introduced refer to the occurrence of unequal masses, as indicated by

$$e:\quad m,\tfrac{1}{2}p\rightarrow m_e,\frac{m_e}{M}p;\qquad \mu:\quad m,\tfrac{1}{2}p\rightarrow m_\mu,\frac{m_\mu}{M}p, \tag{5–5.191}$$

where

$$M=m_\mu+m_e. \tag{5–5.192}$$

The earlier discussion can be followed to Eq. (5–5.170), but now $(1/m^2)(I_1 + I_2)$ is replaced by

$$\frac{4}{3}\int\frac{(dk)}{(2\pi)^4}\frac{[(pk)^2/M^2]-2k^2}{\left(k^2-\frac{2m_\mu}{M}pk\right)\left(k^2-\frac{2m_e}{M}pk\right)\left(k^2+\frac{2m_\mu}{M}pk\right)\left(k^2+\frac{2m_e}{M}pk\right)}. \qquad (5–5.193)$$

The representations

$$\frac{1}{k^2\mp\frac{2m_\mu}{M}pk}\,\frac{1}{k^2\mp\frac{2m_e}{M}pk}$$

$$=-\int ds_\pm\, s_\pm \tfrac{1}{2}\, dv_\pm \exp\left\{-is_\pm\left[k^2\mp\left(1+\frac{m_\mu-m_e}{M}v_\pm\right)pk\right]\right\} \qquad (5–5.194)$$

are combined in

$$\frac{1}{k^2-\frac{2m_\mu}{M}pk}\,\frac{1}{k^2-\frac{2m_e}{M}pk}\,\frac{1}{k^2+\frac{2m_\mu}{M}pk}\,\frac{1}{k^2+\frac{2m_e}{M}pk}$$

$$=\int_0^\infty ds\, s^3\int_{-1}^{1}\frac{1}{2}\,dv_+\int_{-1}^{1}\frac{1}{2}\,dv_-\int_{-1}^{1}\frac{1}{2}\,dv\,\frac{1-v^2}{4}\exp\{-is[k^2-Vpk]\}, \qquad (5–5.195)$$

with

$$V=v+\frac{m_\mu-m_e}{M}\left(v_+\frac{1+v}{2}-v_-\frac{1-v}{2}\right). \qquad (5–5.196)$$

The integral (5–5.176) continues to apply, with $4m^2\to M^2$, $v\to V$, and the structure of $(1/m^2)I$, as inferred from (5–5.177), is replaced by ($-i\varepsilon$ is omitted)

$$-\frac{i}{(4\pi)^2}\frac{4}{M^2}\int\tfrac{1}{2}\,dv_+\,\tfrac{1}{2}dv_-\,\tfrac{1}{2}dv\,\frac{1-v^2}{2}\left[\frac{3}{V^2}-\frac{2V^2}{(V^2)^2}\right]. \qquad (5–5.197)$$

Now we notice that

$$\frac{d}{dv}\frac{V}{V^2}=\left(1+\frac{m_\mu-m_e}{2M}(v_++v_-)\right)\left(\frac{1}{V^2}-\frac{2V^2}{(V^2)^2}\right), \qquad (5–5.198)$$

and a partial integration converts (5–5.197) into

$$-\frac{i}{(4\pi)^2}\frac{4}{M^2}\int\tfrac{1}{2}\,dv_+\,\tfrac{1}{2}dv_-\,\tfrac{1}{2}dv\left[\frac{1-v^2}{V^2}+\frac{1}{1+\frac{m_\mu-m_e}{2M}(v_++v_-)}\frac{vV}{V^2}\right]; \qquad (5–5.199)$$

the combination in brackets can also be presented as

$$\frac{1}{V^2}+\frac{[(m_\mu-m_e)/2M](v_+-v_-)}{1+[(m_\mu-m_e)/2M](v_++v_-)}\frac{v}{V^2}. \tag{5–5.200}$$

The $v$ integration is performed first. The computation rule of Eq. (5–5.181) is introduced by writing

$$\int\frac{1}{2}dv\,\frac{1}{V^2}=\frac{1}{1+[(m_\mu-m_e)/2M](v_++v_-)}P\int\frac{1}{2}dV\,\frac{1}{V^2-\varepsilon^2}, \tag{5–5.201}$$

where the limits of $V$ integration are $1+[(m_\mu-m_e)/M]v_+$ and $-(1+[(m_\mu-m_e)/M]v_-)$. This gives

$$\int_{-1}^{1}\frac{1}{2}dv\,\frac{1}{V^2}=-\frac{1}{1+[(m_\mu-m_e)/M]v_+}\,\frac{1}{1+[(m_\mu-m_e)/M]v_-}. \tag{5–5.202}$$

The other needed integral, appearing in

$$\int\frac{1}{2}dv\,\frac{v}{V^2}=\frac{1}{[1+[(m_\mu-m_e)/2M](v_++v_-)]^2}\int\frac{1}{2}dV\,\frac{V-[(m_\mu-m_e)/2M](v_+-v_-)}{V^2}, \tag{5–5.203}$$

is

$$P\int\frac{1}{2}dV\,\frac{V}{V^2-\varepsilon^2}=\tfrac{1}{2}\log\frac{1+[(m_\mu-m_e)/M]v_+}{1+[(m_\mu-m_e)/M]v_-}. \tag{5–5.204}$$

They are combined in

$$\int_{-1}^{1}\frac{1}{2}dv\,[\ ]=\frac{1}{2}\frac{[(m_\mu-m_e)/2M](v_+-v_-)}{[1+[(m_\mu-m_e)/2M](v_++v_-)]^3}\log\frac{1+[(m_\mu-m_e)/M]v_+}{1+[(m_\mu-m_e)/M]v_-}-\frac{1}{[1+[(m_\mu-m_e)/2M](v_++v_-)]^2}, \tag{5–5.205}$$

where the bracket on the left side represents the combination of Eq. (5–5.200). Now we have only to observe that the right-hand side of (5–5.205) is reproduced by performing the differentiations in

$$\frac{1}{2}\frac{M}{m_\mu-m_e}\frac{\partial}{\partial v_+}\frac{\partial}{\partial v_-}\frac{v_+-v_-}{1+[(m_\mu-m_e)/2M](v_++v_-)}\log\frac{1+[(m_\mu-m_e)/M]v_+}{1+[(m_\mu-m_e)/M]v_-}. \tag{5–5.206}$$

This immediately yields

$$\int\tfrac{1}{2}dv_+\,\tfrac{1}{2}dv_-\,\tfrac{1}{2}dv\,[\ ]=-\frac{1}{2}\frac{M}{m_\mu-m_e}\log\frac{m_\mu}{m_e}, \tag{5–5.207}$$

and (5-5.199) becomes

$$\frac{i}{(4\pi)^2}\frac{2}{m_\mu^2 - m_e^2}\log\frac{m_\mu}{m_e}. \tag{5-5.208}$$

The muonium energy displacement can be inferred from the positronium result of Eq. (5-5.184) by the substitution

$$\frac{1}{m^2} \to \frac{2}{m_\mu^2 - m_e^2}\log\frac{m_\mu}{m_e} \cong \frac{2}{m_\mu^2}\log\frac{m_\mu}{m_e}, \tag{5-5.209}$$

together with the appropriate form of $|\psi(0)|^2$:

$$|\psi(0)|^2 = \frac{1}{\pi}\alpha^3\left(\frac{m_\mu m_e}{m_\mu + m_e}\right)^3. \tag{5-5.210}$$

This contribution to the muonium hyperfine splitting is, therefore,

$$-\frac{16}{\pi}\alpha^3\left(\frac{m_e}{m_\mu}\right)^2\left(\log\frac{m_\mu}{m_e}\right)\left(1+\frac{m_e}{m_\mu}\right)^{-3}\text{Ry}. \tag{5-5.211}$$

On comparison with the energy splitting of the elementary theory, as derived from (4-17.16) by the substitutions $M = M_p \to m_\mu$, $m \to m_e$, $Z = 1$, $S = \frac{1}{2}$, $g_S = 2$:

$$\Delta E_{\text{n.rel.}} = \frac{16}{3}\alpha^2\frac{m_e}{m_\mu}\left(1+\frac{m_e}{m_\mu}\right)^{-3}\text{Ry}, \tag{5-5.212}$$

we can characterize (5-5.211) as a fractional modification given by

$$-\frac{3}{\pi}\alpha\frac{m_e}{m_\mu}\log\frac{m_\mu}{m_e} = -1.797\times 10^{-4}. \tag{5-5.213}$$

The numerical value implied by (5-5.212) is

$$\Delta E_{\text{n.rel.}} = 4453.8\text{ MHz}, \tag{5-5.214}$$

to be compared with an experimental value:

$$\Delta E_{\text{meas.}} = 4463.3\text{ MHz}, \tag{5-5.215}$$

both of which are quoted only with sufficient accuracy for our present purpose. Most of the discrepancy of 9.5 MHz is removed by incorporating the $\alpha/2\pi$ modification of the two magnetic moments, which increases (5-5.214) to

$$\Delta E_{\text{n.rel.}+\alpha} = 1.00232\times 4453.8\text{ MHz} = 4464.2\text{ MHz}. \tag{5-5.216}$$

The residual discrepancy, which is now in the opposite sense, is 0.9 MHz. As described in Eq. (4–17.110), the effects discussed in Section 4–17 imply a decrease in the theoretical value of

$$-[1.64 \times 10^{-5}][4.45 \times 10^{3}\,\mathrm{MHz}] = -0.073\,\mathrm{MHz}, \tag{5–5.217}$$

thereby reducing the discrepancy to 0.8 MHz. But this is just what is produced by the mass effect described in Eq. (5–5.213),

$$-[1.8 \times 10^{-4}][4.45 \times 10^{3}\,\mathrm{MHz}] = -0.8\,\mathrm{MHz}. \tag{5–5.218}$$

Thus, in contrast with the hydrogen hyperfine structure, purely electrodynamic mechanisms suffice to give excellent agreement with experiment in the positronium and muonium systems.

H.: I am disturbed by one thing in this comparison with experiment. You have taken into account effects of order $\alpha^2$ that arise from the interaction between the particles, as in (5–5.217), but not modifications of the same order of magnitude in individual particle properties—the magnetic moments. Isn't this inconsistent?

S.: You are right, in principle. But, in practice, the numerical coefficients in the effects you mention are sufficiently small that our limited comparison with experiment, at the level of some tens of parts per million, is not significantly affected. Nevertheless, direct measurements of an accuracy to detect the $\alpha^2$ modifications in the electron and muon magnetic moments do exist, as we have already noted in Section 4–3, and one of our next tasks will be to develop the corresponding theory of the electron magnetic moment. But, although it is greatly simplified by the use of source theory, this is still a rather lengthy calculation. Nevertheless, it might be helpful to pause and fill a gap in the treatment of quantum electrodynamic effects of order $\alpha$. As in the forthcoming discussion of the electron magnetic moment, this topic refers to the effect of external electromagnetic fields. In the next section (Section 5–6), the focus is on strong fields. This general treatment will be useful in understanding Section 5.9.

## 5–6 STRONG MAGNETIC FIELDS

A major objective of these concluding sections on electrodynamics is an improved treatment of the electron magnetic moment, a quantity that is defined in weak magnetic fields. But, first, we shall explore some effects of strong magnetic fields. These include the strong-field modification of the $\alpha/2\pi$ moment, the existence of an induced moment—a magnetic polarizability—and the properties of the photon radiation emitted while moving in the magnetic field. In addition, we shall develop and apply a useful variant of the non-causal computational method.

The starting point is the action term associated with the exchange of a single photon accompanying the particle,

$$-\int (dx)(dx')\tfrac{1}{2}\psi(x)\gamma^0 M(x,x')\psi(x'), \tag{5–6.1}$$

a contribution that alters the spin-$\frac{1}{2}$ Green's-function equation into

$$(\gamma\Pi + m + M)\overline{G}_+ = 1. \tag{5–6.2}$$

Here [cf. Eq. (4–16.1)],

$$\begin{aligned} M &= ie^2\int\frac{(dk)}{(2\pi)^4}\gamma^\nu\frac{1}{k^2}\frac{1}{\gamma(\Pi-k)+m}\gamma_\nu \\ &= ie^2\int\frac{(dk)}{(2\pi)^4}\gamma^\nu\frac{1}{k^2}\frac{m-\gamma(\Pi-k)}{(\Pi-k)^2-eq\sigma F+m^2}\gamma_\nu, \end{aligned} \tag{5–6.3}$$

where contact terms are left implicit. The exponential representation used in Eq. (4–16.2),

$$\frac{1}{k^2}\frac{1}{(\Pi-k)^2-eq\sigma F+m^2} = -\int_0^\infty ds\, s\int_0^1 du\, e^{-is\chi(u)},$$

$$\begin{aligned} \chi(u) &= u\left[(\Pi-k)^2-eq\sigma F+m^2\right]+(1-u)k^2 \\ &= (k-u\Pi)^2+u(1-u)\Pi^2+u(m^2-eq\sigma F), \end{aligned} \tag{5–6.4}$$

and the $\xi$-device of Section 4–14 (with $\langle = \langle\xi'=0|,\ \rangle = |\xi'=0\rangle$), convert $M$ into the expression

$$M = -ie^2\int ds\, s\, du\, \gamma^\nu\left\langle e^{-is\chi}[m-\gamma(\Pi-k)]\right\rangle\gamma_\nu. \tag{5–6.5}$$

We are going to exploit the analogy between $e^{-is\chi}$ and the unitary operator that describes the development during the time interval $s$, under the action of the energy operator $\chi$. Thus, with definitions such as

$$\xi(s) = e^{is\chi}\xi e^{-is\chi}, \tag{5–6.6}$$

we derive equations of motion,

$$\frac{d\xi(s)}{ds} = \frac{1}{i}[\xi(s),\chi]. \tag{5–6.7}$$

The procedure resembles that of Section 4–8, but is here applied to the system of charged particle and photon. Some of the equations of motion are

$$\frac{dk(s)}{ds} = 0, \qquad \frac{d\xi(s)}{ds} = 2[k - u\Pi(s)], \tag{5–6.8}$$

and

$$\frac{d\Pi(s)}{ds} = 2ueqF[\Pi(s) - k]. \tag{5–6.9}$$

The latter uses the commutator [Eq. (4–8.44)]

$$[\Pi, \Pi] = ieqF, \tag{5–6.10}$$

and assumes the constancy of the field $F$. It is the last circumstance that enables us to solve these equations of motion, which then form a linear system. Thus, (5–6.9) is solved by the matrix statement [it is Eq. (4–8.48), with $us$ replacing $s$]

$$\Pi(s) - k = e^{2ueqFs}(\Pi - k), \tag{5–6.11}$$

and then (5–6.8) yields

$$\xi(s) = \xi + 2(1-u)sk - \frac{e^{2ueqFs} - 1}{eqF}(\Pi - k). \tag{5–6.12}$$

We can also present the latter as

$$eqF(\xi(s) - \xi) = Dk - A\Pi, \tag{5–6.13}$$

where

$$\begin{aligned} A &= e^{2ueqFs} - 1, \\ D &= A + 2(1-u)eqFs. \end{aligned} \tag{5–6.14}$$

As a first indication of the method to be followed here, let us relate the expectation value $\langle e^{-isX}k\rangle$, appearing in (5–6.5), to the basic expectation value $\langle e^{-isX}\rangle$. For that, we employ the statement of time evolution [Eq. (5–6.6)]

$$\xi e^{-isX} = e^{-isX}\xi(s), \tag{5–6.15}$$

and the null eigenvalue ($\xi = 0$) reference of the expectation value, to deduce that

$$\begin{aligned} \langle [\xi, e^{-isX}]\rangle &= \langle e^{-isX}[\xi(s) - \xi]\rangle \\ &= \frac{1}{eqF}\langle e^{-isX}(Dk - A\Pi)\rangle = 0, \end{aligned} \tag{5–6.16}$$

which gives

$$\langle e^{-is\chi}k\rangle = \langle e^{-is\chi}\rangle\frac{A}{D}\Pi. \tag{5–6.17}$$

Another useful relation appears on employing two commutators,

$$\begin{aligned} 0 &= \left\langle\left[\xi_\mu,\left[\xi_\nu, e^{-is\chi}\right]\right]\right\rangle \\ &= \left\langle e^{-is\chi}\left(\xi_\mu(s)\xi_\nu(s) - \xi_\mu(s)\xi_\nu - \xi_\nu(s)\xi_\mu + \xi_\mu\xi_\nu\right)\right\rangle \\ &= \left\langle e^{-is\chi}\left(\xi_\mu(s) - \xi_\mu\right)\left(\xi_\nu(s) - \xi_\nu\right)\right\rangle + \left\langle e^{-is\chi}\left[\xi_\mu, \xi_\nu(s)\right]\right\rangle. \end{aligned} \tag{5–6.18}$$

The commutator appearing here is evaluated in convenient form as

$$\begin{aligned} \left[(eqF\xi)_\mu, (eqF\xi(s))_\nu\right] &= \left[(eqF\xi)_\mu, (Dk)_\nu\right] \\ &= i\left(eqFD^T\right)_{\mu\nu}, \end{aligned} \tag{5–6.19}$$

on applying the commutation relation

$$\left[\xi_\mu, k_\nu\right] = ig_{\mu\nu}, \tag{5–6.20}$$

and introducing the transposed $D$-matrix. The latter is obtained from $D$ by reversing the sign of $F$, according to the antisymmetry of $F_{\mu\nu}$. The statement in (5–6.18) now becomes

$$\left\langle e^{-is\chi}(Dk - A\Pi)_\mu(Dk - A\Pi)_\nu\right\rangle + \langle e^{-is\chi}\rangle i\left(eqFD^T\right)_{\mu\nu} = 0 \tag{5–6.21}$$

and then, applying (5–6.17),

$$\left\langle e^{-is\chi}k_\mu k_\nu\right\rangle = \langle e^{-is\chi}\rangle\left[\left(\frac{A}{D}\Pi\right)_\nu\left(\frac{A}{D}\Pi\right)_\mu - i\left(\frac{eqF}{D}\right)_{\mu\nu}\right]. \tag{5–6.22}$$

Despite the unsymmetrical appearance of the right-hand side in (5–6.22), this structure is indeed symmetrical in $\mu$ and $\nu$. The necessary algebraic property of the matrices is

$$AA^T + D + D^T = 0. \tag{5–6.23}$$

It is confirmed, first by noting that

$$(A + 1)(A^T + 1) = e^{2ueqFs}e^{-2ueqFs} = 1 \tag{5–6.24}$$

implies

$$AA^T + A + A^T = 0, \tag{5–6.25}$$

and then by applying the antisymmetry of $D - A$,

$$A + A^T = D + D^T. \tag{5–6.26}$$

The main problem, the evaluation of $\langle e^{-is\chi}\rangle$, is now solved by devising a differential equation:

$$\begin{aligned} i\frac{\partial}{\partial s}\langle e^{-is\chi}\rangle &= \langle e^{-is\chi}\chi\rangle \\ &= \langle e^{-is\chi}\rangle u\Pi^2 - \langle e^{-is\chi}k\rangle 2u\Pi \\ &\quad + \langle e^{-is\chi}k^2\rangle + \langle e^{-is\chi}\rangle u(m^2 - eq\sigma F). \end{aligned} \tag{5–6.27}$$

With the aid of Eqs. (5–6.17, 22), it immediately follows that

$$\begin{aligned} i\frac{\partial}{\partial s}\log\langle e^{-is\chi}\rangle &= u\Pi^2 - 2u\Pi\left(\frac{A}{D}\right)^T\Pi + \Pi\left(\frac{A}{D}\right)^T\frac{A}{D}\Pi \\ &\quad - i\,\mathrm{tr}'\left(\frac{eqF}{D}\right) + u(m^2 - eq\sigma F), \end{aligned} \tag{5–6.28}$$

where the prime on the trace is a reminder that only the vector indices are involved. In order to have a symmetrical matrix in the $\Pi$ quadratic form, we rewrite the right side of (5–6.28), apart from the last term, as

$$u\Pi^2 + \Pi\left[\left(\frac{A}{D}\right)^T\frac{A}{D} - u\left(\frac{A}{D} + \left(\frac{A}{D}\right)^T\right)\right]\Pi - i\,\mathrm{tr}'\left(eqF\frac{uA+1}{D}\right), \tag{5–6.29}$$

which applies the commutator (5–6.10). Now, if we use the relation (5–6.23), in which $A$ and $A^T$ are commutative, we get

$$\begin{aligned} \left(\frac{A}{D}\right)^T\frac{A}{D} - u\left[\frac{A}{D} + \left(\frac{A}{D}\right)^T\right] &= -\frac{uA+1}{D} - \left(\frac{uA+1}{D}\right)^T \\ &= -\frac{1}{2eqF}\left[\frac{\partial D/\partial s}{D} - \frac{\partial D^T/\partial s}{D^T}\right], \end{aligned} \tag{5–6.30}$$

where the last version incorporates the properties

$$\frac{\partial D}{\partial s} = 2eqF(uA + 1), \qquad \frac{\partial D^T}{\partial s} = -2eqF(uA^T + 1). \tag{5–6.31}$$

This puts (5–6.29) into the form

$$u\Pi^2 + \Pi\left(-\frac{1}{2eqF}\right)\frac{\partial}{\partial s}\log\left(-\frac{D}{D^T}\right)\Pi - \tfrac{1}{2}i\frac{\partial}{\partial s}\,\mathrm{tr}'(\log D), \qquad (5\text{–}6.32)$$

and the resulting integral of (5–6.28) is

$$\langle e^{is\chi}\rangle = C\left(\det'\frac{D}{2eqF}\right)^{-1/2} e^{-is\Phi}, \qquad (5\text{–}6.33)$$

where

$$\Phi = u(\Pi^2 + m^2 - eq\sigma F) + \Pi\left(-\frac{1}{2eqFs}\right)\log\left(-\frac{D}{D^T}\right)\Pi. \qquad (5\text{–}6.34)$$

In introducing the determinant we have made use of the differential property [Eq. (4–8.23)]

$$\delta\log\det' X = \mathrm{tr}'(X^{-1}\delta X) = \delta\,\mathrm{tr}'(\log X), \qquad (5\text{–}6.35)$$

and provided a multiplicative factor to simplify the form of the integration constant $C$.

To evaluate $C$, we consider the limit of small $s$, where

$$\frac{D}{2eqF} = s + u^2eqFs^2 + \cdots, \qquad -\frac{D}{D^T} = 1 + 2u^2eqFs + \cdots. \qquad (5\text{–}6.36)$$

Then, (5–6.33) exhibits the dominant behavior

$$\langle e^{-is\chi}\rangle \sim C\frac{1}{s^2}, \qquad (5\text{–}6.37)$$

since the dimensionality of the determinant is 4. The singularity at $s = 0$ arises from the increasingly large values of $k$ that are demanded, as $s \to 0$, by complementarity with $\xi = 0$. Accordingly, the limiting structure is given by the elementary integral [Eq. (4–8.57)],

$$\int\frac{(dk)}{(2\pi)^4}e^{-isk^2} = \frac{1}{(4\pi)^2}\frac{1}{is^2}, \qquad (5\text{–}6.38)$$

and

$$C = -\frac{i}{(4\pi)^2}. \qquad (5\text{–}6.39)$$

The complete result can be presented as

$$\langle e^{-is\chi}\rangle = -\frac{i}{(4\pi)^2}\frac{1}{s^2}\left(\det{}'\,\frac{2eqFs}{D}\right)^{1/2} e^{-is\Phi}. \tag{5–6.40}$$

The known zero-field form [cf. Eqs. (4–16.9, 10)] emerges on using the expansions of (5–6.36), since the determinant reduces to unity, and

$$F = 0: \qquad \Phi = u(1-u)\Pi^2 + m^2 u = m^2u^2 + u(1-u)\left[m^2 - (\gamma\Pi)^2\right]. \tag{5–6.41}$$

To facilitate the next step, which is concerned with the Dirac matrices, we write

$$\Phi = \Phi_1 - ueq\sigma F, \tag{5–6.42}$$

thereby isolating the spin matrices, and present our results to this point as

$$M = -\frac{\alpha}{4\pi}\int\frac{ds}{s}\,du\left(\det{}'\,\frac{2eqFs}{D}\right)^{1/2} e^{-is\Phi_1}$$
$$\times\gamma^\nu e^{isueq\sigma F}\left[m - \gamma\frac{2(1-u)eqFs}{D}\Pi\right]\gamma_\nu + \text{c.t.}, \tag{5–6.43}$$

where the contact terms (c.t.) must still be made explicit. Now we again examine equations of motion, this time of the matrix quantities

$$\gamma(s) = e^{-isueq\sigma F}\gamma e^{isueq\sigma F}. \tag{5–6.44}$$

These equations are

$$\frac{d}{ds}\gamma(s) = 2ueqF\gamma(s), \tag{5–6.45}$$

since

$$[\gamma, \sigma F] = -2iF\gamma, \tag{5–6.46}$$

and the solution is

$$\gamma(s) = (1 + A)\gamma = \gamma(1 + A^T). \tag{5–6.47}$$

It is used to write

$$\gamma e^{isueq\sigma F} = e^{isueq\sigma F}\gamma(1 + A^T), \tag{5–6.48}$$

which is then combined with the rearrangement

$$\left(m-\gamma\frac{2(1-u)eqFs}{D}\Pi\right)\gamma=\gamma\left(m+\gamma\frac{2(1-u)eqFs}{D}\Pi\right)+2\frac{2(1-u)eqFs}{D}\Pi. \tag{5–6.49}$$

In doing this, one encounters

$$\gamma(1+A^T)\gamma=-4-\mathrm{tr}' A+2i\sigma A, \tag{5–6.50}$$

where

$$\sigma A=\tfrac{1}{2}\sigma^{\mu\nu}A_{\mu\nu}=\tfrac{1}{2}\sigma^{\mu\nu}\left(2ueqFs\frac{\sinh 2ueqFs}{2ueqFs}\right)_{\mu\nu}. \tag{5–6.51}$$

The combination is

$$\begin{aligned}\gamma^\nu e^{isueq\sigma F}&\left[m-\gamma\frac{2(1-u)eqFs}{D}\Pi\right]\gamma_\nu\\ &=e^{isueq\sigma F}\left\{(-4-\mathrm{tr}' A+2i\sigma A)\left(m+\gamma\frac{2(1-u)eqFs}{D}\Pi\right)\right.\\ &\qquad\left.+2\gamma(1+A^T)\frac{2(1-u)eqFs}{D}\Pi\right\}.\end{aligned} \tag{5–6.52}$$

In the absence of the homogeneous field, the latter reduces to

$$-4[m+(1-u)\gamma\Pi]+2(1-u)\gamma\Pi=-2[2m+(1-u)\gamma\Pi], \tag{5–6.53}$$

and (5–6.43) becomes

$$M_0=\frac{\alpha}{2\pi}\int\frac{ds}{s}\,du\,e^{-ism^2u^2}e^{-isu(1-u)[m^2-(\gamma\Pi)^2]}[2m+(1-u)\gamma\Pi]+\text{c.t.}. \tag{5–6.54}$$

The contact term is chosen to make $M_0$, and its first derivative with respect to $\gamma\Pi$, vanish at $\gamma\Pi+m=0$. These are the normalization conditions. Accordingly,

$$\text{c.t.}=-m_c-\zeta_c(\gamma\Pi+m), \tag{5–6.55}$$

where

$$m_c=\frac{\alpha}{2\pi}m\int\frac{ds}{s}\,du\,(1+u)e^{-ism^2u^2} \tag{5–6.56}$$

and

$$\zeta_c = \frac{\alpha}{2\pi}\int \frac{ds}{s}\,du\,(1-u)e^{-ism^2u^2}$$

$$-i\frac{\alpha}{\pi}m^2\int ds\,du\,u(1-u^2)e^{-ism^2u^2}. \tag{5–6.57}$$

Of course, only the combination of the two parts of $M_0$, and of $M$, is physically significant. Writing out the contact terms separately is a fiction, which is given mathematical meaning by stopping all the $s$-integrals at some arbitrarily small lower limit. (The photon mass is another such fiction, used in connection with the $u$-integral.)

Now let us specialize to a pure magnetic field:

$$F_{12} = -F_{21} = H. \tag{5–6.58}$$

In this situation the matrix $F$ has the eigenvalues $iH$, $-iH$, 0, 0, as follows from the components

$$(F\Pi)_1 = H\Pi_2, \qquad (F\Pi)_2 = -H\Pi_1, \qquad (F\Pi)_3 = 0, \qquad (F\Pi)_0 = 0. \tag{5–6.59}$$

One can then evaluate the determinant of (5–6.43)—or, rather, its inverse—as

$$\Delta = \det{}'\,\frac{D}{2eqFs}$$

$$= \left(1 - u + \frac{e^{2iueHs}-1}{2ieHs}\right)\left(1 - u + \frac{1-e^{-2iueHs}}{2ieHs}\right), \tag{5–6.60}$$

where the particular sign of $q = \pm 1$ is irrelevant. On introducing the variable

$$(eH > 0): \qquad x = ueHs, \tag{5–6.61}$$

this is written as

$$\Delta = \left[1 - u + u\frac{\sin x}{x}e^{ix}\right]\left[1 - u + u\frac{\sin x}{x}e^{-ix}\right]$$

$$= (1-u)^2 + 2u(1-u)\frac{\sin x\cos x}{x} + u^2\left(\frac{\sin x}{x}\right)^2, \tag{5–6.62}$$

which quantity varies from 1, at $x = 0$, to $(1-u)^2$, at $x = \infty$. The related combination appearing in (5-6.52), $2(1-u)eqFs/D$, has the twofold eigenvalue $1-u$, associated with the 03 plane, and the conjugate pair of eigenvalues

$$(1-u)\left(1-u+\frac{e^{\pm 2iueHs}-1}{\pm 2ieHs}\right)^{-1} = \frac{1-u}{\Delta}\left(1-u+\frac{e^{\mp 2iueHs}-1}{\mp 2ieHs}\right)$$
$$= \frac{1-u}{\Delta}\left(1-u+u\frac{\sin 2x}{2x} \mp iu\frac{\sin^2 x}{x}\right). \quad (5\text{-}6.63)$$

The various possibilities are united in

$$\gamma\frac{2(1-u)eqFs}{D}\Pi = (1-u)\gamma\Pi + (1-u)\left(\frac{1-u+u\dfrac{\sin 2x}{2x}}{\Delta} - 1\right)\gamma\Pi_H$$
$$-\frac{u^2(1-u)}{\Delta}s\frac{\sin^2 x}{x^2}\gamma eqF\Pi, \quad (5\text{-}6.64)$$

where $\Pi_H$ is the projection of $\Pi$ onto the plane (12) defined by the magnetic field. Alternatively, one can remark that

$$\gamma F\Pi = i\boldsymbol{\sigma}\cdot\mathbf{H}\gamma\Pi_H = iH\sigma_3\gamma\Pi_H, \quad (5\text{-}6.65)$$

according to the algebraic properties of $\sigma_3 = i\gamma_1\gamma_2$,

$$\sigma_3\gamma_1 = i\gamma_2, \qquad \sigma_3\gamma_2 = -i\gamma_1, \quad (5\text{-}6.66)$$

which enables us to present (5-6.64) as

$$\gamma\frac{2(1-u)eqFs}{D}\Pi = (1-u)\gamma\Pi + (1-u)\left(\frac{1-u+u\dfrac{\sin x}{x}e^{-i\zeta x}}{\Delta} - 1\right)\gamma\Pi_H,$$
$$(5\text{-}6.67)$$

where

$$\zeta = q\sigma_3. \quad (5\text{-}6.68)$$

The analogous combination of (5-6.52), with the additional factor

$$1 + A^T = e^{-2ueqFs}, \quad (5\text{-}6.69)$$

also has the twofold eigenvalue $1 - u$, while the other eigenvalues are

$$e^{\mp 2ix}\frac{1-u}{\Delta}\left[1 - u + u\frac{\sin x}{x}e^{\mp ix}\right]. \tag{5–6.70}$$

Accordingly,

$$\begin{aligned}\gamma(1 + A^T)&\frac{2(1-u)eqFs}{D}\Pi \\ &= (1-u)\gamma\Pi + (1-u)\left(e^{-2i\zeta x}\frac{1 - u + u\dfrac{\sin x}{x}e^{-i\zeta x}}{\Delta} - 1\right)\gamma\Pi_H.\end{aligned} \tag{5–6.71}$$

Also appearing in (5–6.52) are

$$\operatorname{tr}' A = 2(\cos 2x - 1) = -4\sin^2 x \tag{5–6.72}$$

and

$$\sigma A = \zeta \sin 2x, \tag{5–6.73}$$

which occur as

$$\begin{aligned}-4 - \operatorname{tr}' A + 2i\sigma A &= -4\cos^2 x + 2i\zeta \sin 2x \\ &= -4\cos x\, e^{-i\zeta x}.\end{aligned} \tag{5–6.74}$$

Utilizing these results, we present the brace of (5–6.52) as

$$\begin{aligned}&-4\cos x\, e^{-i\zeta x}\left[m + (1-u)\gamma\Pi + (1-u)\left(\frac{1-u}{\Delta} - 1\right)\gamma\Pi_H \right.\\ &\quad\left. + \frac{u(1-u)}{\Delta}\frac{\sin x}{x}e^{-i\zeta x}\gamma\Pi_H\right] \\ &\quad + 2(1-u)\gamma\Pi - 2(1-u)\gamma\Pi_H + 2\frac{(1-u)^2}{\Delta}e^{-2i\zeta x}\gamma\Pi_H \\ &\quad + 2\frac{u(1-u)}{\Delta}\frac{\sin x}{x}e^{-3i\zeta x}\gamma\Pi_H \\ &\quad = -2m - 2e^{-2i\zeta x}[m + (1-u)\gamma\Pi] \\ &\quad\quad -2(1-u)\left[\frac{1-u}{\Delta} + \frac{u}{\Delta}\frac{\sin x}{x}e^{-i\zeta x} - e^{-2i\zeta x}\right]\gamma\Pi_H.\end{aligned} \tag{5–6.75}$$

The spin factor standing to the left of the brace in (5–6.52) can be recombined with $\Phi_1$ to form $\Phi$ (Eq. (5–6.42)], which we exhibit as

$$\Phi = u(1-u)\left(\Pi^2 + m^2 - eq\sigma F\right) + u^2\left(m^2 - eq\sigma F\right)$$
$$+\Pi\left[\left(-\frac{1}{2eqFs}\right)\log\left(-\frac{D}{D^T}\right) + u^2\right]\Pi. \qquad (5\text{–}6.76)$$

The last term vanishes for $F = 0$, according to (5–6.36), and therefore only involves the $\Pi_H$ components. The quantity in brackets has a unique value in that subspace, namely

$$-\frac{1}{2ieHs}\log\frac{e^{2ix} - 1 + 2(1-u)ieHs}{1 - e^{-2ix} + 2(1-u)ieHs} + u^2 = \frac{\beta}{eHs} - u(1-u), \qquad (5\text{–}6.77)$$

where the angle $\beta$ is specified by

$$\tan\beta = \frac{(1-u)\sin x}{(1-u)\cos x + u\dfrac{\sin x}{x}}. \qquad (5\text{–}6.78)$$

As a consequence,

$$\Pi\left[\left(-\frac{1}{2eqFs}\right)\log\left(-\frac{D}{D^T}\right) + u^2\right]\Pi = \frac{u}{x}\left(\beta - (1-u)x\right)\Pi_H^2. \qquad (5\text{–}6.79)$$

Note that $\beta$ begins as $(1-u)x$, for small $x$, and approaches $x$ for large values of $x$.

The material for the general construction of $M$ is now available. For most applications, however, it suffices to use the fact that the particle field, to a good first approximation, obeys the equations

$$(\gamma\Pi + m)\psi = 0, \qquad \left(\Pi^2 - eq\sigma F + m^2\right)\psi = 0, \qquad (5\text{–}6.80)$$

and thereby to simplify the structure of $M$ as it contributes to the action. Doing this, we find that

$$M = \frac{\alpha}{2\pi}m\int\frac{ds}{s}\,du\,e^{-ism^2u^2}\Bigg\{\Delta^{-1/2}\exp\left[-i\frac{us}{x}\left(\beta - (1-u)x\right){\Pi_H}^2\right]e^{i\zeta ux}$$
$$\times\left[1 + ue^{-2i\zeta x} + \frac{1-u}{m}\left(\frac{1-u}{\Delta} + \frac{u}{\Delta}\frac{\sin x}{x}e^{-i\zeta x} - e^{-2i\zeta x}\right)\gamma\Pi_H\right]$$
$$-(1+u)\Bigg\}, \qquad (5\text{–}6.81)$$

in which the contact term has now been made explicit (as simplified by $\gamma\Pi + m = 0$). Observe that in the absence of the magnetic field, when $x = 0$ and $\Delta = 1$, the $\gamma\Pi_H$ term disappears and $M$ vanishes as it should. A further simplification will result from the remark that

$$i\zeta\gamma\Pi_H = \frac{1}{eH}eq\gamma F\Pi = \frac{i}{2eH}[eq\sigma F, \gamma\Pi + m] \tag{5–6.82}$$

effectively vanishes when set between $\psi$-fields obeying (5–6.80). Under these circumstances, where spin-dependent exponentials of the form

$$e^{i\lambda\zeta} = \cos\lambda + i\zeta\sin\lambda, \tag{5–6.83}$$

multiply $\gamma\Pi_H$, the $\zeta$-term can be omitted, which is equivalent to replacing the exponential function by the average of its value for $\zeta = \pm 1$. But, before we can apply this observation to (5–6.81), in which an exponential function of $\Pi_H^2$ also appears, it is necessary to study the energy spectrum and the associated eigenfunctions, as they are implied by the field equations of (5–6.80).

It is convenient, and involves no loss in generality, to specialize the coordinate system by choosing $\Pi_3$, the component of momentum along the magnetic field direction, to be zero. The field $\psi$ will be projected onto subspaces of intrinsic parity, labeled by the eigenvalues $\gamma^{0\prime} = \pm 1$. Since $\gamma^0$ and $i\gamma_5$ anticommute, the matrices

$$\gamma^0\gamma = i\gamma_5\sigma, \qquad (i\gamma_5)^2 = 1 \tag{5–6.84}$$

only couple different subspaces. Accordingly, the Dirac equation referring to an energy eigenvalue $\Pi^{0\prime} = E$, when presented as

$$(E - m\gamma^0 - i\gamma_5\sigma\cdot\Pi)\psi = 0, \tag{5–6.85}$$

decomposes into the pair of equations ($\Pi_3 = 0$)

$$(E - m)\psi_+ = \sigma\cdot\Pi\psi_-, \qquad (E + m)\psi_- = \sigma\cdot\Pi\psi_+. \tag{5–6.86}$$

On eliminating fields between these equations, we infer the system

$$\begin{aligned}(E^2 - m^2)\psi_+ &= (\sigma\cdot\Pi)^2\psi_+ = (\Pi_H^2 - eHq\sigma_3)\psi_+,\\ (E^2 - m^2)\psi_- &= (\sigma\cdot\Pi)^2\psi_- = (\Pi_H^2 - eHq\sigma_3)\psi_-,\end{aligned} \tag{5–6.87}$$

which would also follow directly from the second-order form of the Dirac equation presented in (5–6.80). Evidently the energy eigenvalues are obtained by assigning to

$$\zeta = q\sigma_3, \tag{5-6.88}$$

an eigenvalue, $\zeta' = \pm 1$, and, independently, introducing an eigenvalue for

$$\Pi_H{}^2 = \Pi_1{}^2 + \Pi_2{}^2, \qquad [\Pi_1, \Pi_2] = ieqH. \tag{5-6.89}$$

The familiar one-dimensional oscillator problem provides the latter spectrum,

$$(\Pi_H{}^2)' = (2n+1)eH, \qquad n = 0, 1, 2, \ldots, \tag{5-6.90}$$

and one infers the energy values

$$E^2 = m^2 + (2n + 1 - \zeta')eH. \tag{5-6.91}$$

Note that the ground state of the system, with energy $E = m$, is uniquely characterized by the quantum numbers

$$E = m: \qquad n = 0, \quad \zeta' = 1. \tag{5-6.92}$$

All other energy levels are doubly degenerate with the same energy, $(m^2 + 2n'eH)^{1/2}$, being realized by the two sets of quantum numbers

$$n = n', \quad \zeta' = +1 \quad \text{and} \quad n = n' - 1, \quad \zeta' = -1.$$

No distinction has been drawn in this account between the quantum numbers assigned to $\psi_+$ and to $\psi_-$. But, since $\sigma_3$ anticommutes with

$$\boldsymbol{\sigma} \cdot \boldsymbol{\Pi} = \sigma_1 \Pi_1 + \sigma_2 \Pi_2, \tag{5-6.93}$$

the eigenvalues assigned to $\zeta$ in the two subspaces, for a state of given energy, must be of opposite sign, with corresponding differences in the eigenvalues of $\Pi_H{}^2$. Thus, a more precise description of the eigenvalues associated with the energy (5-6.91) is given by

$$(\zeta - \zeta')\psi_+ = 0, \qquad \left[\Pi_H{}^2 - (2n+1)eH\right]\psi_+ = 0 \tag{5-6.94}$$

and

$$(\zeta + \zeta')\psi_- = 0, \qquad \left[\Pi_H{}^2 - (2n + 1 - 2\zeta')eH\right]\psi_- = 0, \tag{5-6.95}$$

which are summarized in the following characteristics of the complete $\psi$-field:

$$\gamma^0 \zeta \psi = \zeta' \psi, \qquad \Pi_H{}^2 \psi = (2n + 1 - \zeta' + \zeta)eH\psi. \tag{5-6.96}$$

That $\gamma^0\zeta$ provides an exact quantum number can be seen directly from the Dirac equation of (5–6.85). For the ground state, $n = 0$, $\zeta' = +1$, the negative value that would be assigned to $\Pi_H^2$ in the $\gamma^{0\prime} = -1$ subspace shows that $\psi_-$ vanishes. The special properties of the ground state are conveyed by the eigenvector statements

$$E = m: \qquad (\gamma^0 - 1)\psi = 0, \quad \gamma\Pi_H\psi = 0. \tag{5–6.97}$$

For excited states, $\gamma\Pi_H$ does not have a definite eigenvalue. But what is required is a kind of expectation value

$$\langle\gamma\Pi_H\rangle = \big\langle(\gamma^0\Pi^{0\prime} - m)\big\rangle = \langle\gamma^0\rangle E - m, \tag{5–6.98}$$

referring to the field structure that appears in the action [Eq. (5–6.1)]. To evaluate $\langle\gamma^0\rangle$ we use $m$ as a variable parameter, deducing that

$$(\gamma\Pi + m)\frac{\partial\psi}{\partial m} = \left(\gamma^0\frac{\partial E}{\partial m} - 1\right)\psi, \tag{5–6.99}$$

and thereby

$$\langle\gamma^0\rangle\frac{\partial E}{\partial m} = 1. \tag{5–6.100}$$

According to the energy expression (5–6.91),

$$\frac{\partial E}{\partial m} = \frac{m}{E}, \tag{5–6.101}$$

and thus

$$\langle\gamma^0\rangle = E/m, \tag{5–6.102}$$

which yields

$$\langle\gamma\Pi_H\rangle = \frac{E^2 - m^2}{m} = (2n + 1 - \zeta')\frac{eH}{m}. \tag{5–6.103}$$

Harold looks bewildered.

H.: How can the expectation value of $\gamma^0$, which has eigenvalues of unit magnitude, be greater than unity?

S.: I remind you of the $\gamma^0$ factor that appears in the action. If we make the meaning of $\langle\gamma^0\rangle$ explicit for the contribution of a particular eigenfunction $\psi$ and its complex conjugate $\psi^*$, it reads

$$\langle\gamma^0\rangle = \frac{\int(d\mathbf{x})\psi^*\psi}{\int(d\mathbf{x})\psi^*\gamma^0\psi}, \tag{5–6.104}$$

which is certainly greater than one, in general. When we introduce the intrinsic parity decomposition, this expectation value becomes

$$\langle\gamma^0\rangle = \frac{\int(d\mathbf{x})[\psi_+^*\psi_+ + \psi_-^*\psi_-]}{\int(d\mathbf{x})[\psi_+^*\psi_+ - \psi_-^*\psi_-]}, \tag{5–6.105}$$

where, according to (5–6.86, 87),

$$\int(d\mathbf{x})\psi_-{}^*\psi_- = \int(d\mathbf{x})\psi_+{}^*\frac{(\boldsymbol{\sigma}\cdot\boldsymbol{\Pi})^2}{(E+m)^2}\psi_+ = \frac{E-m}{E+m}\int(d\mathbf{x})\psi_+{}^*\psi_+, \tag{5–6.106}$$

and, indeed,

$$\langle\gamma^0\rangle = \frac{1+\dfrac{E-m}{E+m}}{1-\dfrac{E-m}{E+m}} = \frac{E}{m}. \tag{5–6.107}$$

While we are about it, let us note another useful expectation value, that of $\zeta$. We have only to use the eigenvalue stated for $\gamma^0\zeta$ in Eq. (5–6.96),

$$\langle\zeta\rangle = \langle\gamma^0\gamma^0\zeta\rangle = \zeta'\frac{E}{m}. \tag{5–6.108}$$

Let us also inquire about the energy spectrum when the $\alpha/2\pi$ magnetic moment is introduced into the Dirac equation:

$$\gamma\Pi_H\psi = \left(\gamma^0\Pi^{0\prime} - m + \frac{\alpha}{2\pi}\frac{eq}{2m}\sigma_3 H\right)\psi. \tag{5–6.109}$$

The equivalent pair of equations is ($\Pi_3 = 0$)

$$\left(E - m + \zeta'\frac{\alpha}{2\pi}\frac{eH}{2m}\right)\psi_+ = \boldsymbol{\sigma}\cdot\boldsymbol{\Pi}\psi_-, \qquad \left(E + m + \zeta'\frac{\alpha}{2\pi}\frac{eH}{2m}\right)\psi_- = \boldsymbol{\sigma}\cdot\boldsymbol{\Pi}\psi_+, \tag{5–6.110}$$

where $\zeta'$ replaces $q\gamma^0\sigma_3$. Since these equations only differ from the set (5–6.86) in the substitution

$$E \to E + \zeta' \frac{\alpha}{2\pi} \frac{eH}{2m}, \tag{5–6.111}$$

the energy spectrum is predicted to be

$$E = -\zeta' \frac{\alpha}{2\pi} \frac{eH}{2m} + \left[m^2 + (2n + 1 - \zeta')eH\right]^{1/2}. \tag{5–6.112}$$

For the ground state, in particular,

$$n = 0, \quad \zeta' = +1: \qquad E = m - \frac{\alpha}{2\pi} \frac{eH}{2m}. \tag{5–6.113}$$

With increasing magnetic field strength the total energy decreases monotonically and, if this formula continued to apply, would vanish at the field strength

$$H = \frac{4\pi}{\alpha} \frac{m^2}{e} = 8 \times 10^{16} \text{ G}. \tag{5–6.114}$$

While extraordinarily large, this magnitude might be approached under the astrophysical circumstances encountered in neutron stars. The actual situation concerning the formula of (5–6.113) is quite different, however, as we now proceed to explain.

We exhibit $M$ [Eq. (5–6.81)] in the ground state, where

$$n = 0, \quad \zeta' = +1: \qquad (\Pi_H{}^2)' = eH, \quad \gamma\Pi_H \to 0. \tag{5–6.115}$$

For this circumstance,

$$\begin{aligned} \exp\left[-i\frac{us}{x}[\beta - (1-u)x]\Pi_H{}^2\right] &= e^{-i\beta}e^{i(1-u)x} \\ &= \Delta^{-1/2}\left(1 - u + u\frac{\sin x}{x}e^{ix}\right)e^{-iux}, \end{aligned} \tag{5–6.116}$$

and ($\gamma^{0\prime} = +1$, $\zeta \to \zeta' = +1$)

$$M = \frac{\alpha}{2\pi} m \int \frac{dx}{x}\, du \exp\left[-i\frac{m^2 u}{eH}x\right]\left\{\frac{1 + ue^{-2ix}}{1 - u + u\dfrac{\sin x}{x}e^{-ix}} - 1 - u\right\}, \tag{5–6.117}$$

which makes use of (5–6.62). We give this expression another form by the transformation

$$x \to -iy, \tag{5–6.118}$$

which yields

$$M = \frac{\alpha}{2\pi} m \int_0^\infty \frac{dy}{y} \int_0^1 du \exp\left[-\frac{m^2 u}{eH} y\right] \left\{ \frac{1 + ue^{-2y}}{1 - u + u\dfrac{\sinh y}{y} e^{-y}} - 1 - u \right\}. \tag{5–6.119}$$

The substitution (5–6.118) is a rotation of the integration path to the lower imaginary axis. Its justification involves the absence of a singularity at the origin, and throughout the quadrant

$$x = \xi - i\eta, \qquad \xi > 0, \quad \eta > 0. \tag{5–6.120}$$

In particular, a zero of the denominator in (5–6.117) would require that

$$(1 - u)2\eta + u(1 - e^{-2\eta} \cos 2\xi) = 0, \tag{5–6.121}$$

which obviously cannot be satisfied for $\eta > 0$. The reality of $M$ thus made explicit was to be expected—the ground state is stable against radiative decay.

For weak magnetic fields, which are characterized by

$$\frac{eH}{m^2} \ll 1, \tag{5–6.122}$$

only correspondingly small values of $y$ contribute in (5–6.119), provided $u \gg eH/m^2$. The initial term in the $y$-expansion of the brace in (5–6.119) is

$$\{\ \} = -u(1 - u)y + \cdots, \tag{5–6.123}$$

and

$$M \cong -\frac{\alpha}{2\pi} m \int_0^1 du\, u(1 - u) \int_0^\infty dy \exp\left[-\frac{m^2 u}{eH} y\right]$$

$$= -\frac{\alpha}{2\pi}\frac{eH}{m} \int_0^1 du\,(1 - u) = -\frac{\alpha}{2\pi}\frac{eH}{2m}, \tag{5–6.124}$$

which, as the weak-field magnetic moment term for $\zeta' = +1$, restates (5–6.113). The next power in the expansion of the brace is displayed in

$$\{\ \} = -u(1 - u)y + \left(\tfrac{4}{3}u - \tfrac{5}{3}u^2 + u^3\right)y^2 + \cdots. \tag{5–6.125}$$

Its consequence is

$$M \cong -\frac{\alpha}{2\pi}\frac{eH}{2m} + \frac{\alpha}{2\pi}m\left(\frac{eH}{m^2}\right)^2 \int du\left(\frac{4}{3}\frac{1}{u} - \frac{5}{3} + u\right)$$

$$= -\frac{\alpha}{2\pi}\frac{eH}{2m} + \frac{\alpha}{2\pi}\frac{(eH)^2}{m^3}\left[\frac{4}{3}\int_{\to 0}^{1}\frac{du}{u} - \frac{7}{6}\right], \tag{5–6.126}$$

and here we meet an infra-red problem. But the mathematical origin of this 'problem' is already evident; the expansion (5–6.125) can only be used for values of $u$ such that

$$1 > u > u_0, \tag{5–6.127}$$

where

$$1 \gg u_0 \gg \frac{eH}{m^2}. \tag{5–6.128}$$

To deal with the remaining interval, $u < u_0$, the brace is expanded in powers of $u$:

$$\{\ \} = u\left(e^{-2y} - \frac{\sinh y}{y}e^{-y}\right) + \cdots, \tag{5–6.129}$$

where only the term linear in $u$, which connects with the logarithm of (5–6.126), need be retained. This contribution to $M$ is

$$\frac{\alpha}{2\pi}m\int_0^{u_0} du\, u\int_0^{\infty}\frac{dy}{y}\exp\left[-\frac{m^2}{eH}uy\right]\left(e^{-2y} - \frac{\sinh y}{y}e^{-y}\right)$$

$$= \frac{\alpha}{2\pi}m\int_0^{u_0} du\, u\left[\frac{m^2}{2eH}u\log\left(1 + \frac{2eH}{m^2 u}\right) - 1\right]$$

$$= \frac{\alpha}{2\pi}m\left[\frac{m^2}{6eH}u_0^3\log\left(1 + \frac{2eH}{m^2 u_0}\right) - \tfrac{1}{3}u_0^2\right.$$

$$\left. - \frac{2}{3}\frac{eH}{m^2}u_0 + \frac{4}{3}\left(\frac{eH}{m^2}\right)^2\log\left(1 + \frac{m^2}{2eH}u_0\right)\right]. \tag{5–6.130}$$

Its expansion up to quadratic field terms, which is based upon $u_0 \gg eH/m^2$, is given by

$$\frac{\alpha}{2\pi}m\left[-\frac{eH}{m^2}u_0+\frac{4}{9}\left(\frac{eH}{m^2}\right)^2+\frac{4}{3}\left(\frac{eH}{m^2}\right)^2\log\left(\frac{m^2}{2eH}u_0\right)\right]. \tag{5–6.131}$$

The term linear in $H$ is needed to restore the piece missing from the integral of $1-u$ [cf. Eq. (5–6.124)] because one has now stopped that evaluation at the lower limit $u_0$. On adding the rest to (5–6.126), the logarithmic dependence on $u_0$ disappears, yielding

$$M \cong -\frac{\alpha}{2\pi}\frac{eH}{2m}+\frac{\alpha}{2\pi}\frac{(eH)^2}{m^3}\left[\tfrac{4}{3}\log\left(\frac{m^2}{2eH}\right)+\frac{4}{9}-\frac{7}{6}\right]$$

$$= -\frac{\alpha}{2\pi}\frac{eH}{2m}\left[1-\frac{8}{3}\frac{eH}{m^2}\left(\log\left(\frac{m^2}{2eH}\right)-\frac{13}{24}\right)\right]. \tag{5–6.132}$$

Here is an indication that, with increasing magnetic field strength, the energy of the ground state does not continue to decrease below $m$ at the rate suggested by the weak-field moment. The preceding calculation referred to a definite spin orientation, $\zeta'=+1$, which prevents any further physical identification of individual terms. We shall soon see, however, that the term quadratic in $H$ (apart from a logarithmic dependence) is actually spin-independent. It therefore represents an induced magnetic moment, a magnetic polarization of the particle, which, being opposed to the direction of the field, is diamagnetic in character.

The question now naturally arises about the strong-field behavior of $M$, where the inequality of (5–6.122) is reversed,

$$\frac{eH}{m^2} \gg 1. \tag{5–6.133}$$

To answer it, we divide the $y$-integration domain in (5–6.119) at $y_0$, where

$$eH/m^2 \gg y_0 \gg 1. \tag{5–6.134}$$

The contribution to $M$ from $y<y_0$ is independent of $H$. For $y>y_0$, we can simplify the double integral of (5–6.119) to

$$\int_{y_0}^{\infty}dy\int_0^1 du\,\exp\left[-\frac{m^2}{eH}uy\right]\left[\frac{1}{(1-u)y+\frac{1}{2}u}-\frac{1+u}{y}\right], \tag{5–6.135}$$

which is dominated by values of $y\sim eH/m^2\gg 1$, and of $1-u\sim m^2/eH\ll 1$.

The performance of the $u$-integral, under these circumstances, gives

$$\sim \int_{y_0}^{\infty} dy \exp\left[-\frac{m^2}{eH}y\right]\frac{1}{y}\log 2y \sim \frac{1}{2}\left(\log\frac{2eH}{m^2}\right)^2. \quad (5\text{–}6.136)$$

This leading asymptotic term is quite sufficient to indicate that, far from vanishing at the magnetic field strength of (5–6.114), the energy of the ground state in very strong fields:

$$E = m\left[1 + \frac{\alpha}{4\pi}\left\{\left(\log\frac{2eH}{m^2}\right)^2 + \cdots\right\}\right] \quad (5\text{–}6.137)$$

has increased above $m$. The two limiting forms indicate that, at a value of $H$ in the neighborhood of the characteristic value

$$H = \frac{m^2}{e} = 4 \times 10^{16}\ \mathrm{G}, \quad (5\text{–}6.138)$$

the total energy reaches a minimum value, which is only less than $m$ by a fractional amount of the order $\alpha$. Incidentally, the latter observation is essential to justify this treatment of the strong-field situation, since it is still based on the simplifications of Eq. (5–6.80).

In order to facilitate writing the general expression for $M$ that refers to a state with quantum numbers $n$ and $\zeta'$, we introduce the symbols

$$D_{\pm} = 1 - u + u\frac{\sin x}{x}e^{\pm ix}. \quad (5\text{–}6.139)$$

Accordingly, we have [Eq. (5–6.62)]

$$\Delta = D_+ D_-, \quad (5\text{–}6.140)$$

while the coefficient of $\gamma\Pi_H$ in (5–6.81) acquires the form presented by

$$\frac{1 - u + u\dfrac{\sin x}{x}e^{-i\zeta x}}{\Delta} - e^{-2i\zeta x} = D_+^{-\frac{1}{2}(1+\zeta)}D_-^{-\frac{1}{2}(1-\zeta)} - e^{-2i\zeta x}, \quad (5\text{–}6.141)$$

and the ground-state combination of (5–6.116) reads

$$\exp\left[-i\frac{us}{x}[\beta - (1-u)x]eH\right] = \left(\frac{D_+}{D_-}\right)^{1/2} e^{-iux}. \quad (5\text{–}6.142)$$

For the general situation where [Eq. (5–6.96)]

$$\Pi_H^2 \to (2n + 1 - \zeta' + \zeta)eH = (2n' + \zeta)eH, \; n' = n + \tfrac{1}{2}(1 - \zeta'), \quad (5\text{–}6.143)$$

the structure appearing in (5–6.81) becomes

$$\Delta^{-1/2} \exp\left[-i\frac{us}{x}(\beta - (1-u)x)\Pi_H^2\right] e^{i\zeta ux} \to \left(\frac{D_+}{D_-}e^{-2iux}\right)^{n'} D_+^{-\frac{1}{2}(1-\zeta)} D_-^{-\frac{1}{2}(1+\zeta)}. \quad (5\text{–}6.144)$$

Then, if one uses the projection matrices $\frac{1}{2}(1 \pm \zeta)$ to express the $\zeta$-dependence of functions, and recalls that the function of $\zeta$ multiplying $\gamma\Pi_H$ is to be averaged over its $\zeta = \pm 1$ values, we infer the following effective form of $M$ in a state of energy quantum number $n'$:

$$M = \frac{\alpha}{2\pi} m \int_0^\infty \frac{dx}{x} \int_0^1 du \exp\left[-i\frac{m^2}{eH}ux\right]$$
$$\times \left\{\left(\frac{D_+}{D_-}e^{-2iux}\right)^{n'} \left[\tfrac{1}{2}(1+\zeta)\frac{1 + ue^{-2ix}}{D_-} + \tfrac{1}{2}(1-\zeta)\frac{1 + ue^{2ix}}{D_+}\right.\right.$$
$$\left.\left. + (1-u)\frac{eH}{m^2}n'\left(\frac{2}{D_+D_-} - \frac{e^{-2ix}}{D_-} - \frac{e^{2ix}}{D_+}\right)\right] - 1 - u\right\}. \quad (5\text{–}6.145)$$

Now, the transformation $x \to -iy$ cannot be used, as expected from the radiative instability of all the levels above the ground state. Nevertheless, for weak magnetic fields and $u > u_0$, small values of $x$ should still dominate. We shall first proceed to the same accuracy as in the ground-state discussion, retaining only terms quadratic in $x$. With that limitation,

$$\frac{D_+}{D_-}e^{-2iux} = 1 - \tfrac{2}{3}iu(1-u)^2x^3 + \ldots \quad (5\text{–}6.146)$$

is replaced by unity, which appears to remove this part of the $n'$-dependence. But clearly a restriction on $n'$ is implied, such that

$$n'u(1-u)^2x^3 \ll 1, \quad (5\text{–}6.147)$$

or

$$n' \ll \left(\frac{m^2}{eH}\right)^3 u_0^2 = \frac{m^2}{eH}\left(\frac{u_0}{eH/m^2}\right)^2, \quad (5\text{–}6.148)$$

which excludes $E^2 - m^2$ being large in comparison with $m^2$. This is an essentially non-relativistic situation. To the required $H^2$ accuracy, the term explicitly linear in $H$ does not contribute, and the expansion

$$\frac{1 + ue^{\pm 2ix}}{D_\pm} = 1 + u \pm iu(1-u)x - u(\tfrac{4}{3} - \tfrac{5}{3}u + u^2)x^2 + \cdots \quad (5\text{–}6.149)$$

produces the following $u > u_0$ contribution to $M$:

$$-\zeta\frac{\alpha}{2\pi}\frac{eH}{2m}(1 - 2u_0) + \frac{\alpha}{2\pi}\frac{(eH)^2}{m^3}\left[\tfrac{4}{3}\log\frac{1}{u_0} - \frac{7}{6}\right]. \quad (5\text{–}6.150)$$

Note that this expression is real, and that it coincides with the corresponding ground-state result on placing $\zeta = +1$.

For $u < u_0$, we expand in powers of $u$:

$$\left(\frac{D_+}{D_-}e^{-2iux}\right)^{n'} = 1 + 2iun'\left(\frac{\sin^2 x}{x} - x\right) + \cdots,$$

$$\frac{1 + ue^{\pm 2ix}}{D_\pm} = 1 + u\left(e^{\pm 2ix} + 1 - \frac{\sin x}{x}e^{\pm ix}\right) + \cdots, \quad (5\text{–}6.151)$$

and evaluate the explicit $H$-term at $u = 0$. The resulting contribution to $M$ is

$$\frac{\alpha}{2\pi}m\int_0^{u_0} du\int_0^\infty \frac{dx}{x}\exp\left[-i\frac{m^2}{eH}ux\right]$$
$$\times\left\{2iun'\left(\frac{\sin^2 x}{x} - x\right) + u\left(\cos 2x - \frac{\sin 2x}{2x}\right)\right.$$
$$\left. + i\zeta u\left(\frac{\sin^2 x}{x} - \sin 2x\right) + \frac{eH}{m^2}2n'(1 - \cos 2x)\right\}. \quad (5\text{–}6.152)$$

On placing $n' = 0$, $\zeta = \zeta' = +1$, this reduces to the already evaluated ground-state expression, where $x$ can be replaced by $-iy$ [Eq. (5–6.130)]. Accordingly, we remove the ground-state form to get the additional terms

$$\frac{\alpha}{2\pi}m\int_0^{u_0} du\int_0^\infty \frac{dx}{x}\exp\left[-i\frac{m^2}{eH}ux\right]$$
$$\times\left\{2iun'\left(\frac{\sin^2 x}{x} - x\right) + i(\zeta - 1)u\left(\frac{\sin^2 x}{x} - \sin 2x\right) + \frac{eH}{m^2}2n'(1 - \cos 2x)\right\}.$$
$$(5\text{–}6.153)$$

Since this is the entire source of the imaginary part of $M$, we have

$$\operatorname{Im} M = \frac{\alpha}{2\pi} m \int_0^{u_0} du$$

$$\times \left\{ u \int_0^\infty dx \cos\left(\frac{m^2}{eH} ux\right) \left[ 2n'\left(\frac{\sin^2 x}{x^2} - 1\right) + (\zeta - 1)\left(\frac{\sin^2 x}{x^2} - \frac{\sin 2x}{x}\right)\right]\right.$$

$$\left. - 4n' \frac{eH}{m^2} \int_0^\infty dx \sin\left(\frac{m^2}{eH} ux\right) \frac{\sin^2 x}{x} \right\}. \quad (5\text{–}6.154)$$

Apart from vanishing delta-function integrals, the three closely related integrals encountered here are $(\lambda > 0)$

$$\int_0^\infty dx \cos \lambda x \frac{\sin 2x}{x} = 2\int_0^\infty dx \sin \lambda x \frac{\sin^2 x}{x} = \frac{\pi}{2}\eta(2 - \lambda),$$

$$\int_0^\infty dx \cos \lambda x \frac{\sin^2 x}{x^2} = \frac{\pi}{2}(1 - \tfrac{1}{2}\lambda)\eta(2 - \lambda), \quad (5\text{–}6.155)$$

where $\eta(x)$ is the Heaviside unit step function

$$\eta(x) = \begin{cases} x > 0: & 1, \\ x < 0: & 0. \end{cases} \quad (5\text{–}6.156)$$

Thus, all the integrals that constitute $\operatorname{Im} M$ vanish for $u > 2eH/m^2$. This property is clarified on noting that, in such circumstances, the substitution $x \to -iy$ is permissible in (5–6.153), as evidenced by the existence of the resulting $y$-integral, which is then explicitly real. To the limited accuracy that we are working at, no distinction need be made here between $\zeta'$ and $\zeta$. The immediate outcome is

$$\operatorname{Im} M = -\tfrac{1}{4}\alpha m n \int_0^{2eH/m^2} du \left[ \frac{m^2}{eH}\left(u - \frac{eH}{m^2}\right)^2 + \frac{eH}{m^2} \right]$$

$$+ \tfrac{1}{4}\alpha m(1 - \zeta') \int_0^{2eH/m^2} du \left(u - \frac{eH}{m^2}\right) = -\tfrac{2}{3}\alpha m \left(\frac{eH}{m^2}\right)^2 n, \quad (5\text{–}6.157)$$

which is independent of $\zeta'$. With the notation

$$\operatorname{Im} M = -\tfrac{1}{2}\gamma_0, \quad (5\text{–}6.158)$$

this result is expressed as

$$\gamma_0 = \tfrac{4}{3}\alpha m\left(\frac{eH}{m^2}\right)^2 n. \tag{5–6.159}$$

The imaginary part of $M$ effectively produces the replacement of $m$ by $m - \frac{1}{2}i\gamma_0$. What this implies for the energy of the system is contained in the differential relation (5–6.101), namely

$$E \to E + \frac{\partial E}{\partial m}\left(-\tfrac{1}{2}i\gamma_0\right) = E - \tfrac{1}{2}i\gamma, \tag{5–6.160}$$

with

$$\gamma = (m/E)\gamma_0. \tag{5–6.161}$$

The latter quantity is identified as the decay constant of the system, the inverse of the mean lifetime, as expressed by the probability time factor

$$\left|\exp\left[-i\left(E - \tfrac{1}{2}i\gamma\right)t\right]\right|^2 = e^{-\gamma t}. \tag{5–6.162}$$

For non-relativistic states of motion, $\gamma \cong \gamma_0$. It is in these circumstances that an elementary semi-classical calculation of $\gamma$ can be performed, with concordant results. The classical formula for the power radiated by an accelerated electron is

$$P = \tfrac{2}{3}\alpha(\dot{\mathbf{v}})^2, \tag{5–6.163}$$

where, according to the classical equation of motion,

$$\dot{\mathbf{v}} = \frac{e}{m}\mathbf{v} \times \mathbf{H}. \tag{5–6.164}$$

Considering motion in the plane perpendicular to the magnetic field, we have

$$P = \tfrac{4}{3}\alpha m\left(\frac{eH}{m^2}\right)^2 \tfrac{1}{2}m\mathbf{v}^2. \tag{5–6.165}$$

Now, the quantum expression for the non-relativistic energy $\frac{1}{2}m\mathbf{v}^2$, as implied by (5–6.90), is

$$\tfrac{1}{2}m\mathbf{v}^2 = \frac{\left(\Pi_H{}^2\right)'}{2m} = \omega\left(n + \tfrac{1}{2}\right), \qquad \omega = \frac{eH}{m}, \tag{5–6.166}$$

in which $\omega$ is the classical orbital rotational frequency. To produce the quantum transcription of this semiclassical result, one divides the radiated power by the energy of a quantum, which is $\omega$, to get the emission probability per unit time:

$$\gamma = P/\omega, \tag{5–6.167}$$

and also replaces the total non-relativistic energy by the excitation energy above the ground state, in order to incorporate the stability of the latter. The result for $\gamma \cong \gamma_0$ is just (5–6.159).

This limited treatment is completed by evaluating the real part of the expression (5–6.153):

$$\frac{\alpha}{2\pi} m \int_0^{u_0} du \biggl\{ u \int_0^\infty dx \sin\left(\frac{m^2}{eH} ux\right) \times \left[ 2n'\left(\frac{\sin^2 x}{x^2} - 1\right) + (\zeta - 1)\left(\frac{\sin^2 x}{x^2} - \frac{\sin 2x}{x}\right)\right] + 4n' \frac{eH}{m^2} \int_0^\infty dx \cos\left(\frac{m^2}{eH} ux\right) \frac{\sin^2 x}{x} \biggr\}. \tag{5–6.168}$$

Here we meet the integrals

$$\begin{aligned} \int_0^\infty \frac{dx}{x} \cos \lambda x \sin^2 x &= \tfrac{1}{4} \log\left|1 - \frac{4}{\lambda^2}\right|, \\ \int_0^\infty \frac{dx}{x} \sin \lambda x \sin 2x &= \tfrac{1}{2} \log\left|\frac{\lambda + 2}{\lambda - 2}\right|, \end{aligned} \tag{5–6.169}$$

which are combined in the partial integration evaluation of

$$\int_0^\infty \frac{dx}{x^2} \sin \lambda x \sin^2 x = \tfrac{1}{4}\lambda \log\left|1 - \frac{4}{\lambda^2}\right| + \tfrac{1}{2} \log\left|\frac{\lambda + 2}{\lambda - 2}\right|. \tag{5–6.170}$$

[Note too that the $\lambda$-derivative of the last statement reproduces the first entry of (5–6.169).] There is also the elementary integral

$$\int_0^\infty dx \sin \lambda x = \frac{1}{\lambda}. \tag{5–6.171}$$

At this stage, (5–6.168) can be presented as

$$\frac{\alpha}{2\pi}(1 - \zeta)\frac{eH}{m} u_0 + \frac{\alpha}{\pi} m \left(\frac{eH}{m^2}\right)^2 \biggl\{ 2n \int_0^\infty dt\, t^2 \left[\log\left|1 - \frac{1}{t^2}\right| + \frac{1}{t^2}\right] + n' \int_0^\infty dt \log\left|1 - \frac{1}{t^2}\right| + 2n' \int_0^\infty dt\, t \left[\log\left|\frac{t+1}{t-1}\right| - \frac{2}{t}\right]\biggr\}, \tag{5–6.172}$$

where

$$t = \tfrac{1}{2}\lambda = \frac{m^2}{2eH}u, \tag{5-6.173}$$

and the integrals have been so arranged that, with negligible error, infinity can replace the actual upper limit at

$$t_0 = \frac{m^2}{2eH}u_0 \gg 1. \tag{5-6.174}$$

We have also substituted $\zeta'$ for $\zeta$ in the quadratic $H$-term.

The three integrals that appear here are evaluated as real parts of corresponding complex integrals,

$$\int_0^\infty dt \log\left|1 - \frac{1}{t^2}\right| = \operatorname{Re}\int_0^\infty dt \log\left(1 - \frac{1}{t^2}\right) = \operatorname{Re}\log\frac{t+1}{t-1}\bigg|_0^\infty = 0,$$

$$\int_0^\infty dt\left(t^2 \log\left|1 - \frac{1}{t^2}\right| + 1\right) = \operatorname{Re}\int_0^\infty dt\left[t^2 \log\left(1 - \frac{1}{t^2}\right) + 1\right]$$

$$= \operatorname{Re}\tfrac{1}{3}\log\frac{t+1}{t-1}\bigg|_0^\infty = 0, \tag{5-6.175}$$

and

$$\int_0^\infty dt\left[t\log\left|\frac{t+1}{t-1}\right| - 2\right] = \operatorname{Re}\int_0^\infty dt\left[t\log\frac{t+1}{t-1} - 2\right]$$

$$= \operatorname{Re}\tfrac{1}{2}\log\frac{t-1}{t+1}\bigg|_0^\infty = 0. \tag{5-6.176}$$

All that remains, then, is the term linear in $H$ and proportional to $u_0$, which restores the piece removed from the related integral that had been stopped at the lower limit $u_0$. The net result for the real part of $M$ is to replace the logarithmic lower limit $u_0$ in (5–6.150) by the same value that appeared in the ground-state calculation. The complete statement of $M$, to the present accuracy, is therefore

$$M = -\zeta\frac{\alpha}{2\pi}\frac{eH}{2m} + \frac{2\alpha}{3\pi}m\left(\frac{eH}{m^2}\right)^2\left[\log\left(\frac{m^2}{2eH}\right) - \frac{13}{24}\right] - i\frac{2\alpha}{3}m\left(\frac{eH}{m^2}\right)^2 n, \tag{5-6.177}$$

confirming the earlier remark about the generality of the diamagnetic term. One should observe how simply the diamagnetic and radiation damping terms are related, as expressed by the combination

$$\log\left(\frac{m^2}{2eH}\right) - \pi i n = \log\left(\frac{m^2}{2eH}e^{-\pi i n}\right). \tag{5-6.178}$$

The absence of spin dependence in the decay constant (5-6.159) is to be expected, according to an elementary non-relativistic calculation of the magnetic dipole radiation associated with the spin transition $\zeta = -1 \to +1$. It is based on the electric dipole formula of Eq. (3-15.69), which is converted by the substitution

$$e\mathbf{x} \to \frac{e}{2m}\boldsymbol{\sigma} \tag{5-6.179}$$

into the decay-constant expression

$$\gamma_0 = \tfrac{4}{3}\alpha\omega^3\frac{1}{(2m)^2}|\langle + |\boldsymbol{\sigma}| - \rangle|^2 = \tfrac{2}{3}\alpha\left(\frac{eH}{m^2}\right)^3 m. \tag{5-6.180}$$

Here, we have used the matrix property

$$|\langle + |\boldsymbol{\sigma}| - \rangle|^2 = \langle + |\boldsymbol{\sigma}\cdot\boldsymbol{\sigma}| + \rangle - |\langle + |\boldsymbol{\sigma}| + \rangle|^2 = 3 - 1 = 2 \tag{5-6.181}$$

and recognized that the spin transition frequency $\omega$ equals that of an orbital transition $n \to n - 1$ (recall $2n + 1 - \zeta'$), which is the classical rotation frequency of (5-6.166). Being cubic in the magnetic field strength, this process has escaped a treatment that halts at the second power of an expansion. We shall now proceed to include these cubic terms.

We first attempt to reproduce the decay constant of (5-6.180) by examining a state that decays entirely by magnetic dipole radiation. This is the first excited level with $n = 0$, $\zeta' = -1$. The starting point, then, is the $n' = 1$ form of $M$:

$$\begin{aligned} M = \frac{\alpha}{2\pi}m\int_0^\infty \frac{dx}{x}\int_0^1 du \exp\left[-i\frac{m^2}{eH}ux\right] \\ \times\Bigg\{e^{-2iux}\left[\tfrac{1}{2}(1+\zeta)\frac{D_+}{D_-^{\,2}}(1+ue^{-2ix}) + \tfrac{1}{2}(1-\zeta)\frac{1}{D_-}(1+ue^{2ix})\right. \\ \left. +(1-u)\frac{eH}{m^2}\left(\frac{2}{D_-^{\,2}} - \frac{D_+}{D_-^{\,2}}e^{-2ix} - \frac{1}{D_-}e^{2ix}\right)\right] - 1 - u\Bigg\}. \end{aligned} \tag{5-6.182}$$

Since we are interested here only in the imaginary part of $M$, it is well to recognize that some of these terms do permit the $x \to -iy$ transformation that leads to a purely real expression. They are the ones involving only $D_-$, where

$$D_- \to 1 - u + u\frac{\sinh y}{y}e^{-y}, \tag{5–6.183}$$

and the ones containing

$$\frac{D_+}{D_-^{\,2}}e^{-2ix} \to \frac{(1-u)e^{-2y} + u\dfrac{\sinh y}{y}e^{-y}}{\left(1 - u + u\dfrac{\sinh y}{y}e^{-y}\right)^2}. \tag{5–6.184}$$

What remains is

$$\operatorname{Im} M = \operatorname{Im}\frac{\alpha}{2\pi}m\int_0^\infty \frac{dx}{x}\int_0^1 du \exp\left[-i\left(\frac{m^2}{eH} + 2\right)ux\right]$$
$$\times\left\{\tfrac{1}{2}(1+\zeta)u\left(\frac{\dfrac{\sin x}{x}e^{ix}}{D_-^{\,2}} - 1\right) + \left[\tfrac{1}{2}(1-\zeta)u - (1-u)\frac{eH}{m^2}\right]\left(\frac{e^{2ix}}{D_-} - 1\right)\right\}, \tag{5–6.185}$$

where we have applied the contact term in individual expressions to remove any singularity at $x = 0$.

It will be seen that the transformation $x \to -iy$, with its implied reality, is permitted for $u > 2eH/m^2$. Accordingly, only small values of $u$ contribute to $\operatorname{Im} M$, and we employ an expansion. For the second of the two structures that compose (5–6.185), the presence of $u$ or $eH/m^2$ as a factor implies that no more than the linear term in $u$ of an expansion is needed,

$$\frac{e^{2ix}}{D_-} - 1 = e^{2ix} - 1 - u\left(\frac{\sin x}{x}e^{ix} - e^{2ix}\right) + \cdots. \tag{5–6.186}$$

A similar remark applies to the first term, and it is strengthened by the observation that

$$\langle\zeta\rangle = \zeta'\frac{E}{m} = -\left(1 + 2\frac{eH}{m^2}\right)^{1/2}, \tag{5–6.187}$$

or

$$\left\langle \tfrac{1}{2}(1+\zeta)\right\rangle \cong -\frac{1}{2}\frac{eH}{m^2}; \tag{5–6.188}$$

it suffices to evaluate its factor at $u = 0$. Introducing the expectation value

$$\left\langle \tfrac{1}{2}(1-\zeta)\right\rangle \cong 1+\frac{1}{2}\frac{eH}{m^2}, \tag{5–6.189}$$

we get the following expression for Im $M$:

$$\begin{aligned}\operatorname{Im}\frac{\alpha}{2\pi}m\int_0^\infty \frac{dx}{x}\int_0^1 du\,\exp\left[-i\left(\frac{m^2}{eH}+2\right)ux\right]\\ \times\left\{\left(\frac{1}{2}\frac{eH}{m^2}u-u^2\right)\left(\frac{\sin x}{x}e^{ix}-1\right)+\left(u-\frac{eH}{m^2}+\frac{1}{2}\frac{eH}{m^2}u+u^2\right)(e^{2ix}-1)\right\}.\end{aligned} \tag{5–6.190}$$

The two $x$-integrals that appear here are of the forms

$$\int_0^\infty \frac{dx}{x}[\sin\lambda x-\sin(\lambda-2)x]=\pi\eta(2-\lambda) \tag{5–6.191}$$

and

$$\int_0^\infty \frac{dx}{x}\left[\sin\lambda x-\frac{\sin x}{x}\sin(\lambda-1)x\right]=\pi(1-\tfrac{1}{2}\lambda)\eta(2-\lambda). \tag{5–6.192}$$

For the contribution with only one $u$ or $H$ factor, one must be careful to note that

$$\eta(2-\lambda)=\eta\left[2-\left(\frac{m^2}{eH}+2\right)u\right]\cong\eta\left(2-\frac{m^2}{eH}u\right)-2u\,\delta\left(2-\frac{m^2}{eH}u\right). \tag{5–6.193}$$

The immediate outcome is

$$\begin{aligned}\operatorname{Im}M&=\tfrac{1}{2}\alpha m\left\{\int_0^{2eH/m^2}du\left[\frac{eH}{m^2}u-\tfrac{1}{4}u^2+\frac{m^2}{2eH}u^3\right]-4\left(\frac{eH}{m^2}\right)^3\right\}\\ &=-\tfrac{1}{3}\alpha\left(\frac{eH}{m^2}\right)^3 m,\end{aligned} \tag{5–6.194}$$

and

$$n = 0, \quad \zeta' = -1: \qquad \gamma_0 = \tfrac{2}{3}\alpha\left(\frac{eH}{m^2}\right)^3 m, \tag{5-6.195}$$

as expected.

Now let us perform the expansion up to cubic terms in $H$ for the general expression. We begin with the $x$-expansion for $u > u_0$:

$$\begin{aligned}
\left(\frac{D_+}{D_-}e^{-2iux}\right)^{n'} &= 1 - \tfrac{2}{3}in'u(1-u)^2x^3 + \cdots, \\
\frac{1 + ue^{\pm 2ix}}{D_\pm} &= 1 + u \pm iu(1-u)x - u\left(\tfrac{4}{3} - \tfrac{5}{3}u + u^2\right)x^2 \\
&\quad \mp iu(1-u)\left(1 - \tfrac{4}{3}u + u^2\right)x^3 + \cdots, \\
\frac{2}{D_+D_-} - \frac{e^{-2ix}}{D_-} - \frac{e^{2ix}}{D_+} &= 4\left(1 - \tfrac{2}{3}u\right)x^2 + \cdots.
\end{aligned} \tag{5-6.196}$$

These are combined into the following expanded form of the brace in Eq. (5–6.145):

$$\begin{aligned}
\{\ \} = &-i\zeta u(1-u)\left[x - \left(1 - \tfrac{4}{3}u + u^2\right)x^3\right] - u\left(\tfrac{4}{3} - \tfrac{5}{3}u + u^2\right)x^2 \\
&- \tfrac{2}{3}in'u(1-u)(1-u^2)x^3 + 4n'\frac{eH}{m^2}(1-u)\left(1 - \tfrac{2}{3}u\right)x^2.
\end{aligned} \tag{5-6.197}$$

After performing the $x$-integration, we have

$$\begin{aligned}
M(u > u_0) = \frac{\alpha}{2\pi}m\int_{u_0}^1 du\Bigg\{ &-\zeta\frac{eH}{m^2}(1-u) + \left(\frac{eH}{m^2}\right)^2\frac{1}{u}\left(\tfrac{4}{3} - \tfrac{5}{3}u + u^2\right) \\
&-2\zeta\left(\frac{eH}{m^2}\right)^3\frac{1-u}{u^2}\left(1 - \tfrac{4}{3}u + u^2\right) \\
&-\tfrac{8}{3}n'\left(\frac{eH}{m^2}\right)^3\frac{1-u}{u^2}\left(1 - u + \tfrac{1}{2}u^2\right)\Bigg\},
\end{aligned} \tag{5-6.198}$$

and then

$$\begin{aligned}
M(u > u_0) = \frac{\alpha}{2\pi}m\Bigg\{ &-\zeta\frac{eH}{2m^2} + \left(\frac{eH}{m^2}\right)^2\left(\tfrac{4}{3}\log\frac{1}{u_0} - \frac{7}{6}\right) \\
&-2\zeta\left(\frac{eH}{m^2}\right)^3\left(\frac{1}{u_0} - \tfrac{7}{3}\log\frac{1}{u_0} + \frac{5}{6}\right) - \tfrac{8}{3}n'\left(\frac{eH}{m^2}\right)^3\left(\frac{1}{u_0} - 2\log\frac{1}{u_0} + \frac{1}{4}\right)\Bigg\}.
\end{aligned} \tag{5-6.199}$$

Since no distinction need be made between $\zeta'$ and $\zeta$ in $H^3$ terms, one can also present the last two terms of the brace as

$$\tfrac{8}{3}(2n+1)\left(\frac{eH}{m^2}\right)^3\left(\log\frac{1}{u_0}-\frac{1}{8}-\frac{1}{2u_0}\right)+2\zeta\left(\frac{eH}{m^2}\right)^3\left(\log\frac{1}{u_0}-\frac{2}{3}-\frac{1}{3u_0}\right). \quad (5\text{–}6.200)$$

In writing the last forms, we have not troubled to keep positive powers of $u_0$, which we know will be cancelled eventually by related terms in $M(u < u_0)$. But we have temporarily retained another type of $u_0$-dependence, the $H^3/u_0$ structure. To see how these terms are cancelled, consider the special example of the ground-state evaluation of the $u < u_0$ contribution, as presented in Eq. (5–6.130). Carrying the expansion in $eH/m^2u_0$ one step beyond that given in (5–6.131) supplies the additional term

$$\frac{\alpha}{2\pi}m\left\{2\left(\frac{eH}{m^2}\right)^3\frac{1}{u_0}\right\}, \quad (5\text{–}6.201)$$

which does indeed cancel the similar term in (5–6.199) when the latter is evaluated for $n' = 0$, $\zeta = +1$. In short, the $H^3/u_0$ terms are a residue of the $H^2$ calculation, and will not be considered further.

Now we proceed to the $u$-expansion for $u < u_0$. The ingredients are:

$$\left(\frac{D_+}{D_-}e^{-2iux}\right)^{n'} = 1+2in'u\left(\frac{\sin^2 x}{x}-x\right)-2in'u^2\frac{\sin^2 x}{x}\left(\frac{\sin 2x}{2x}-1\right)$$
$$-2n'^2u^2\left(\frac{\sin^4 x}{x^2}-2\sin^2 x+x^2\right)+\cdots,$$

$$\frac{1+ue^{\pm 2ix}}{D_\pm} = 1+u\left(e^{\pm 2ix}+1-\frac{\sin x}{x}e^{\pm ix}\right)$$
$$+u^2\left(\frac{\sin^2 x}{x^2}e^{\pm 2ix}-2\frac{\sin x}{x}e^{\pm ix}+1-\frac{\sin x}{x}e^{\pm 3ix}+e^{\pm 2ix}\right)+\cdots, \quad (5\text{–}6.202)$$

and

$$\frac{2}{D_+D_-}-\frac{e^{-2ix}}{D_-}-\frac{e^{2ix}}{D_+} = 2(1-\cos 2x)$$
$$+2u\left(\frac{\sin x}{x}\cos 3x-\cos 2x-2\frac{\sin x}{x}\cos x+2\right)+\cdots. \quad (5\text{–}6.203)$$

We shall record only the $u^2$ and $Hu$ terms in the structure of $M(u < u_0)$:

$$\frac{\alpha}{2\pi}m\int_0^{u_0} du\int_0^\infty \frac{dx}{x}\exp\left[-i\frac{m^2}{eH}ux\right]$$

$$\times\left\{u^2A_0 + 2in'u^2A_1 - 2n'^2u^2A_2\right.$$

$$\left.-i\zeta u^2B_0 + 2n'\zeta u^2B_1 + 2n'\frac{eH}{m^2}uC_1 + 2in'^2\frac{eH}{m^2}uC_2\right\}. \quad (5\text{–}6.204)$$

The following symbols have been introduced for various functions of $x$:

$$A_0 = \frac{\sin^2 x}{x^2}\cos 2x - \frac{\sin 2x}{x} + 1 - \frac{\sin x}{x}\cos 3x + \cos 2x,$$

$$A_1 = \frac{\sin^2 x}{x}\left(\cos 2x + 2 - \frac{\sin 2x}{x}\right) - x\left(\cos 2x + 1 - \frac{\sin 2x}{2x}\right), \quad (5\text{–}6.205)$$

$$A_2 = \frac{\sin^4 x}{x^2} - 2\sin^2 x + x^2$$

and

$$B_0 = \frac{\sin^2 x}{x^2}\sin 2x - 2\frac{\sin^2 x}{x} - \frac{\sin x}{x}\sin 3x + \sin 2x,$$

$$B_1 = \left(\frac{\sin^2 x}{x} - x\right)\left(\sin 2x - \frac{\sin^2 x}{x}\right); \quad (5\text{–}6.206)$$

also

$$C_1 = \frac{\sin x}{x}\cos 3x - 2\frac{\sin x}{x}\cos x + 1,$$

$$C_2 = 4\left(\frac{\sin^2 x}{x} - x\right)\sin^2 x. \quad (5\text{–}6.207)$$

We have already seen, in the discussion of the $H^2$ portion of Re $M$, that no contribution appears for $u < u_0$ unless there is a related term in the $u > u_0$ structure. Specifically, this expresses the vanishing of the terms linear in $n'$ and $\zeta$ [Eqs. (5–6.168–176)], which do not appear in (5–6.150). A similar situation occurs in the $H^3$ calculation. The $u > u_0$ form of (5–6.199) only has $H^3$ terms proportional to $\zeta$ and $n'$. Correspondingly, there are vanishing contributions in (5–6.204) for the terms with coefficients $n'^2$, $n'\zeta$, as well as the one lacking any quantum number dependence. The integrals involved in demonstrating this are of the general types already encountered in the $H^2$ discussion, and we refrain from giving further details. The remaining structure in (5–6.204) can be presented as ($\zeta' \cong \zeta$)

$$\operatorname{Re} \frac{\alpha}{2\pi} m \int_0^{u_0} du \int_0^{\infty} \frac{dx}{x} \exp\left[-i\frac{m^2}{eH} ux\right]$$

$$\times\left\{(2n+1)\left[iu^2 A_1 + \frac{eH}{m^2} uC_1\right] - \zeta\left[iu^2(A_1 + B_0) + \frac{eH}{m^2} uC_1\right]\right\}, \quad (5\text{–}6.208)$$

where

$$A_1 + B_0 = \tfrac{3}{2}\sin 2x - 2x\cos^2 x - \frac{1}{2x}\sin^2 2x. \quad (5\text{–}6.209)$$

The $x$-integrals appearing here are effectively given by

$$\int_0^{\infty} \frac{dx}{x} \sin\lambda x\,(A_1 + B_0) = -\frac{1}{\lambda} - \frac{\lambda}{\lambda^2 - 4} - \tfrac{1}{8}\lambda \log\left|1 - \frac{16}{\lambda^2}\right|$$

$$+ \tfrac{3}{4}\log\left|\frac{\lambda+2}{\lambda-2}\right| - \tfrac{1}{2}\log\left|\frac{\lambda+4}{\lambda-4}\right|,$$

$$\int_0^{\infty} \frac{dx}{x} \sin\lambda x\, B_0 = -\tfrac{3}{4}\lambda \log\left|1 - \frac{4}{\lambda^2}\right| + \tfrac{1}{4}\lambda \log\left|1 - \frac{16}{\lambda^2}\right|$$

$$-\tfrac{1}{2}\left(1 + \tfrac{1}{4}\lambda^2\right)\log\left|\frac{\lambda+2}{\lambda-2}\right| + \tfrac{1}{16}\lambda^2 \log\left|\frac{\lambda+4}{\lambda-4}\right|, \quad (5\text{–}6.210)$$

and

$$\int_0^{\infty} \frac{dx}{x} \cos\lambda x\, C_1 = -1 + \tfrac{3}{2}\log\left|1 - \frac{4}{\lambda^2}\right| - \log\left|1 - \frac{16}{\lambda^2}\right|$$

$$+ \tfrac{3}{4}\lambda \log\left|\frac{\lambda+2}{\lambda-2}\right| - \tfrac{1}{4}\lambda \log\left|\frac{\lambda+4}{\lambda-4}\right|. \quad (5\text{–}6.211)$$

The subsequent integrals over the variable

$$\lambda = \frac{m^2}{eH} u, \quad (5\text{–}6.212)$$

in which positive powers of $\lambda_0 \sim u_0$ are omitted, appear as

$$\int_0^{\lambda_0} d\lambda\, \lambda^2 \int_0^\infty \frac{dx}{x} \sin \lambda x\,(A_1 + B_0) + \int_0^{\lambda_0} d\lambda\, \lambda \int_0^\infty \frac{dx}{x} \cos \lambda x\, C_1$$

$$\to -2 \log \frac{\lambda_0}{2} + 1 \qquad (5\text{–}6.213)$$

and

$$\int_0^{\lambda_0} d\lambda\, \lambda^2 \int_0^\infty \frac{dx}{x} \sin \lambda x\, B_0 \to -\tfrac{14}{3} \log \frac{\lambda_0}{2} + \tfrac{32}{5}(\log 2 - \tfrac{1}{2}) + \tfrac{11}{18}. \qquad (5\text{–}6.214)$$

This gives the following result for (5–6.204):

$$\frac{\alpha}{2\pi} m \left(\frac{eH}{m^2}\right)^3 \left\{(2n+1)\left[\tfrac{8}{3} \log\left(\frac{m^2}{2eH} u_0\right) - \tfrac{32}{5}(\log 2 - \tfrac{1}{2}) + \tfrac{7}{18}\right]\right.$$

$$\left. + \zeta\left[2 \log\left(\frac{m^2}{2eH} u_0\right) - 1\right]\right\}, \qquad (5\text{–}6.215)$$

which, added to (5–6.199, 200), together with the analogous $H^2$ terms [Eq. (5–6.131)], yields

$$\operatorname{Re} M = \frac{\alpha}{2\pi} m \left\{ -\zeta \frac{eH}{2m^2} + \left(\frac{eH}{m^2}\right)^2 \left[\tfrac{4}{3} \log\left(\frac{m^2}{2eH}\right) - \frac{13}{18}\right] + \zeta \left(\frac{eH}{m^2}\right)^3 \left[2 \log\left(\frac{m^2}{2eH}\right) - \frac{7}{3}\right]\right.$$

$$\left. + (2n+1)\left(\frac{eH}{m^2}\right)^3 \left[\tfrac{8}{3} \log\left(\frac{m^2}{2eH}\right) - \tfrac{32}{5}(\log 2 - \tfrac{1}{2}) + \tfrac{1}{18}\right]\right\}. \qquad (5\text{–}6.216)$$

We see here the first strong-field modifications of the $\alpha/2\pi$ spin moment and of the magnetic polarizability, which work in opposite directions on these two properties.

*A bit of history*: A more complicated-appearing but equivalent result was produced some time ago by R. Newton, Phys. Rev. 96, 523 (1954). His method was a related one, in its use of the mass operator $M$, but was sufficiently more cumbersome that only these initial terms of an expansion in $H$ were exhibited.

One can, of course, compute the $H^3$ terms in $\operatorname{Im} M$ also, but we shall not trouble to do so. The physical processes that would be included here involve relativistic modifications in the electric dipole process and electric quadrupole contributions. These additional effects must remain small, however, for weak fields and low energies. The situation is quite different at ultra-relativistic energies where very high multipole moments dominate, and we proceed directly to that calculation.

Let us return to the $x$-expansion of (5–6.146) and remove the $n'$ restriction of (5–6.147, 148). Now, we write

$$\left(\frac{D_+}{D_-}e^{-2iux}\right)^{n'} \cong \exp\left[-\tfrac{2}{3}in'u(1-u)^2x^3\right] \qquad (5\text{–}6.217)$$

as an approximate expression of the regime

$$n' \gg 1, \qquad u(1-u)^2x^3 \ll 1. \qquad (5\text{–}6.218)$$

The integral formula for $M$, Eq. (5–6.145), will be dominated by the two exponential functions with large coefficients in the argument, (5–6.217) and $\exp[-i(m^2/eH)ux]$. The important range of $x$ occurs where the two arguments are roughly comparable,

$$\frac{m^2}{eH} \sim n'(1-u)^2x^2. \qquad (5\text{–}6.219)$$

Under the high-energy circumstances expressed by

$$E^2 - m^2 = 2n'eH \gg m^2, \qquad (5\text{–}6.220)$$

these dominant values of $x$,

$$x \sim \frac{1}{1-u}\frac{m}{E}, \qquad (5\text{–}6.221)$$

will indeed be small compared with unity, apart from a narrow range of $u$ near one. Accordingly, we retain only the leading terms in an $x$-expansion of the bracket in (5–6.145),

$$[\ ] \cong 1 + u - i\zeta u(1-u)x + 2\frac{E^2}{m^2}(1-u)\left(1-\tfrac{1}{3}u\right)x^2. \qquad (5\text{–}6.222)$$

Having in mind the estimate of (5–6.221), we introduce a new integration variable $y$, such that

$$x = \frac{1}{1-u}\frac{m}{E}y. \qquad (5\text{–}6.223)$$

This converts the bracket of (5–6.222) into

$$[\ ] \cong 1 + u - i\zeta' uy + 2\frac{1-\frac{2}{3}u}{1-u}y^2, \qquad (5\text{–}6.224)$$

where we have also replaced $\zeta(m/E)$ by its effective value $\zeta'$, and the product of the two exponential functions becomes

$$\exp\left[-i\tfrac{3}{2}\xi\left(y + \tfrac{1}{3}y^3\right)\right], \qquad (5\text{–}6.225)$$

in which

$$\xi = \frac{2}{3}\frac{m^2}{eH}\frac{m}{E}\frac{u}{1-u}. \qquad (5\text{–}6.226)$$

Since our concern here is with Im $M$, we write, directly,

$$\operatorname{Im} M = -\frac{\alpha}{2\pi}m\int_0^1 du\int_0^\infty \frac{dy}{y}\left\{(1+u)\left[\sin\tfrac{3}{2}\xi\left(y+\tfrac{1}{3}y^3\right) - \sin\tfrac{3}{2}\xi y\right]\right.$$

$$\left. + \zeta' uy\cos\tfrac{3}{2}\xi\left(y+\tfrac{1}{3}y^3\right) + 2\frac{1-\frac{2}{3}u}{1-u}y^2\sin\tfrac{3}{2}\xi\left(y+\tfrac{1}{3}y^3\right)\right\}. \qquad (5\text{–}6.227)$$

The basic integral that appears is

$$\int_0^\infty dy\cos\tfrac{3}{2}\xi\left(y+\tfrac{1}{3}y^3\right) = 3^{-1/2}K_{1/3}(\xi), \qquad (5\text{–}6.228)$$

which uses an example of the Bessel function of imaginary argument,

$$K_\nu(z) = \tfrac{1}{2}\pi i\exp\left[i\frac{\pi}{2}\nu\right]H_\nu^{(1)}(iz). \qquad (5\text{–}6.229)$$

(For all needed information on such functions, the classic reference, G. N. Watson, *Bessel Functions*, Cambridge University Press, can be consulted.) On using the property

$$\int_0^\infty dy\left(1+y^2\right)\cos\tfrac{3}{2}\xi\left(y+\tfrac{1}{3}y^3\right) = \frac{2}{3\xi}\int_0^\infty d\sin\tfrac{3}{2}\xi\left(y+\tfrac{1}{3}y^3\right) = 0, \qquad (5\text{–}6.230)$$

we infer that

$$\frac{d}{d\xi}\int_0^\infty \frac{dy}{y}\sin\tfrac{3}{2}\xi\left(y+\tfrac{1}{3}y^3\right) = 3^{-1/2}K_{1/3}(\xi), \qquad (5\text{–}6.231)$$

and then

$$\tfrac{1}{2}\pi - \int_0^\infty \frac{dy}{y} \sin \tfrac{3}{2}\xi\left(y + \tfrac{1}{3}y^3\right) = 3^{-1/2} \int_\xi^\infty d\eta\, K_{1/3}(\eta). \qquad (5\text{–}6.232)$$

The integral that produced $\frac{1}{2}\pi$ was identified through the limit of large $\xi$, where $y$ becomes correspondingly small,

$$\lim_{\xi \gg 1} \int_0^\infty \frac{dy}{y} \sin \tfrac{3}{2}\xi\left(y + \tfrac{1}{3}y^3\right) = \int_0^\infty \frac{dy}{y} \sin \tfrac{3}{2}\xi y = \tfrac{1}{2}\pi. \qquad (5\text{–}6.233)$$

Combining the result of a partial integration,

$$\tfrac{3}{2}\xi \int_0^\infty dy\left(y + y^3\right) \sin \tfrac{3}{2}\xi\left(y + \tfrac{1}{3}y^3\right) = 3^{-1/2} K_{1/3}(\xi), \qquad (5\text{–}6.234)$$

with the recurrence relation

$$\left(\frac{d}{d\xi} + \frac{1}{3\xi}\right) K_{1/3}(\xi) = -K_{2/3}(\xi), \qquad (5\text{–}6.235)$$

we also deduce that

$$\int_0^\infty dy\, y \sin \tfrac{3}{2}\xi\left(y + \tfrac{1}{3}y^3\right) = 3^{-1/2} K_{2/3}(\xi). \qquad (5\text{–}6.236)$$

A final property of interest, which exploits another recurrence relation,

$$2\frac{d}{d\xi} K_{2/3}(\xi) + K_{1/3}(\xi) = -K_{5/3}(\xi), \qquad (5\text{–}6.237)$$

is

$$2K_{2/3}(\xi) - \int_\xi^\infty d\eta\, K_{1/3}(\eta) = \int_\xi^\infty d\eta\, K_{5/3}(\eta). \qquad (5\text{–}6.238)$$

The various integral evaluations are inserted in (5–6.227) to give

$$\operatorname{Im} M = -\frac{\alpha}{2\pi} m\, 3^{-1/2} \int_0^1 du \left\{ -(1 + u) \int_\xi^\infty d\eta\, K_{1/3}(\eta) \right.$$

$$\left. + \zeta' u K_{1/3}(\xi) + 2\frac{1 - \frac{2}{3}u}{1 - u} K_{2/3}(\xi) \right\}, \qquad (5\text{–}6.239)$$

or [Eqs. (5–6.158, 161)]

$$\gamma = \frac{\alpha}{\pi} m \frac{m}{E} 3^{-1/2} \int_0^1 du \left\{ \int_\xi^\infty d\eta\, K_{5/3}(\eta) - u \int_\xi^\infty d\eta\, K_{1/3}(\eta) + \frac{2}{3} \frac{u}{1-u} K_{2/3}(\xi) + \zeta' u K_{1/3}(\xi) \right\}. \tag{5–6.240}$$

For all presently attainable artificial magnetic field strengths and electron energies,

$$\frac{eH}{m^2} \frac{E}{m} \cong \left(\tfrac{1}{2}\, 10^{-10}\, \mathrm{G}^{-1}\, \mathrm{GeV}^{-1}\right) HE \tag{5–6.241}$$

is very small compared to unity. Since the functions $K_\nu(\xi)$ of direct concern decrease exponentially for $\xi \gg 1$, the important values of $u$ are such that [cf. (5–6.226)]

$$u \sim \frac{eH}{m^2} \frac{E}{m} \ll 1. \tag{5–6.242}$$

Accordingly, a leading approximation to $\gamma$ is given by

$$\gamma \cong \frac{\alpha}{\pi} m \frac{m}{E} 3^{-1/2} \int_0^1 du \int_\xi^\infty d\eta\, K_{5/3}(\eta), \tag{5–6.243}$$

or

$$\gamma \cong \frac{\alpha}{2\pi} \frac{eH}{m^2} m 3^{1/2} \int_0^\infty d\xi\, \xi K_{5/3}(\xi) = \tfrac{5}{2}\, 3^{-1/2} \alpha \frac{eH}{m^2} m, \tag{5–6.244}$$

which evaluation uses an example of the integral

$$\int_0^\infty d\xi\, \xi^{\mu-1} K_\nu(\xi) = 2^{\mu-2} \Gamma\left(\frac{\mu-\nu}{2}\right) \Gamma\left(\frac{\mu+\nu}{2}\right). \tag{5–6.245}$$

The radiation emitted under the circumstances being considered, as expressed by (5–6.242), has a special character—it is classical radiation. To appreciate this, it may be simplest to turn matters about and inquire concerning the classical radiation that surely is emitted by an electron moving in a macroscopic orbit under the control of a macroscopic magnetic field; this is the experimentally well-known synchrotron radiation. In this classical limit the exponential function $\exp[-is\chi(u)]$ of Eq. (5–6.4) will be rapidly oscillatory, and the major contribution to the $k$-integral is concentrated in the neighborhood of the stationary phase point,

$$k = u\Pi. \tag{5–6.246}$$

This is not the momentum of a real photon

$$-k^2 = -u^2\Pi^2 \simeq u^2m^2, \tag{5–6.247}$$

but becomes so if $u$ is sufficiently small. Accordingly, to an accuracy that neglects $u^2$, we identify the energy of the emitted photon, $\omega = k^0$, with

$$\omega = uE, \tag{5–6.248}$$

and we recognize in the classical criterion,

$$\omega \ll E, \tag{5–6.249}$$

a consistent restriction to small values of $u$.

We can now present (5–6.243) as the spectral integral

$$\gamma \simeq \frac{\alpha}{\pi}m\frac{m}{E}3^{-1/2}\int_0^\infty \frac{d\omega}{E}\int_\xi^\infty d\eta\, K_{5/3}(\eta),$$
$$\xi \simeq \frac{2}{3}\frac{m^2}{eH}\frac{m}{E}\frac{\omega}{E}, \tag{5–6.250}$$

and pick out of the integrand the classical power spectrum,

$$\gamma = \int d\omega \frac{1}{\omega}P(\omega), \tag{5–6.251}$$

where

$$P(\omega) = \frac{\alpha}{\pi}m\frac{m}{E}\frac{\omega}{E}3^{-1/2}\int_\xi^\infty d\eta\, K_{5/3}(\eta). \tag{5–6.252}$$

With this identification, the total radiated power is

$$P = \int d\omega\, P(\omega) = \frac{9}{4}\frac{\alpha}{\pi}\left(\frac{eH}{m}\right)^2\left(\frac{E}{m}\right)^2 3^{-1/2}\int_0^\infty d\xi\, \xi\int_\xi^\infty d\eta\, K_{5/3}(\eta)$$
$$= \tfrac{2}{3}\alpha\left(\frac{eH}{m}\right)^2\left(\frac{E}{m}\right)^2, \tag{5–6.253}$$

which applies the integral [(5–6.245)]

$$\frac{1}{2}\int_0^\infty d\xi\, \xi^2 K_{5/3}(\xi) = \tfrac{8}{9}\, 3^{-1/2}\pi. \tag{5–6.254}$$

To verify that (5–6.253) is indeed the classical expression for radiated power, at high energies, we return to the classical, non-relativistic formula, Eq. (5–6.163), and remove the low-energy limitation. To do that we remark that power, the coefficient in the linear relation between emitted energy and elapsed time, must be a relativistic invariant. We therefore replace time by proper time, and non-relativistic momentum $\mathbf{p} = m\mathbf{v}$ by the four-vector of momentum, to produce an invariant:

$$P = \frac{2}{3}\frac{\alpha}{m^2}(\dot{\mathbf{p}})^2 \to \frac{2}{3}\frac{\alpha}{m^2}\frac{dp^\mu}{ds}\frac{dp_\mu}{ds}, \tag{5–6.255}$$

where [cf. Eq. (1–3.76)]

$$(ds)^2 = -dx^\mu\, dx_\mu = (1 - \mathbf{v}^2)(dt)^2 = (m/E)^2(dt)^2. \tag{5–6.256}$$

Since the classical equations of motion read [they are the first line of Eq. (1–3.77), written in three-dimensional notation]

$$\frac{d\mathbf{p}}{dt} = e\mathbf{v} \times \mathbf{H}, \qquad \frac{dE}{dt} = 0, \tag{5–6.257}$$

we have, for motion in the plane perpendicular to the field,

$$P = \frac{2}{3}\frac{\alpha}{m^2}\left(\frac{E}{m}\right)^2 (eH)^2\mathbf{v}^2. \tag{5–6.258}$$

At ultra-relativistic energies, where $\mathbf{v}^2 \cong 1$, this is indeed the same as (5–6.253).

The nature of the spectrum, in which the important frequencies [$\xi \sim 1$] are of the order

$$\omega \sim \frac{eH}{m^2}\left(\frac{E}{m}\right)^2 m, \tag{5–6.259}$$

is also a classical result. The rotational frequency of the electron in the magnetic field, as inferred from (5–6.257) and the relation $\mathbf{p} = E\mathbf{v}$, is

$$\omega_0 = \frac{eH}{E} = \frac{eH}{m^2}\frac{m}{E}m. \tag{5–6.260}$$

This is the fundamental frequency of the classical radiation, generalizing the non-relativistic result deduced from (5–6.164). But, unlike the non-relativistic situation, most of the radiation appears at very high harmonics of the fundamental. There are two reasons for this. First, owing to the high energy of the particle, the radiation is concentrated near the instantaneous direction of motion of the particle, appearing in a narrow cone with an opening angle of the order

$$\theta \sim m/E. \tag{5–6.261}$$

Only that fraction of the orbit is actually effective in directing radiation toward an observer, and on this account the important harmonic numbers would be $\sim E/m$. The second point is that, in consequence of the Doppler effect, the detected frequency differs markedly from the emitted frequency. A signal generated at the point $\mathbf{r}(t)$, at the time $t$, is received, at the point $\mathbf{r}'$, at the time

$$t' = t + |\mathbf{r}' - \mathbf{r}(t)|. \tag{5–6.262}$$

Hence,

$$dt' = \left[1 - \frac{\mathbf{r}' - \mathbf{r}}{|\mathbf{r}' - \mathbf{r}|} \cdot \mathbf{v}(t)\right] dt = (1 - v\cos\theta)\, dt, \tag{5–6.263}$$

where $\theta$ is the angle between the emission direction and the instantaneous velocity. Then, since

$$v \sim \cos\theta \sim 1 - \tfrac{1}{2}(m/E)^2, \tag{5–6.264}$$

we have

$$dt' \sim (m/E)^2\, dt, \tag{5–6.265}$$

and the significant detected frequencies are of the order

$$\omega \sim \left(\omega_0 \frac{E}{m}\right)\left(\frac{E}{m}\right)^2, \tag{5–6.266}$$

which is the content of (5–6.259).

In the strict classical limit, where $u \ll 1$ is considered negligible in comparison with unity, all reference to the spin naturally disappears. It is interesting, then, to proceed to the level of first quantum corrections, where the quantum number $\zeta'$ does make an appearance. For that purpose, we return to the expression for $\gamma$, Eq.

(5–6.240), and approximate it as

$$\gamma \cong \frac{\alpha}{\pi} m \frac{m}{E} 3^{-1/2} \int_0^1 du$$
$$\times \left\{ \int_\xi^\infty d\eta\, K_{5/3}(\eta) + u\left[\tfrac{2}{3}K_{2/3}(\xi) - \int_\xi^\infty d\eta\, K_{1/3}(\eta)\right] + \zeta' u K_{1/3}(\xi)\right\}, \quad (5\text{–}6.267)$$

where, in the leading term,

$$\xi \cong \frac{2}{3}\frac{m^2}{eH}\frac{m}{E}u(1+u) = \frac{\omega}{\omega_c}\left(1 + \frac{\omega}{E}\right), \quad (5\text{–}6.268)$$

and we have now introduced the characteristic classical frequency

$$\omega_c = \frac{3}{2}\frac{eH}{m^2}\left(\frac{E}{m}\right)^2 m. \quad (5\text{–}6.269)$$

It suffices to write $\xi = \omega/\omega_c$ in the terms that have an explicit $u$-factor. Concerning the quantum correction terms that are independent of $\zeta'$, we note that

$$\left(\frac{E}{\omega_c}\right)^2 \int_0^1 du\, u\left[\tfrac{2}{3}K_{2/3}(\xi) - \int_\xi^\infty d\eta\, K_{1/3}(\eta)\right]$$
$$= \frac{2}{3}\int_0^\infty d\xi\, K_{2/3}(\xi) - \frac{1}{2}\int_0^\infty d\xi\, \xi^2 K_{1/3}(\xi) = 0, \quad (5\text{–}6.270)$$

as a consequence of the integral evaluations deduced from (5–6.245). One aspect of this fact refers to unpolarized particles, where $\zeta' = +1$ and $-1$ occur with equal probabilities. Then the spectral density of $\gamma$, which is $\omega^{-1}P(\omega)$, has the classical form, but with the substitution

$$\omega \to \omega\left(1 + \frac{\omega}{E}\right). \quad (5\text{–}6.271)$$

All first quantum corrections in $\gamma$ are made explicit by writing

$$\gamma = \frac{\alpha}{\pi} m \frac{m}{E} 3^{-1/2} \int_0^\infty \frac{d\omega}{E}\left\{\int_\xi^\infty d\eta\, K_{5/3}(\eta) - \frac{\omega}{E}\xi K_{5/3}(\xi) + \zeta'\frac{\omega}{E}K_{1/3}(\xi)\right\}, \quad (5\text{–}6.272)$$

where $\xi$ now reverts to its classical form, $\omega/\omega_c$. The corrected decay rate is

$$\gamma = \tfrac{5}{2}\, 3^{-1/2}\alpha \frac{eH}{m^2} m\left[1 - \tfrac{8}{5}\, 3^{-1/2}\frac{eH}{m^2}\frac{E}{m}\left(1 - \frac{3^{3/2}}{16}\zeta'\right)\right]. \quad (5\text{–}6.273)$$

Introducing an additional factor of $\omega$ in (5–6.272) then gives the radiation power,

$$P = \tfrac{2}{3}\alpha\left(\frac{eH}{m}\right)^2\left(\frac{E}{m}\right)^2\left[1 - \tfrac{55}{16}\,3^{1/2}\,\frac{eH}{m^2}\frac{E}{m}\left(1 - \tfrac{8}{55}\,3^{1/2}\zeta'\right)\right]. \quad (5\text{–}6.274)$$

Harold looks skeptical.

H.: That last calculation, of the first quantum correction to radiated power, is not very convincing. Suppose you had not recognized that two of the terms in $\gamma$ cancel [Eq. (5–6.270)], and had inserted the additional factor of $\omega$ in their integrands as well. That would give a different answer for $P$.

S.: You are quite right. Put more generally, one cannot infer a unique integrand, $\omega^{-1}P(\omega)$, given only the integral expression for $\gamma$. For that, an additional argument is needed, which we shall now develop.

Let us return to the expression for $M$, Eq. (5–6.3) or subsequent forms, and insert a unit factor,

$$1 = \int_{-\infty}^{\infty} d\omega\,\delta(\omega - k^0) = \int_{-\infty}^{\infty} d\omega \int_{-\infty}^{\infty} d\tau\,\frac{1}{2\pi}e^{i(\omega - k^0)\tau}. \quad (5\text{–}6.275)$$

When applied to Im $M$, where only real processes contribute, the inferred spectral distribution in $\omega$ will be that of the radiated photon energy, thus supplying the desired photon spectrum without ambiguity. The $k$-integration symbolized by $\langle\ \rangle$ is now modified by the presence of the factor

$$e^{-ik^0\tau} = e^{ik\lambda}, \qquad \lambda^0 = \tau, \quad \lambda = 0. \quad (5\text{–}6.276)$$

This effectively induces the following substitution in $\chi(u)$ [Eq. (5–6.4):

$$(k - u\Pi)^2 \to (k - u\Pi)^2 - \frac{1}{s}k\lambda = \left(k - u\Pi - \frac{\lambda}{2s}\right)^2 - \frac{u}{s}\Pi\lambda - \frac{\lambda^2}{4s^2}. \quad (5\text{–}6.277)$$

The subsequent $k$-transformation

$$k \to k + \frac{\lambda}{2s}, \quad (5\text{–}6.278)$$

together with the last two $\lambda$-dependent terms of (5–6.277), modifies Eq. (5–6.43), for example, by the additional factor (this procedure depends upon the circumstance $\lambda = 0$)

$$\exp\left[iu\Pi\lambda + i\frac{\lambda^2}{4s}\right] = \exp\left[-iuE\tau - i\frac{\tau^2}{4s}\right], \quad (5\text{–}6.279)$$

and by the substitution

$$\frac{2(1-u)eqFs}{D}\Pi \rightarrow \frac{2(1-u)eqFs}{D}\Pi - \frac{\lambda}{2s}. \tag{5–6.280}$$

The additional term that the latter produces in the brace of (5–6.52), or its rearrangement in (5–6.75), is

$$\begin{aligned}(-4\cos x\, e^{-i\zeta x})\left(-\frac{\gamma\lambda}{2s}\right) + 2\gamma(1+A^T)\left(-\frac{\lambda}{2s}\right) &= -\gamma^0\frac{\tau}{s}(2\cos x\, e^{-i\zeta x}-1)\\ &= -2m\left[\gamma^0\frac{\tau}{2ms}e^{-2i\zeta x}\right].\end{aligned} \tag{5–6.281}$$

In the last version, we have also made explicit the factor of $-2m$ that is removed in producing the bracket of (5–6.81). Thus, the additional term appears in the related bracket of (5–6.145) as

$$\left[\tfrac{1}{2}(1+\zeta)\frac{e^{-2ix}}{D_-} + \tfrac{1}{2}(1-\zeta)\frac{e^{2ix}}{D_+}\right]\frac{\tau u}{x}\frac{eH}{2m}\frac{E}{m}, \tag{5–6.282}$$

which also involves the effective substitution of Eq. (5–6.102). One should not overlook the necessity for a related supplement to the contact term $-(1+u)$. Recall that the latter is designed to produce a null result for zero field strength, where $x = eHus$, but not $x/H$, vanishes. Hence this additional contact term is

$$-\frac{\tau u}{x}\frac{eH}{2m}\frac{E}{m}. \tag{5–6.283}$$

Leaving these extra terms aside for the moment, we see that the explicitly $\tau$-dependent factors of (5–6.275) and (5–6.279) combine into

$$\frac{1}{2\pi}\int_{-\infty}^{\infty} d\tau \exp[i(\omega - uE)\tau]\exp\left[-\frac{i}{6}\frac{(\tau uE)^2}{y\xi}\left(\frac{m}{E}\right)^2\right], \tag{5–6.284}$$

where we have introduced the variable $y$ of (5–6.223), and $\xi$ [Eq. (5–6.226)]. The omission of the Gaussian function of $\tau$ would instantly reduce this integral to $\delta(\omega - uE)$, which is the classical identification of (5–6.248). Since the important ranges of the several variables are $y \sim 1$, $\xi \sim 1$, $\tau uE \sim 1$, the Gaussian function is indeed close to unity under ultra-relativistic circumstances. If that were the whole story, we should arrive at (5–6.267), with the additional factor

$$\int d\omega\, \delta(\omega - uE) \tag{5–6.285}$$

serving to effectively replace $u$ with $\omega/E$, and the terms discussed in (5–6.270) would be retained in the spectral distribution.

But there are the additional contributions of (5–6.282) and (5–6.283). Keeping only the leading term in the $x$-expansion replaces the bracket of (5–6.282) by unity, and produces the following added term in $\gamma$:

$$-\frac{\alpha}{2\pi}\frac{eH}{m^2}m\int_0^\infty d\omega\int_0^1 du\, u\frac{d}{du}\delta(\omega-uE)$$

$$\times\int_0^\infty\frac{dy}{y^2}\left[\cos\tfrac{3}{2}\xi\left(y+\tfrac{1}{3}y^3\right)-\cos\tfrac{3}{2}\xi y\right]. \qquad (5\text{–}6.286)$$

A partial integration in $u$, followed by one in $y$, replaces this by

$$-\frac{\alpha}{\pi}m\frac{m}{E}\int_0^\infty d\omega\int_0^1 du\, u\delta(\omega-uE)$$

$$\times\int_0^\infty\frac{dy}{y}\left[\left(1+\tfrac{2}{3}y^2\right)\sin\tfrac{3}{2}\xi\left(y+\tfrac{1}{3}y^3\right)-\sin\tfrac{3}{2}\xi y\right]$$

$$=-\frac{\alpha}{\pi}m\frac{m}{E}3^{-1/2}\int_0^\infty d\omega\int_0^1 du\, u\delta(\omega-uE)$$

$$\times\left[\tfrac{2}{3}K_{2/3}(\xi)-\int_\xi^\infty d\eta\, K_{1/3}(\eta)\right], \qquad (5\text{–}6.287)$$

which precisely cancels those spin-independent quantum correction terms of (5–6.267) that had been discarded in (5–6.270). We conclude that the identification of the spectral distribution leading to the power calculation of Eq. (5–6.274) is correct.

This already overlong section will be closed here. While additional topics remain to be explored, in the areas of very strong fields, and high-energy radiation processes, they are sufficiently tied in with other considerations, of astrophysics, and accelerator design, that further discussion would lead us too far from the main line of development.

## 5–7 ELECTRON MAGNETIC MOMENT

It is our intention now to find the coefficient of $(\alpha/2\pi)^2$ in a power-series expansion of the electron magnetic moment,

$$\mu_e=\tfrac{1}{2}g_e=1+c_1\frac{\alpha}{2\pi}+c_2\left(\frac{\alpha}{2\pi}\right)^2+\cdots, \qquad (5\text{–}7.1)$$

where it is known that $c_1 = 1$. One contribution to $c_2$ has been available for some time, and we begin by evaluating it. It is the effect of modifying the photon propagation function, associated with electron pairs, which was discussed in Section 4–3 in connection with the difference between the electron and muon moments. The relevant form is (4–3.107), with $m' = m$ as noted in the text. Since this is a multiplicative correction to $\alpha/2\pi$, the vacuum polarization correction to $c_2$ is

$$(c_2)_{\text{v.pol.}} = 4\int_0^1 du\, u(1-u)^2 \int_0^1 dv\, v^2\left(1 - \tfrac{1}{3}v^2\right)\frac{1}{(1+u)^2 - (1-u)^2 v^2}. \quad (5\text{–}7.2)$$

The result of performing the $v$-integration is already given in Eqs. (4–3.112, 113), where now

$$x = 2\frac{u^{1/2}}{1-u}, \quad (5\text{–}7.3)$$

and we present it as

$$(c_2)_{\text{v.pol.}} = \frac{4}{3}\int_0^1 du\, u\left[\left(1 - \frac{2u}{(1-u)^2}\right)\frac{1+u}{1-u}\log\frac{1}{u} + \frac{4u}{(1-u)^2} - \frac{5}{3}\right]. \quad (5\text{–}7.4)$$

Partial integration and the substitution $1 - u = t$ reduce the integral to

$$10 - \tfrac{1}{12} - 6\int_0^1 dt\,\frac{1}{t}\log\frac{1}{1-t} = \tfrac{119}{12} - \pi^2, \quad (5\text{–}7.5)$$

according to (5–4.107), and

$$(c_2)_{\text{v.pol.}} = \tfrac{4}{3}\left(\tfrac{119}{12} - \pi^2\right) = 0.06275. \quad (5\text{–}7.6)$$

This is a rather small contribution, if one anticipates that $c_2$ should be of order unity.

The formula of Eq. (4–3.107) was derived in a causal manner, as a modification of the technique developed in Section 4–2. Before continuing, let us note the corresponding non-causal derivation, as a modification of the work in Section 4–16. For that, we replace the null photon mass by the variable mass $M$, with

$$M^2 = \frac{4m^2}{1-v^2}, \quad (5\text{–}7.7)$$

the distribution of which is described by the weight factor [Eq. (4–3.105)]

$$\frac{\alpha}{\pi}\, dv\,\frac{v^2\left(1 - \tfrac{1}{3}v^2\right)}{1-v^2}. \quad (5\text{–}7.8)$$

The replacement of $k^2$ by $k^2 + M^2$ alters the function $\chi(u)$ of (4-16.3) by adding

$$4m^2\frac{1-u}{1-v^2}. \tag{5-7.9}$$

This influences the right-hand side of Eq. (4-16.17), which is changed to

$$\frac{\alpha}{2\pi}\frac{m}{s}u^2(1-u)\exp\left[-ism^2\left(u^2+4\frac{1-u}{1-v^2}\right)\right]eq\sigma F. \tag{5-7.10}$$

The consequence for Eq. (5-16.18) is indicated by the substitution

$$\frac{\alpha}{2\pi}2\int_0^1 du\,(1-u) \rightarrow \frac{\alpha}{2\pi}2\int_0^1 du\,\frac{u^2(1-u)}{u^2+4\dfrac{1-u}{1-v^2}}. \tag{5-7.11}$$

When the latter is multiplied by (5-7.8) and integrated with respect to $v$, the resulting coefficient of $(\alpha/2\pi)^2$ is

$$(c_2)_{\text{v.pol.}} = 4\int_0^1 du\,u^2(1-u)\int_0^1 dv\,v^2\left(1-\tfrac{1}{3}v^2\right)\frac{1}{(2-u)^2-u^2v^2}; \tag{5-7.12}$$

the substitution $u \rightarrow 1-u$ confirms the equivalence with (5-7.2).

Paralleling the use of the modified photon propagation function in Eq. (4-16.1) is the introduction of the modified electron propagation function. The structure of $\overline{G}_+$ appropriate to a weak, homogeneous electromagnetic field is contained in Eqs. (4-2.31, 40). Employing the form given in Eq. (4-2.44) for the explicit field dependence of $M(F)$, and the expansion

$$\overline{G}_+ = \frac{1}{\gamma\Pi+m} - \frac{1}{\gamma\Pi+m}M(F)\frac{1}{\gamma\Pi+m}, \tag{5-7.13}$$

which is sufficiently accurate for our purposes, we get

$$\begin{aligned}\overline{G}_+ = {} & \frac{1}{\gamma\Pi+m} + \frac{1}{\gamma\Pi+m}\frac{\alpha}{2\pi}\frac{eq}{2m}\sigma F\frac{1}{\gamma\Pi+m} \\ & + \frac{\alpha}{2\pi}\int_m^\infty \frac{dM}{M}\frac{m^2}{M^4}\frac{1}{\gamma\Pi+m}eq\sigma F\left(\frac{M-m}{\gamma\Pi+M} - \frac{M+m}{\gamma\Pi-M}\right) \\ & + \frac{\alpha}{4\pi}\int_{\rightarrow m}^\infty \frac{dM}{M}\left(1-\frac{m^2}{M^2}\right)\left[\frac{1-\dfrac{2mM}{(M-m)^2}}{\gamma\Pi+M} + \frac{1+\dfrac{2mM}{(M+m)^2}}{\gamma\Pi-M}\right],\end{aligned} \tag{5-7.14}$$

where symmetrization of the factors multiplying $\sigma F$ is understood in the third term of $\overline{G}_+$. The lower limit of integration in the last term is a reminder of an infra-red singularity, which is non-physical. Indeed, this term will be cancelled completely by another contribution to be introduced later, and we set it aside to examine the implications of the explicitly field-dependent parts of (5–7.14).

The non-causal method of Section 4–16 will be used. The vacuum amplitude of (4–16.1) is modified to

$$e^2\int\frac{(dk)}{(2\pi)^4}\psi_1\gamma^0\gamma^\mu\frac{1}{k^2}\overline{G}_+(\Pi-k)\gamma_\mu\psi_2. \qquad (5\text{–}7.15)$$

The terms of interest in $\overline{G}_+$ then produce matrix combinations of the type

$$\gamma^\mu[m-\gamma(\Pi-k)]\sigma F[m'-\gamma(\Pi-k)]\gamma_\mu, \qquad (5\text{–}7.16)$$

where $m'$ may be $m$, $M$, or $-M$, and a symmetrization between $m$ and $m'$ is applied. On using such relations as

$$\gamma^\mu[m-\gamma(\Pi-k)]=[m+\gamma(\Pi-k)]\gamma^\mu+2(\Pi-k)^\mu, \qquad (5\text{–}7.17)$$

and the familiar property

$$\gamma^\mu\sigma F\gamma_\mu=0, \qquad (5\text{–}7.18)$$

we find that (5–7.16), with its implicit symmetrization, reduces to

$$(m+m')[\sigma F\gamma(\Pi-k)+\gamma(\Pi-k)\sigma F]. \qquad (5\text{–}7.19)$$

We then encounter momentum integrals of the form ($\Pi\to p$)

$$\int\frac{(dk)}{(2\pi)^4}\frac{1}{k^2}\frac{1}{(p-k)^2+m^2}\frac{1}{(p-k)^2+m'^2}(p-k)$$

$$=-i\int_0^\infty ds\,s^2\int_0^1 du\,u\int_{-1}^1\frac{1}{2}\,dv\int\frac{(dk)}{(2\pi)^4}(p-k)$$

$$\times\exp\left\{-is\left[u(p-k)^2+(1-u)k^2\right.\right.$$

$$\left.\left.+u\left(\frac{1+v}{2}m^2+\frac{1-v}{2}m'^2\right)+(1-u)\mu^2\right]\right\}, \qquad (5\text{–}7.20)$$

in which a photon mass $\mu$ has also been introduced. The redefinition $k - up \to k$, and the real-particle property $-p^2 = m^2$, convert this into

$$-i\int ds\, s^2\, du\, u\, \tfrac{1}{2}\, dv \int \frac{(dk)}{(2\pi)^4} p(1-u)$$

$$\times \exp\left[-is\left(k^2 + u^2m^2 + u\frac{1-v}{2}(m'^2 - m^2) + (1-u)\mu^2\right)\right]$$

$$= \frac{i}{(4\pi)^2} p \int_0^1 du\, u(1-u) \int_0^1 dw \frac{1}{u^2m^2 + uw(m'^2 - m^2) + (1-u)\mu^2}, \quad (5\text{-}7.21)$$

where

$$w = \tfrac{1}{2}(1 - v). \quad (5\text{-}7.22)$$

Observe that an infra-red singularity does appear for $m' = m$.

With the aid of these results we find that the second term of (5–7.14), describing the additional magnetic moment of the electron, produces the following contribution to the vacuum amplitude (5–7.15):

$$e^2\frac{\alpha}{2\pi}\int \frac{(dk)}{(2\pi)^4} \psi_1\gamma^0 eq \frac{\sigma F\gamma(\Pi - k) + \gamma(\Pi - k)\sigma F}{k^2\left[(\Pi - k)^2 + m^2\right]^2}\psi_2$$

$$= i\left(\frac{\alpha}{2\pi}\right)^2 \psi_1\gamma^0\frac{eq}{2m}\sigma F\psi_2(-2)\int_0^1 du\, u(1-u)\frac{1}{u^2 + (\mu/m)^2}. \quad (5\text{-}7.23)$$

This exhibits another piece of $c_2$,

$$(c_2)'_{\bar{G}} = -2\int_0^1 du \frac{u(1-u)}{u^2 + (\mu/m)^2} = -2\left(\log\frac{m}{\mu} - 1\right). \quad (5\text{-}7.24)$$

Concerning the third term of (5–7.14), we remark that the two parts associated with $M$ and $-M$ give equal contributions, leading to the vacuum amplitude

$$e^2\frac{\alpha}{2\pi}\int\frac{(dk)}{(2\pi)^4}\int\frac{dM^2}{M^2}\frac{m^2}{M^4}(M^2 - m^2)$$

$$\times\psi_1\gamma^0 eq\frac{\sigma F\gamma(\Pi - k) + \gamma(\Pi - k)\sigma F}{k^2\left[(\Pi - k)^2 + m^2\right]\left[(\Pi - k)^2 + M^2\right]}\psi_2$$

$$= i\left(\frac{\alpha}{2\pi}\right)^2\psi_1\gamma^0\frac{eq}{2m}\sigma F\psi_2(c_2)''_{\bar{G}}, \quad (5\text{-}7.25)$$

with

$$(c_2)''_{\bar{G}} = -2\int_{m^2}^{\infty}\frac{dM^2}{M^2}\left(\frac{m^2}{M^2}\right)^2(M^2-m^2)\int_0^1 du\int_0^1 dw\,\frac{1-u}{um^2+w(M^2-m^2)}$$

$$= -2\int_{m^2}^{\infty}\frac{dM^2}{M^2}\left(\frac{m^2}{M^2}\right)^2\int_0^1 du\,(1-u)\log\frac{um^2+M^2-m^2}{um^2}$$

$$= -\int_1^{\infty}dx\int_0^1 du\,(1-u)\frac{1}{x^2}\frac{1}{x-1+u}. \tag{5-7.26}$$

The last version was produced by a partial integration on the variable

$$x = M^2/m^2. \tag{5-7.27}$$

Now

$$\int_1^{\infty}dx\,\frac{1}{x^2}\frac{1}{x-1+u} = \frac{1}{(1-u)^2}\log\frac{1}{u} - \frac{1}{1-u}, \tag{5-7.28}$$

and, using the substitution $1 - u = t$, we get

$$(c_2)''_{\bar{G}} = -\int_0^1 dt\left(\frac{1}{t}\log\frac{1}{1-t} - 1\right) = -\left(\frac{\pi^2}{6} - 1\right), \tag{5-7.29}$$

again invoking (5–4.107). The complete contribution associated with the explicitly field-dependent terms in the modified particle propagation function is therefore

$$(c_2)_{\bar{G}} = -\left(2\log\frac{m}{\mu} + \frac{\pi^2}{6} - 3\right). \tag{5-7.30}$$

We have begun the discussion of the magnetic moment problem by computing some obvious contributions. Now we must examine the whole picture. The initial causal arrangement referred to the exchange of an electron and a photon in a homogeneous magnetic field. At the next level of description, the characterization of the two-particle process is modified, and three-particle exchange takes place. The corrections to the two-particle mechanism involve the introduction of modified propagation functions for the photon and electron (effects that have already been considered) and, associated with the interaction of the electron and photon, of form factors for the two-particle emission and absorption acts. The three-particle processes are brought in by considering the emission and absorption, not of two real particles, but of one real and one virtual particle. A virtual electron decays into a real electron and a real photon. The subsequent recombination of these particles is the mechanism for producing $\overline{G}_+$, which has already been discussed. But there is a second possibility in which it is the initially emitted photon that later combines with the electron to produce the virtual particle that is detected, along with the other photon. [Aside to the reader: Draw the causal diagram! It can be presented as a rectangle with heavy, virtual-electron lines constituting the narrow top and bottom, while wavy, thin, real-photon lines form the sides. A real-electron line traces one diagonal, and the initial and final virtual-electron lines are attached at the other two vertices.] Similarly, a virtual photon decays into two real electrons, of opposite charge. The recombination of these particles generates $\overline{D}_+$, which effect has already been computed. But the exchange of the roles of the two like charges at the absorption end produces a new process. [The causal diagram here is also a rectangle, with virtual-photon lines forming the top and bottom, and real-electron lines the sides. A real-electron line occupies one diagonal, and the initial and final virtual-electron lines are placed at the other vertices.]

Before continuing, let us review the machinery that introduces form factors for the two-particle process. It is a consequence of an interaction that contributes to Compton scattering. The latter is produced in either of two ways that are related by photon crossing symmetry. The first one involves the recombination of the initial electron and photon to form a virtual electron that decays into the final electron and photon. We do not consider this mechanism explicitly, since it is an iteration of the two-particle exchange that produces the modified particle propagation function. The action principle handles it automatically. (Compare the discussion with Harold at the end of Section 5-4.) In the second possibility, the roles of the initial and final photon are interchanged. The initial electron emits the final photon to become a virtual particle which, on absorbing the initial photon, produces the final electron. [A causal diagram can be drawn in lozenge (diamond) shape, with real-photon lines forming one set of parallel lines and real-particle lines the other set. A horizontal virtual-electron line connects two vertices, and the initial and final virtual-electron lines are tied to the other vertices.]

The artistic reader is now in a position to recognize that the topologies of these three causal diagrams are the same. That is, with the distinction between real and virtual particles ignored, and on performing deformations that maintain the connectivity of the lines, the three diagrams can be made identical. (The resulting non-causal diagram is what is known as a Feynman diagram. A simple version of it is produced by drawing out the electron pictorial representations into a single straight line, with the two photon graphic symbols traced as intersecting arcs.)

Thus, we have the option of evaluating three distinct causal processes, or of performing one non-causal calculation. It is the latter strategy that will be adopted here.

What has just been described can be demonstrated analytically, of course. However, we shall not trouble to consider all three causal arrangements (while urging the reader to do so), but just select one to produce the common space-time form of the coupling. For the two-particle process, the vacuum amplitude representing the partial Compton scattering is

$$i\int(dx)(dx')\psi_1(x)\gamma^0 eq\gamma A_2(x)G_+(x,x')eq\gamma A_1(x')\psi_2(x'), \tag{5-7.31}$$

where the fields are those of the real particles that enter and leave the collision, and the presence of the homogeneous magnetic field influences the form of the electron propagation function. The sources of these particles are

$$i\eta_2(x)J_2^\mu(x')\big|_{\text{eff.}} = \delta(x-x')eq\gamma^\mu\psi_2(x) \tag{5-7.32}$$

and

$$iJ_1^\mu(x')\eta_1(x)\gamma^0\big|_{\text{eff.}} = \psi_1(x)\gamma^0 eq\gamma^\mu\delta(x-x'), \tag{5-7.33}$$

in which $\psi_{1,2}(x)$ here refer to the extended particle sources. Putting these elements together gives the desired vacuum amplitude, apart from contact terms,

$$-i\int(dx)\cdots(dx''')\psi_1(x)\gamma^0 eq\gamma^\mu G_+(x,x')eq\gamma^\nu G_+(x',x'')$$
$$\times eq\gamma_\mu G_+(x'',x''')eq\gamma_\nu\psi(x''')D_+(x-x'')D_+(x'-x'''). \tag{5-7.34}$$

It is written more compactly as

$$-i\int(dx)(dx')\psi_1(x)\gamma^0 M^{(2)}(x,x')\psi_2(x'), \tag{5-7.35}$$

where

$$M^{(2)} = e^4\int\frac{(dk)}{(2\pi)^4}\frac{(dk')}{(2\pi)^4}\frac{1}{k^2}\frac{1}{k'^2}\gamma^\mu\frac{1}{\gamma(\Pi-k)+m}\gamma^\nu$$
$$\times\frac{1}{\gamma(\Pi-k-k')+m}\gamma_\mu\frac{1}{\gamma(\Pi-k')+m}\gamma_\nu + \text{c.t.} \tag{5-7.36}$$

is the mass operator contribution characterizing this two-photon exchange process.

There is a close relationship with the structure of $M$ [Eq. (5–6.3)], which describes single photon exchange,

$$M = ie^2 \int \frac{(dk)}{(2\pi)^4} \gamma^\nu \frac{1}{k^2} \frac{1}{\gamma(\Pi - k) + m} \gamma_\nu + \text{c.t.}, \tag{5–7.37}$$

where the contact terms appearing here,

$$\text{c.t.} = -m_c - \zeta_c(\gamma\Pi + m), \tag{5–7.38}$$

are specified in Eqs. (5–6.56, 57). This example illustrates the completely local nature of contact terms, in contrast with the non-locality that is characteristic of a multi-particle exchange process. Now consider the effect of an infinitesimal alteration of the vector field $A$ in (5–7.37),

$$\delta M = ie^2 \int \frac{(dk)}{(2\pi)^4} \gamma^\nu \frac{1}{k^2} \frac{1}{\gamma(\Pi - k) + m} eq\gamma\,\delta A \frac{1}{\gamma(\Pi - k) + m} \gamma_\nu + \zeta_c eq\gamma\,\delta A. \tag{5–7.39}$$

On writing

$$\delta A(x) = \int \frac{(dk')}{(2\pi)^4} e^{ik'x}\,\delta A(k') \tag{5–7.40}$$

and

$$\delta M = \int \frac{(dk')}{(2\pi)^4} \frac{\delta_r M}{\delta A(k')} e^{ik'x}\,\delta A(k')$$

$$= \int \frac{(dk')}{(2\pi)^4} e^{ik'x}\,\delta A(k') \frac{\delta_l M}{\delta A(k')}, \tag{5–7.41}$$

we can present (5–7.39) as

$$\frac{\delta_r M}{\delta A_\nu(k')} = ie^2 \int \frac{(dk)}{(2\pi)^4} \gamma^\mu \frac{1}{k^2} \frac{1}{\gamma(\Pi - k) + m} eq\gamma^\nu \frac{1}{\gamma(\Pi - k - k') + m} \gamma_\mu + \zeta_c eq\gamma^\nu, \tag{5–7.42}$$

or, interchanging $k$ and $k'$, as

$$\frac{\delta_l M}{\delta A_\mu(-k)} = ie^2 \int \frac{(dk')}{(2\pi)^4} \gamma^\nu \frac{1}{k'^2} \frac{1}{\gamma(\Pi - k - k') + m} eq\gamma^\mu \frac{1}{\gamma(\Pi - k') + m} \gamma_\nu + \zeta_c eq\gamma^\mu. \tag{5–7.43}$$

Both of these integral structures can be recognized in (5–7.36). But, before introducing them as component parts of $M^{(2)}$, we must understand better the role of the contact terms.

The $\gamma^\mu \cdots \gamma_\mu$ structure in $M^{(2)}$ represents the exchange of a photon with momentum $k$, in which the system is probed by a photon of momentum $k'$. This is described by the combination (5–7.42), which necessitates the following contact term in $M^{(2)}$:

$$-ie^2\int\frac{(dk')}{(2\pi)^4}\frac{1}{k'^2}\zeta_c\gamma^\nu\frac{1}{\gamma(\Pi-k')+m}\gamma_\nu = -\zeta_c M. \tag{5–7.44}$$

(Note that contact terms added to $M^{(2)}$ as a whole are without effect in this non-causal situation, since $\psi_1$ and $\psi_2$ do not overlap.) And the $\gamma^\nu \cdots \gamma_\nu$ structure in $M^{(2)}$ represents the exchange of a photon with momentum $k'$, in which the system is probed by a photon of momentum $k$. That is described by the combination (5–7.43), which requires the following contact term in $M^{(2)}$:

$$-ie^2\int\frac{(dk)}{(2\pi)^4}\frac{1}{k^2}\gamma^\mu\frac{1}{\gamma(\Pi-k)+m}\zeta_c\gamma_\mu = -\zeta_c M. \tag{5–7.45}$$

The essential point is that, since the two photon exchanges are independent, both of these contact terms are needed, and the complete contact term structure in (5–7.36) is

$$\text{c.t.} = -2\zeta_c M. \tag{5–7.46}$$

In view of the importance of this conclusion, we add another consideration in which the effects of the two photon exchange processes are more clearly separated. Let us examine how $M^{(2)}$ responds to an arbitrary infinitesimal variation of the electromagnetic field. We write this response, without contact terms, as

$$\begin{aligned}\delta M^{(2)} = e^4\int\frac{(dk)}{(2\pi)^4}\frac{(dk')}{(2\pi)^4}\frac{1}{k^2}\frac{1}{k'^2}\\ \times\{\gamma^\mu\,\delta G_+(\Pi-k)\,\gamma^\nu G_+(\Pi-k-k')\gamma_\mu G_+(\Pi-k')\gamma_\nu\\ +\gamma^\mu G_+(\Pi-k)\gamma^\nu\,\delta G_+(\Pi-k-k')\,\gamma_\mu G_+(\Pi-k')\gamma_\nu\\ +\gamma^\mu G_+(\Pi-k)\gamma^\nu G_+(\Pi-k-k')\gamma_\mu\,\delta G_+(\Pi-k')\gamma_\nu\}.\end{aligned} \tag{5–7.47}$$

Considering the first of these three terms, we see that the $\gamma^\mu \cdots \gamma_\mu$ structure now involves the differential action of two fields. An inspection of (5–7.38) shows that no contact term appears for such circumstances. But the $\gamma^\nu \cdots \gamma_\nu$ combination in the same first term has the form of (5–7.43), which demands the contact term exhibited there. The addition thus implied to (5–7.47) is

$$-ie^2\int\frac{(dk)}{(2\pi)^4}\frac{1}{k^2}\gamma^\mu\,\delta G_+(\Pi-k)\,\zeta_c eq\gamma_\mu=-\zeta_c\,\delta M. \tag{5–7.48}$$

For the second of the three terms, both of the single-photon exchange structures involve two differentiations with respect to fields, with the consequent absence of contact terms. The discussion of the last term in (5–7.47) is analogous to that of the first one, leading to another contact term exactly equal to (5–7.48). In this way, we recognize again, and more explicitly, the existence of the c.t. (5–7.46).

Returning to the structure of $M^{(2)}$ [Eqs. (5–7.36, 46)] we now observe that

$$\begin{aligned}M^{(2)}&=-i\int\frac{(dk')}{(2\pi)^4}\frac{\delta_r M}{\delta A_\nu(k')}\frac{1}{k'^2}\frac{1}{\gamma(\Pi-k')+m}eq\gamma_\nu-\zeta_c M\\&=-i\int\frac{(dk)}{(2\pi)^4}eq\gamma^\mu\frac{1}{k^2}\frac{1}{\gamma(\Pi-k)+m}\frac{\delta_l M}{\delta A^\mu(-k)}-\zeta_c M,\end{aligned} \tag{5–7.49}$$

since the single $M$ structure only incorporates one of the two equal contact terms. Concerning the residual one, we note the form of $\zeta_c$ as the sum of two contributions, given in (5–6.57):

$$\zeta_c=\zeta'+\zeta'', \tag{5–7.50}$$

with

$$\zeta'=\frac{\alpha}{2\pi}\int\frac{ds}{s}\,du\,(1-u)\,e^{-ism^2u^2} \tag{5–7.51}$$

and (introducing a photon mass)

$$\begin{aligned}\zeta''&=-i\frac{\alpha}{\pi}m^2\int ds\,du\,u(1-u^2)\,e^{-is(m^2u^2+\mu^2)}\\&=-\frac{\alpha}{\pi}\int_0^1du\,u\frac{1-u^2}{u^2+(\mu/m)^2}=-\frac{\alpha}{2\pi}\left(2\log\frac{m}{\mu}-1\right).\end{aligned} \tag{5–7.52}$$

Hence the contribution to $c_2$ that is associated with the $\zeta''$-component of the contact term,

$$(c_2)_{\zeta''}=2\log\frac{m}{\mu}-1, \tag{5–7.53}$$

removes the fictitious photon-mass dependence exhibited in (5–7.30),

$$(c_2)_{\bar G}+(c_2)_{\zeta''}=2-\frac{\pi^2}{6}. \tag{5–7.54}$$

Another compensation, which was anticipated in the discussion following Eq. (5–7.14), can now be made explicit. The last term of the expression for $\overline{G}_+$ given in that equation is

$$\int_{\to m}^{\infty} dM\left[\frac{A_+(M)}{\gamma\Pi+M}+\frac{A_-(M)}{\gamma\Pi-M}\right], \tag{5–7.55}$$

with

$$A_\pm(M)=\frac{\alpha}{4\pi}\frac{1}{M}\left(1-\frac{m^2}{M^2}\right)\left(1\mp\frac{2mM}{(M\mp m)^2}\right). \tag{5–7.56}$$

It supplies the following contribution to the vacuum amplitude of (5–7.15):

$$e^2\int\frac{(dk)}{(2\pi)^4}\psi_1\gamma^0\gamma^\mu\frac{1}{k^2}\int dM\left[\frac{A_+(M)}{\gamma(\Pi-k)+M}+\frac{A_-(M)}{\gamma(\Pi-k)-M}\right]\gamma_\mu\psi_2. \tag{5–7.57}$$

Now the mass operator $M$, referring to an arbitrary electromagnetic field, can be decomposed in this way:

$$M=M_0(\gamma\Pi)+M_1, \tag{5–7.58}$$

where $M_1$ depends explicitly on electromagnetic field strengths, and $M_0$ is the gauge-covariant form that applies in the absence of electromagnetic fields. As exhibited in the first term of (4–2.31), the latter is

$$\begin{aligned}M_0(\gamma\Pi)&=-(\gamma\Pi+m)^2\int dM\left[\frac{A_+(M)}{\gamma\Pi+M}+\frac{A_-(M)}{\gamma\Pi-M}\right]\\&=-\int dM\left\{A_+(M)\left[\frac{(M-m)^2}{\gamma\Pi+M}-(M-m)+\gamma\Pi+m\right]\right.\\&\qquad\left.+A_-(M)\left[\frac{(M+m)^2}{\gamma\Pi-M}+(M+m)+\gamma\Pi+m\right]\right\}.\end{aligned} \tag{5–7.59}$$

Next, let us compute, say,

$$\begin{aligned}\frac{\delta_l M_0}{\delta A^\mu(-k)}&=-\int dM\left\{A_+(M)\left[(M-m)^2\frac{1}{\gamma(\Pi-k)+M}eq\gamma_\mu\frac{1}{\gamma\Pi+M}-eq\gamma_\mu\right]\right.\\&\qquad\left.+A_-(M)\left[(M+m)^2\frac{1}{\gamma(\Pi-k)-M}eq\gamma_\mu\frac{1}{\gamma\Pi-M}-eq\gamma_\mu\right]\right\},\end{aligned} \tag{5–7.60}$$

and then proceed to simplify the part of the vacuum amplitude inferred from Eqs. (5–7.35, 49) by performing the reduction $\gamma\Pi + m \to 0$ on the right side of (5–7.60), as effectively expressed by

$$\frac{\delta_l M_0}{\delta A^\mu(-k)} \to -\int dM\left\{A_+(M)\left[(M-m)\frac{1}{\gamma(\Pi-k)+M}-1\right]\right.$$
$$\left.+A_-(M)\left[-(M+m)\frac{1}{\gamma(\Pi-k)-M}-1\right]\right\}eq\gamma_\mu$$
$$=[\gamma(\Pi-k)+m]\int dM\left[\frac{A_+(M)}{\gamma(\Pi-k)+M}+\frac{A_-(M)}{\gamma(\Pi-k)-M}\right]eq\gamma_\mu. \tag{5–7.61}$$

This piece of $M^{(2)}$,

$$-ie^2\int\frac{(dk)}{(2\pi)^4}\gamma^\mu\frac{1}{k^2}\int dM\left[\frac{A_+(M)}{\gamma(\Pi-k)+M}+\frac{A_-(M)}{\gamma(\Pi-k)-M}\right]\gamma_\mu, \tag{5–7.62}$$

yields a vacuum-amplitude contribution that precisely cancels (5–7.57).

Our attention is now concentrated on the remaining part of $M^{(2)}$,

$$M_1^{(2)} = -i\int\frac{(dk')}{(2\pi)^4}\frac{\delta_r M_1}{\delta A_\nu(k')}\frac{1}{k'^2}\frac{1}{\gamma(\Pi-k')+m}eq\gamma_\nu - \zeta' M_1. \tag{5–7.63}$$

Accordingly, we need an expression for

$$\frac{\delta_r M_1}{\delta A_\nu(k')} = \frac{\delta_r M}{\delta A_\nu(k')} - \frac{\delta_r M_0}{\delta A_\nu(k')} \equiv -eq\gamma_1^{\nu}(k') \tag{5–7.64}$$

that is accurate to terms linear in the homogeneous field. We shall use an equivalent form of the construction for $M$ given in (5–6.5),

$$M = -ie^2\int ds\, s\, du\, \gamma^\mu\langle[m-\gamma(\Pi-k)]e^{-is\chi}\rangle\gamma_\mu, \tag{5–7.65}$$

where the contact terms need not be added, since they will cancel between $M$ and $M_0$.

We first construct $M_0$ in this way, by considering the field-free situation. With the now permissible transformation $k\text{-}u\Pi \to k$, and the evaluation [Eq. (5–6.38)]

$$\langle e^{-isk^2}\rangle = \frac{1}{(4\pi)^2}\frac{1}{is^2}, \tag{5–7.66}$$

we immediately get (without contact terms)

$$M_0(\gamma\Pi) = \frac{\alpha}{2\pi}\int\frac{ds}{s}\,du\,[2m + (1-u)\gamma\Pi]\exp[-is(m^2u^2 + \mathscr{H})], \quad (5\text{–}7.67)$$

where

$$\mathscr{H} = u(1-u)[-(\gamma\Pi)^2 + m^2]. \quad (5\text{–}7.68)$$

This reproduces (5–6.54), of course. What is needed, however, is the differential form produced by varying $A$ in $M_0$, now applied to an arbitrary electromagnetic field:

$$\begin{aligned}\delta M_0 = &-\frac{\alpha}{2\pi}\int\frac{ds}{s}\,du\,(1-u)eq\gamma\,\delta A\exp[-is(m^2u^2 + \mathscr{H})]\\ &-i\frac{\alpha}{2\pi}\int ds\,du\,[2m + (1-u)\gamma\Pi]e^{-ism^2u^2}\\ &\times\int_{-1}^{1}\tfrac{1}{2}\,dv\exp\left[-is\frac{1+v}{2}\mathscr{H}\right]\delta\mathscr{H}\exp\left[-is\frac{1-v}{2}\mathscr{H}\right], \quad (5\text{–}7.69)\end{aligned}$$

in which

$$\begin{aligned}\delta\mathscr{H} &= \delta[u(1-u)(\Pi^2 - eq\sigma F)]\\ &= -u(1-u)[\Pi eq\,\delta A + eq\,\delta A\,\Pi + eq\sigma\,\delta F]. \quad (5\text{–}7.70)\end{aligned}$$

Expressing this by a functional derivative, as required for (5–7.64), will induce the substitution

$$\Pi \to \Pi - k' = \Pi' \quad (5\text{–}7.71)$$

for any term standing on the right side of $\delta A$. That gives

$$\begin{aligned}\frac{1}{eq}\frac{\delta_r M_0}{\delta A_\nu(k')} = &-\frac{\alpha}{2\pi}\int\frac{ds}{s}\,du\,(1-u)\gamma^\nu\exp[-is(m^2u^2 + \mathscr{H}')]\\ &+i\frac{\alpha}{2\pi}m\int ds\,du\,\frac{dv}{2}u(1-u^2)(\Pi^\nu + \Pi'^\nu + ik'_\lambda\sigma^{\lambda\nu})\\ &\times\exp\left[-is\left(m^2u^2 + \frac{1-v}{2}\mathscr{H}'\right)\right], \quad (5\text{–}7.72)\end{aligned}$$

where the symbol $\mathscr{H}'$ indicates that the substitution (5–7.71) has been made in

$\mathscr{H}$. In writing this form we have also introduced a simplification associated with its eventual use: where $\gamma\Pi + m$ and $\mathscr{H}$ stand entirely on the left, they have been replaced by zero.

Let us begin the discussion of $M$ by first performing the functional differentiation with respect to $A$. From the differential form

$$\delta M = -ie^2\int ds\, s\, du\, \gamma^\mu\langle eq\gamma\,\delta A\, e^{-is\chi}\rangle\gamma_\mu$$
$$-e^2\int ds\, s^2\, du\,\frac{dv}{2}\gamma^\mu$$
$$\times\left\langle[m-\gamma(\Pi-k)]\exp\left[-is\frac{1+v}{2}\chi\right]\delta\chi\exp\left[-is\frac{1-v}{2}\chi\right]\right\rangle\gamma_\mu \quad (5\text{–}7.73)$$

one infers the functional derivative as

$$-\frac{1}{eq}\frac{\delta_r M}{\delta A_\nu(k')} = ie^2\int ds\, s\, du\,\gamma^\mu\langle\gamma^\nu e^{-is\chi'}\rangle\gamma_\mu$$
$$-e^2\int ds\, s^2\, du\,\frac{dv}{2}\gamma^\mu\left\langle[m-\gamma(\Pi-k)]\exp\left[-is\frac{1+v}{2}\chi\right]\right.$$
$$\left.\times\left(-\frac{1}{eq}\frac{\delta_r\chi}{\delta A_\nu(k')}\right)\exp\left[-is\frac{1-v}{2}\chi'\right]\right\rangle\gamma_\mu. \quad (5\text{–}7.74)$$

According to the structure of $\chi$ [Eq. (5–6.4)], the functional derivative that appears on the right side is

$$-\frac{1}{eq}\frac{\delta_r\chi}{\delta A_\nu(k')} = u\big[(\Pi-k)^\nu + (\Pi'-k)^\nu + ik'_\lambda\sigma^{\lambda\nu}\big]. \quad (5\text{–}7.75)$$

Since our need is only for an evaluation to the first power in the homogeneous field, we shall adopt more elementary methods than those elaborated in the preceding section, for example. Thus, in order to combine the two exponential factors in the second term of (5–7.74), we move intervening factors away by using the approximate expressions

$$\left[\exp\left\{-is\frac{1+v}{2}\chi\right\},(\Pi-k)^\nu\right] \cong is\frac{1+v}{2}[\Pi^\nu,\chi]\exp\left\{-is\frac{1+v}{2}\chi\right\}$$
$$= -su(1+v)eqF^{\nu\lambda}(\Pi-k)_\lambda\exp\left\{-is\frac{1+v}{2}\chi\right\}$$
$$(5\text{–}7.76)$$

and

$$\left[(\Pi' - k)^\nu, \exp\left\{-is\frac{1-v}{2}\chi'\right\}\right] \cong \exp\left\{-is\frac{1-v}{2}\chi'\right\} su(1-v)eqF^{\nu\lambda}(\Pi' - k)_\lambda. \tag{5–7.77}$$

The unification of the exponential factors in turn employs an approximation, which is based on the following theorem applicable to operators $A$ and $B$ such that the commutator

$$[A, B] = iC \tag{5–7.78}$$

is commutative with $A$ and $B$:

$$e^{i(A+B)} = e^{iA}e^{iB}e^{i\frac{1}{2}C} = e^{iB}e^{iA}e^{-i\frac{1}{2}C}. \tag{5–7.79}$$

A proof is immediately supplied by comparing the evaluation

$$e^{i(B^2/2C)}e^{iA}e^{-i(B^2/2C)} = e^{i(A+B)} \tag{5–7.80}$$

with the alternative ones in which $e^{iA}$ is used to effect such transformations. [The same procedure also supplies a short derivation of Eq. (2–1.21), which is a generalization of (5–7.79).] The relevance of this theorem in our situation stems from the remark that the commutator $[\chi, \chi']$, being explicitly linear in the homogeneous field, is effectively inoperative in forming additional commutators, since they would contain higher powers of the field. The same restriction to the first power of the homogeneous field also permits a simplified application of (5–7.79),

$$\exp\left[-is\frac{1+v}{2}\chi\right]\exp\left[-is\frac{1-v}{2}\chi'\right] \cong \exp\left\{-is\left[\frac{1+v}{2}\chi + \frac{1-v}{2}\chi'\right]\right\}\left\{1 - \tfrac{1}{2}s^2\frac{1-v^2}{4}[\chi, \chi']\right\}. \tag{5–7.81}$$

The commutator is evaluated as

$$[\chi, \chi'] = -4u^2(\Pi - k)ieqFk'. \tag{5–7.82}$$

Perhaps it is not too soon to introduce another simplification of which we shall make repeated use. The combination

$$(\Pi F)_\nu = \Pi^\mu F_{\mu\nu}, \tag{5–7.83}$$

appearing in (5–7.82), for example, will persist through to the final calculation, where it can only appear in the form

$$\Pi F\gamma = \tfrac{1}{2}i[\gamma\Pi + m, \sigma F], \qquad (5\text{-}7.84)$$

since the alternative

$$\Pi F\Pi = \tfrac{1}{2}ieqF^{\mu\nu}F_{\mu\nu}, \qquad (5\text{-}7.85)$$

is negligible. But the commutator appearing on the right side of (5–7.84) does not contribute in the application to particle fields that obey $(\gamma\Pi + m)\psi = 0$. Accordingly, the $\Pi F$ structure will be systematically omitted as the calculation proceeds. It is in this sense that the commutator of (5–7.82) is replaced by an equivalent statement,

$$[\chi, \chi'] \to 4u^2 kieqFk'. \qquad (5\text{-}7.86)$$

Incidentally, the combination of exponents on the right side of (5–7.81) is

$$\begin{aligned}\frac{1+v}{2}\chi + \frac{1-v}{2}\chi' &= u\left[\frac{1+v}{2}(\Pi - k)^2 + \frac{1-v}{2}(\Pi' - k)^2\right] \\ &\quad +(1-u)k^2 + u(m^2 - eq\sigma F) \\ &= \left[k - u\left(\frac{1+v}{2}\Pi + \frac{1-v}{2}\Pi'\right)\right]^2 \\ &\quad + u(1-u)\left[\frac{1+v}{2}\Pi^2 + \frac{1-v}{2}\Pi'^2\right] \\ &\quad + u^2\frac{1-v^2}{4}k'^2 + u(m^2 - eq\sigma F), \qquad (5\text{-}7.87)\end{aligned}$$

which employs the relation [cf. Eq. (5–7.71)]

$$2\Pi\Pi' = \Pi^2 + \Pi'^2 - k'^2. \qquad (5\text{-}7.88)$$

Whenever $\Pi^2$ can be replaced by $-m^2$ and the $\sigma F$ term omitted, owing to the presence elsewhere of an electromagnetic field factor, this combination will appear as

$$\frac{1+v}{2}\chi + \frac{1-v}{2}\chi' \to (\bar{k})^2 + D', \qquad (5\text{-}7.89)$$

where

$$D' = m^2u^2 + \frac{1-v}{2}\mathscr{H}' + u^2\frac{1-v^2}{4}k'^2 \qquad (5\text{-}7.90)$$

and

$$\bar{k} = k - u\left(\frac{1+v}{2}\Pi + \frac{1-v}{2}\Pi'\right) = k - u\Pi + u\frac{1-v}{2}k'. \qquad (5\text{–}7.91)$$

It is useful, in any fairly elaborate calculation, to have some independent checks on the algebra. One such check is supplied by the requirement of gauge invariance. By definition, $M_1$ depends explicitly on field strengths. There are two field types of interest here, the weak homogeneous field $F$, and the infinitesimal arbitrary field, say $f$, that is required for the functional derivative of (5–7.64). These are indicated, adequately for our purpose, by the initial terms of an expansion,

$$M_1 = M_1(\Pi, F) + M_1(\Pi, f) + M_1(\Pi, F, f), \qquad (5\text{–}7.92)$$

which contains two linear and one bilinear expression in the various fields. Now consider the gauge transformation

$$\delta A_\nu(x) = \partial_\nu\, \delta\lambda(x), \qquad \delta A_\nu(k') = ik'_\nu\, \delta\lambda(k'), \qquad (5\text{–}7.93)$$

and evaluate

$$\delta M_1 = \int \frac{(dk')}{(2\pi)^4} \frac{\delta_r M_1}{\delta A_\nu(k')} e^{ik'x}\, \delta A_\nu(k')$$

$$= -\int \frac{(dk')}{(2\pi)^4} ik'_\nu eq\gamma_1{}^\nu(k') e^{ik'x}\, \delta\lambda(k'), \qquad (5\text{–}7.94)$$

which we express as

$$-\frac{1}{ieq}\frac{\delta_r M_1}{\delta\lambda(k')} = k'_\nu \gamma_1{}^\nu(k'). \qquad (5\text{–}7.95)$$

Since the field strengths are unaltered by the gauge transformation, and $f$ is set equal to zero after the functional differentiation, only the first of the three terms indicated in (5–7.92) contributes. It is given by [Eq. (4–2.31)]

$$M_1(\Pi, F) = \frac{\alpha}{2\pi}\int_m^\infty \frac{dM}{M}\frac{m^2}{M^2} eq\sigma F \cdot \left[\frac{[1-(m/M)]^2}{\gamma\Pi + M} + \frac{[1+(m/M)]^2}{\gamma\Pi - M}\right]$$

$$= -\frac{\alpha}{2\pi}\int_0^1 du\, u(1-u)\, eq\sigma F$$

$$\cdot [2m + (2-u)\gamma\Pi]\frac{1}{u(1-u)\Pi^2 + m^2 u}, \qquad (5\text{–}7.96)$$

where the dot recalls the necessity for symmetrized multiplication, and

$$u = 1 - \frac{m^2}{M^2}. \tag{5-7.97}$$

The response of this structure to the gauge transformation is given by

$$\begin{aligned} k'_\nu \gamma_1''(k') = & -\frac{\alpha}{2\pi}\int_0^1 du\, u(1-u)(2-u)eq\sigma F \cdot \gamma k' \frac{1}{D'_1} \\ & + \frac{\alpha}{2\pi}\int_0^1 du\, u(1-u)eq\sigma F \cdot [2m + (2-u)\gamma\Pi]\frac{1}{D_1}u(1-u)(\Pi + \Pi')k' \\ & \times \frac{1}{D'_1}, \end{aligned} \tag{5-7.98}$$

in which

$$D_1 = u(1-u)\Pi^2 + m^2 u = \mathscr{H} + m^2u^2, \tag{5-7.99}$$

and similarly $D'_1$, represents a specialization of (5-7.90) to the situation $\frac{1}{2}(1 - v) = 1$. On noticing that

$$u(1-u)(\Pi + \Pi')k' = D_1 - D'_1, \tag{5-7.100}$$

and availing oneself of the simplifications associated with the use of $\gamma_1''(k')$ in (5-7.63), this reduces to

$$\begin{aligned} k'_\nu \gamma_1''(k') \to & -\frac{\alpha}{2\pi}\int_0^1 du\, u(1-u)(2-u)eq\sigma F \cdot \gamma k' \frac{1}{D'_1} \\ & + \frac{\alpha}{2\pi} m \int_0^1 du\, u^2(1-u)eq\sigma F\left(\frac{1}{D'_1} - \frac{1}{m^2u^2}\right), \end{aligned} \tag{5-7.101}$$

which will provide a control on the direct calculation of $\gamma_1''(k')$.

Harold speaks up.

H.: It seems that almost every paper touching on electrodynamics that has appeared recently makes some reference to Ward's identity. Is it related to your last remarks, and what is it?

S.: The answer to your first question is yes. It follows from this affirmative response that "Ward's identity" must also be an expression of gauge invariance. Indeed, consider $M$, $M_0$, or $M_1$, which is to say, any object that contains

$$\Pi = p - eqA \tag{5-7.102}$$

and field strengths. As such, it is invariant under the combined gauge transformation

$$A_\mu \to A_\mu + \partial_\mu \lambda,$$
$$p_\mu \to p_\mu + eq\partial_\mu \lambda = e^{-ieq\lambda} p_\mu e^{ieq\lambda}. \tag{5–7.103}$$

One can express this, using $M(A)$ as an example, by the statement

$$e^{-ieq\lambda} M(A + \partial\lambda) e^{ieq\lambda} = M, \tag{5–7.104}$$

or, in infinitesimal form, as

$$ieq[M, \delta\lambda(x)] + \int (d\xi) \frac{\delta M}{\delta A_\nu(\xi)} \partial_\nu \, \delta\lambda(\xi) = 0. \tag{5–7.105}$$

The momentum version of the latter is

$$-k'_\nu \frac{\delta_\gamma M}{\delta A_\nu(k')} = eq(M - M'), \tag{5–7.106}$$

where, again, the prime on $M$ indicates the substitution of $\Pi'$ for $\Pi$. This is a form of Ward's identity. As we have remarked, the same formula will apply to $M_1$, which statement can be written as

$$k'_\nu \gamma_1{}^\nu(k') = M_1 - M'_1. \tag{5–7.107}$$

We are thereby challenged to show the equivalence with (5–7.101), for example, which does follow from the reductions of $M_1$ and $M'_1$, as they are inferred from (5–7.96):

$$M_1 \to -\frac{\alpha}{2\pi} m \int_0^1 du\, u^2(1-u) eq\sigma F \frac{1}{m^2 u^2},$$
$$M'_1 \to -\frac{\alpha}{2\pi} \int_0^1 du\, u(1-u) eq\sigma F \cdot [mu - (2-u)\gamma k'] \frac{1}{D'_1}. \tag{5–7.108}$$

The list of contributions to $\gamma_1{}^\nu(k')$ leads off with the explicitly field-dependent terms produced by the rearrangements of (5–7.76, 77),

$$\gamma_1{}^\nu(k')\big|_a = 2e^2 \int ds\, s^2\, du \frac{dv}{2}$$
$$\times \langle [2m + \gamma(\Pi - k)] [su^2(1+v) eq(Fk)^\nu$$
$$- su^2(1-v) eq(F(k+k'))^\nu] e^{-isk^2} \rangle e^{-isD'}. \tag{5–7.109}$$

The transformation

$$\bar{k} \to k: \qquad k \to k + u\Pi - u\frac{1-v}{2}k', \tag{5–7.110}$$

combined with the basic integrals [Eqs. (4–14.75), (4–8.57)]

$$\begin{aligned} \langle k_\mu k_\nu e^{-isk^2}\rangle &= -\frac{i}{2s}g_{\mu\nu}\langle e^{-isk^2}\rangle, \\ \langle e^{-isk^2}\rangle &= \frac{1}{(4\pi)^2}\frac{1}{is^2}, \end{aligned} \tag{5–7.111}$$

then yields

$$\gamma_1^{\nu}(k')\big|_a = -i\frac{\alpha}{2\pi}\int ds\, du\,\frac{dv}{2}\left[-s\left(m(1+u) + u\frac{1-v}{2}\gamma k'\right)\right.$$
$$\left. \times u^2(1-v)(1+uv)eq(Fk')^{\nu} + iu^2veq(F\gamma)^{\nu}\right]e^{-isD'}, \tag{5–7.112}$$

or

$$\gamma_1^{\nu}(k')\big|_a = -i\frac{\alpha}{2\pi}\int du\,\frac{dv}{2}\left[\left(m(1+u) + u\frac{1-v}{2}\gamma k'\right)u^2(1-v)(1+uv)\right.$$
$$\left. \times eq(Fk')^{\nu}\frac{1}{D'^2} + u^2veq(F\gamma)^{\nu}\frac{1}{D'}\right]. \tag{5–7.113}$$

In connection with the test provided by (5–7.101), we also record that

$$k'_\nu\gamma_1^{\nu}(k')\big|_a = -i\frac{\alpha}{2\pi}\int du\,\frac{dv}{2}u^2veqk'F\gamma\frac{1}{D'}. \tag{5–7.114}$$

The next contribution involves the commutator introduced in (5–7.81) by combining the exponential functions, which commutator is effectively evaluated in (5–7.86). Setting aside the spin term of (5–7.75) for later consideration, this gives

$$\gamma_1^{\nu}(k')\big|_b = -2e^2i\int ds\, s^4\, du\,\frac{dv}{2}u^3\frac{1-v^2}{2}$$
$$\times\left\langle(2m + \gamma(\Pi - k))(2\Pi - 2k - k')^{\nu}eqkFk'e^{-is\bar{k}^2}\right\rangle e^{-isD'}, \tag{5–7.115}$$

and the transformation (5–7.110) effectively converts it into

$$-2e^2 i\int ds\, s^4\, du\, \frac{dv}{2} u^3 \frac{1-v^2}{2}$$

$$\times\left\langle\left(m(1+u) + u\frac{1-v}{2}\gamma k' - \gamma k\right)\right.$$

$$\left.\times[(1-u)(2\Pi - k') - uvk' - 2k]^{\nu} eqkFk' e^{-isk^2}\right\rangle e^{-isD'}. \quad (5\text{–}7.116)$$

Owing to the factor $kFk'$, only terms quadratic in $k$ survive to give

$$\gamma_1^{\nu}(k')\big|_b = -i\frac{\alpha}{4\pi}\int ds\, s\, du\, \frac{dv}{2} u^3 \frac{1-v^2}{2}\left[2\left(m(1+u) + u\frac{1-v}{2}\gamma k'\right)\right.$$

$$\left.\times eq(Fk')^{\nu} + [(1-u)(2\Pi - k') - uvk']^{\nu} eq\gamma Fk'\right] e^{-isD'}$$

$$= i\frac{\alpha}{4\pi}\int du\, \frac{dv}{2} u^3 \frac{1-v^2}{2}\left[2\left(m(1+u) + u\frac{1-v}{2}\gamma k'\right) eq(Fk')^{\nu}\right.$$

$$\left.+[(1-u)(2\Pi - k') - uvk']^{\nu} eq\gamma Fk'\right]\frac{1}{D'^2}. \quad (5\text{–}7.117)$$

Again, we note the contribution to the product $k'_{\nu}\gamma_1^{\nu}(k')$:

$$k'_{\nu}\gamma_1^{\nu}(k')\big|_b = -i\frac{\alpha}{4\pi}\int du\, \frac{dv}{2} u^2 \frac{1-v^2}{2} eq\gamma Fk' \frac{\mathscr{H}' + u^2 v k'^2}{D'^2}$$

$$= -i\frac{\alpha}{2\pi}\int du\, \frac{dv}{2} u^2 \frac{1-v^2}{2} eq\gamma Fk' \frac{\partial}{\partial v}\frac{1}{D'}$$

$$= -i\frac{\alpha}{2\pi}\int du\, \frac{dv}{2} u^2 v eq\gamma Fk' \frac{1}{D'}, \quad (5\text{–}7.118)$$

which, as it happens, precisely cancels (5–7.114).

Before embarking on our major task, the computation of the rearranged form of the second term in (5–7.74), we consider the first term of that expression:

$$ie^2\int ds\, s\, du\, \gamma^{\mu}\langle\gamma^{\nu} e^{-is\chi'}\rangle\gamma_{\mu},$$

$$\chi' = (k - u\Pi')^2 + \mathscr{H}' + m^2u^2 - ueq\sigma F. \quad (5\text{–}7.119)$$

Here, we have introduced the symbol (omitting the prime)

$$\mathscr{H} = u(1-u)(\Pi^2 + m^2), \tag{5–7.120}$$

as distinguished from

$$\mathscr{H} = u(1-u)(\Pi^2 + m^2 - eq\sigma F). \tag{5–7.121}$$

Using the notation [it is (5–7.91), with $\frac{1}{2}(1-v) = 1$]

$$\bar{k} = k - u\Pi', \tag{5–7.122}$$

we employ the spin expansion

$$e^{-is\chi'} \cong \exp[-is(\bar{k}^2 + \mathscr{H}' + m^2u^2)]\,(1 + isueq\sigma F). \tag{5–7.123}$$

The observations that

$$\gamma^\mu\gamma^\nu\gamma_\mu = 2\gamma^\nu, \qquad \gamma^\mu\gamma^\nu\sigma F\gamma_\mu = -2\sigma F\gamma^\nu \tag{5–7.124}$$

then present (5–7.119) as

$$2ie^2\int ds\, s\, du\left\langle \exp[-is(\bar{k}^2 + \mathscr{H}' + m^2u^2)]\right\rangle(1 - isueq\sigma F)\gamma^\nu. \tag{5–7.125}$$

The exponential function involving the sum $\bar{k}^2 + \mathscr{H}'$ can be decomposed into a product of exponentials with a compensating commutator term, in two different ways [cf. Eq. (5–7.79)]. To the required accuracy, limited to the first power of $F$, an average of the two forms will cancel the additional comutators,

$$\exp[-is\bar{k}^2 + \mathscr{H}'] \cong e^{-is\bar{k}^2}\cdot e^{-is\mathscr{H}'}. \tag{5–7.126}$$

Then, as we have remarked before [cf. the discussion preceding (4–16.13)], the distinction between $\bar{k}$ and $k$ in the required integration can only appear in terms quadratic in $F$, which gives for (5–7.125)

$$\frac{\alpha}{2\pi}\int\frac{ds}{s}\,du\exp[-is(m^2u^2 + \mathscr{H}')]\,(1 - isueq\sigma F)\gamma^\nu$$

$$= \frac{\alpha}{2\pi}\int\frac{ds}{s}\,du\,\gamma^\nu(1 - isu(1-u)eq\sigma F)\exp[-is(m^2u^2 + \mathscr{H}')]$$

$$-i\frac{\alpha}{2\pi}\int ds\,du\,ueq\sigma F\gamma^\nu\exp[-is(m^2u^2 + \mathscr{H}')]. \tag{5–7.127}$$

In the latter version we have reintroduced $\mathscr{H}'$, with the appropriate $\sigma F$ correction term, and where the $\sigma F$ factor is already present, ignored the distinction between $\mathscr{H}'$ and $\mathscr{H}'$. The point of this is to recognize that the initial term of (5–7.127), the one that does not have an explicit $\sigma F$ factor, is cancelled by a piece of (5–7.72), referring to $M_0$, namely, the first of the two terms produced by the factor $1 - u$. Accordingly, we are left with this contribution to $\gamma_1{}^\nu(k')$:

$$\gamma_1{}^\nu(k')\big|_c = -\frac{\alpha}{2\pi}\int_0^1 du\, u(1-u)\gamma^\nu eq\sigma F\frac{1}{m^2u^2+\mathscr{H}'}$$

$$-\frac{\alpha}{2\pi}\int_0^1 du\, ueq\sigma F\gamma^\nu\frac{1}{m^2u^2+\mathscr{H}'}, \qquad (5\text{–}7.128)$$

and [Eq. (5–7.99)]

$$k'_\nu\gamma_1{}^\nu(k')\big|_c = -\frac{\alpha}{2\pi}\int_0^1 du\, u[(1-u)\gamma k' eq\sigma F + eq\sigma F\gamma k']\frac{1}{D_1'}. \qquad (5\text{–}7.129)$$

With the spin term of (5–7.75) still set aside, and conscious of the rearrangements already introduced, we find that the residual form of the second term in (5–7.74) reads

$$-e^2\int ds\, s^2\, du\frac{dv}{2}u\gamma^\mu$$

$$\times\left\langle[m-\gamma(\Pi-k)]\left\{(\Pi-k)^\nu\exp\left[-is\left(\frac{1+v}{2}\chi+\frac{1-v}{2}\chi'\right)\right]\right.\right.$$

$$\left.\left.+\exp\left[-is\left(\frac{1+v}{2}\chi+\frac{1-v}{2}\chi'\right)\right](\Pi'-k)^\nu\right\}\right\rangle\gamma_\mu, \qquad (5\text{–}7.130)$$

where, again,

$$\frac{1+v}{2}\chi+\frac{1-v}{2}\chi' = \bar{k}^2+\frac{1+v}{2}\mathscr{H}+\frac{1-v}{2}\mathscr{H}'+u^2\frac{1-v^2}{4}k'^2+m^2u^2-ueq\sigma F,$$

$$\bar{k} = k-u\Pi+u\frac{1-v}{2}k'. \qquad (5\text{–}7.131)$$

First, we exhibit two explicit spin terms, one of which is evident in (5–7.131), while the other appears on moving $\Pi^2+m^2$ to the left, as can be done without effective change, there to be replaced by $eq\sigma F$. This gives

$$\gamma_1^{\nu}(k')|_d = -2ie^2 \int ds\, s^3\, du \frac{dv}{2} u^2(1-u)\frac{1+v}{2} eq\sigma F$$
$$\times \left\langle [2m + \gamma(\Pi - k)][2(\Pi - k) - k']^{\nu} e^{-is\bar{k}^2} \right\rangle e^{-isD'}$$
$$-2ie^2 \int ds\, s^3\, du \frac{dv}{2} u^2 eq\sigma F \left\langle \gamma(\Pi - k)[2(\Pi - k) - k']^{\nu} e^{-is\bar{k}^2} \right\rangle e^{-isD'}. \tag{5–7.132}$$

The use of the transformation (5–7.110), and of the integrals (5–7.111) then yields, after the $s$-integration,

$$\gamma_1^{\nu}(k')|_d = \frac{\alpha}{2\pi} \int du \frac{dv}{2} u^2(1-u)\frac{1+v}{2} eq\sigma F$$
$$\times \left[\left(m(1+u) + u\frac{1-v}{2}\gamma k'\right)[(1-u)(2\Pi - k') - uvk']^{\nu}\frac{1}{D'^2} + \gamma^{\nu}\frac{1}{D'}\right]$$
$$+ \frac{\alpha}{2\pi} \int du \frac{dv}{2} u^2 eq\sigma F$$
$$\times \left[\left(-m(1-u) + u\frac{1-v}{2}\gamma k'\right)\right.$$
$$\left.\times [(1-u)(2\Pi - k') - uvk']^{\nu}\frac{1}{D'^2} + \gamma^{\nu}\frac{1}{D'}\right]. \tag{5–7.133}$$

As in (5–7.118), the product with $k'_\nu$ contains

$$\frac{\mathscr{H}' + u^2 v k'^2}{D'^2} = 2\frac{\partial}{\partial v}\frac{1}{D'}, \tag{5–7.134}$$

and we observe that

$$k'_\nu \gamma_1^{\nu}(k')|_d = -\frac{\alpha}{\pi} \int du \frac{dv}{2} u(1-u) eq\sigma F \frac{1+v}{2}\frac{\partial}{\partial v}\frac{m(1+u) + u\frac{1-v}{2}\gamma k'}{D'}$$
$$+ \frac{\alpha}{\pi} \int du \frac{dv}{2} u eq\sigma F \frac{\partial}{\partial v}\frac{m(1-u) - u\frac{1-v}{2}\gamma k'}{D'}, \tag{5–7.135}$$

or, with a partial integration,

$$k'_\nu \gamma_1^\nu(k')|_d = -\frac{\alpha}{2\pi} m \int_0^1 du\, u^2(1-u) eq\sigma F \frac{1}{m^2 u^2}$$

$$-\frac{\alpha}{2\pi}\int_0^1 du\, ueq\sigma F[m(1-u) - u\gamma k']\frac{1}{D_1'}$$

$$+\frac{\alpha}{2\pi}\int du \frac{dv}{2} u(1-u) eq\sigma F\left[m(1+u) + \frac{1-v}{2} u\gamma k'\right]\frac{1}{D'}; \quad (5\text{–}7.136)$$

the end-point evaluations have introduced

$$v = +1: \qquad D' = m^2u^2,$$
$$v = -1: \qquad D' = D_1'. \qquad (5\text{–}7.137)$$

After removing the spin terms from (5–7.130) in the manner described, we have an effective reduction of the structure of the exponential function, as described by

$$\frac{1+v}{2}\chi + \frac{1-v}{2}\chi' \to \bar{k}^2 + \mathbf{D}', \qquad (5\text{–}7.138)$$

where

$$\mathbf{D}' = m^2u^2 + \frac{1-v}{2}\mathscr{H}' + u^2\frac{1-v^2}{4}k'^2 \qquad (5\text{–}7.139)$$

is to be distinguished from $D'$, which is constructed by use of $\mathscr{H}'$. In view of the absence of spin matrices in this expression, the implied form of (5–7.130) can be simplified to

$$2e^2\int ds\, s^2\, du \frac{dv}{2} u\big\langle (m-\gamma k)\{(\Pi - k)^\nu \exp[-is(\bar{k}^2 + \mathbf{D}')]$$

$$+\exp[-is(\bar{k}^2 + \mathbf{D}')]\,(\Pi' - k)^\nu\}\big\rangle, \quad (5\text{–}7.140)$$

which has already invoked the possibility of omitting $\gamma\Pi + m$ when it stands entirely to the left. Concerning the integration symbolized by $\langle\ \rangle$, we first note that, as in the discussion based on (5–7.126), we have

$$\big\langle \exp[-is(\bar{k}^2 + \mathbf{D}')]\big\rangle \cong \langle e^{-is\bar{k}^2}\rangle \cdot e^{-is\mathbf{D}'}$$

$$\cong \langle e^{-isk^2}\rangle \cdot e^{-is\mathbf{D}'}. \qquad (5\text{–}7.141)$$

Next, we need to evaluate

$$\left\langle k \exp\left[-is(\bar{k}^2 + \mathrm{D}')\right]\right\rangle \cong \langle k e^{-is\bar{k}^2}\rangle \cdot e^{-is\mathrm{D}'}. \tag{5–7.142}$$

For that, we begin with

$$0 = \frac{1}{2s}\left\langle\left[\xi, e^{-is\bar{k}^2}\right]\right\rangle = \int_{-1}^{1} \tfrac{1}{2}\, dw \,\langle e^{-is\bar{k}^2\frac{1}{2}(1+w)}\bar{k}\, e^{-is\bar{k}^2\frac{1}{2}(1-w)}\rangle$$

$$= \left\langle\left(k + u\frac{1-v}{2}k'\right)e^{-is\bar{k}^2}\right\rangle - \int_{-1}^{1} \tfrac{1}{2}\, dw \,\langle e^{-is\bar{k}^2\frac{1}{2}(1+w)} u\Pi e^{-is\bar{k}^2\frac{1}{2}(1-w)}\rangle, \tag{5–7.143}$$

and then use the approximate rearrangements

$$e^{-is\bar{k}^2\frac{1}{2}(1+w)}\Pi e^{-is\bar{k}^2\frac{1}{2}(1-w)} \cong \Pi e^{-is\bar{k}^2} - is\frac{1+w}{2}\left[\bar{k}^2, \Pi\right]e^{-is\bar{k}^2}$$

$$\cong e^{-is\bar{k}^2}\Pi + is\frac{1-w}{2}\left[\bar{k}^2, \Pi\right]e^{-is\bar{k}^2}, \tag{5–7.144}$$

or

$$e^{-is\bar{k}^2\frac{1}{2}(1+w)}\Pi e^{-is\bar{k}^2\frac{1}{2}(1-w)} \cong \Pi \cdot e^{-is\bar{k}^2} - \tfrac{1}{2}isw\left[\bar{k}^2, \Pi\right]e^{is\bar{k}^2}. \tag{5–7.145}$$

But the latter commutator term disappears after the $w$-integration, from which we learn that

$$\langle k e^{-is\bar{k}^2}\rangle \cong \langle e^{-isk^2}\rangle\left(u\Pi - u\frac{1-v}{2}k'\right). \tag{5–7.146}$$

Finally, we require an evaluation for

$$\left\langle k_\mu k_\nu \exp\left[-is(\bar{k}^2 + \mathrm{D}')\right]\right\rangle \cong \langle k_\mu k_\nu e^{-is\bar{k}^2}\rangle \cdot e^{-is\mathrm{D}'}, \tag{5–7.147}$$

and therefore consider

$$0 = \frac{1}{2s}\left\langle\left[\xi_\mu, k_\nu e^{-is\bar{k}^2}\right]\right\rangle = \frac{i}{2s}g_{\mu\nu}\langle e^{-is\bar{k}^2}\rangle + \frac{1}{2s}\left\langle k_\nu\left[\xi_\mu, e^{-is\bar{k}^2}\right]\right\rangle. \tag{5–7.148}$$

The preceding discussion can again be applied to the last integral, with the additional $k$-factor, which yields

$$0 \cong \frac{i}{2s} g_{\mu\nu}\langle e^{-isk^2}\rangle + \left\langle \left(k + u\frac{1-v}{2}k'\right)_\mu k_\nu e^{-is\bar{k}^2}\right\rangle - u\Pi_\mu \cdot \langle k_\nu e^{-is\bar{k}^2}\rangle, \quad (5\text{–}7.149)$$

and then

$$\langle k_\mu k_\nu e^{-is\bar{k}^2}\rangle \cong \langle e^{-isk^2}\rangle\left[-\frac{i}{2s}g_{\mu\nu} + \left(u\Pi - u\frac{1-v}{2}k'\right)_\mu \cdot \left(u\Pi - u\frac{1-v}{2}k'\right)_\nu\right]. \quad (5\text{–}7.150)$$

Thus, all is as in the absence of the homogeneous field, except for the symmetrization required in the last term above.

In dealing with the double symmetrizations that are occasionally required, it is well to keep in mind the identity

$$A \cdot (B \cdot C) - (A \cdot B) \cdot C = \tfrac{1}{4}[[A, C], B], \quad (5\text{–}7.151)$$

which, through the appearance of a double commutator, implies that all such double symmetrizations are equivalent, to the required accuracy. Then using the reduction $\gamma\Pi + m \to 0$, when this combination stands on the left-hand side, we get the following for (5–7.140):

$$-i\frac{\alpha}{2\pi}\int ds\,du\,\frac{dv}{2}u\Big\{\big[(1-u)(2\Pi - k')^\nu \cdot e^{-is\mathbf{D}'} - uvk'^\nu e^{-is\mathbf{D}'}\big]$$

$$\times\big[m(1+u) - \tfrac{1}{2}u(\gamma\Pi' + m) - \tfrac{1}{2}uv\gamma k'\big] - \frac{i}{s}\gamma^\nu e^{-is\mathbf{D}'}\Big\}. \quad (5\text{–}7.152)$$

But let us quickly detach the last term, and carry out a partial integration,

$$\frac{\alpha}{2\pi}\gamma^\nu\int\frac{ds}{s}\,du\,u\,d\left(\frac{1-v}{2}\right)e^{-is\mathbf{D}'}$$

$$= -\frac{\alpha}{2\pi}\gamma^\nu\int\frac{ds}{s}\,du\,u\,e^{-is\mathbf{D}_1'}$$

$$-i\frac{\alpha}{2\pi}\gamma^\nu\int ds\,du\,\frac{dv}{2}u\frac{1-v}{2}(\mathscr{H}' + u^2vk'^2)e^{-is\mathbf{D}'}, \quad (5\text{–}7.153)$$

where

$$\mathbf{D}_1' = m^2u^2 + \mathscr{H}' = D_1' + \mu(1-u)eq\sigma F. \quad (5\text{–}7.154)$$

Apart from the distinction between $\mathrm{D}_1'$ and $D_1'$, the first term on the right-hand side of (5–7.153) cancels the second piece of (5–7.72), the one arising from the last half of the factor $1-u$. The spin term that survives from this incomplete cancellation is

$$i\frac{\alpha}{2\pi}\int ds\,du\,u^2(1-u)\gamma^\nu eq\sigma F e^{-isD_1'}. \qquad (5\text{–}7.155)$$

We now perform the $s$-integrations in the four expressions that have appeared: (5–7.155), the last term of (5–7.153), the first part of (5–7.152), and, setting aside the explicit spin term, the latter part of (5–7.72). This gives

$$\begin{aligned}\gamma_1^\nu(k')\big|_e = {} & \frac{\alpha}{2\pi}\int_0^1 du\,u^2(1-u)\gamma^\nu eq\sigma F\frac{1}{D_1'}\\ & -\frac{\alpha}{2\pi}\gamma^\nu\int du\frac{dv}{2}u\frac{1-v}{2}\frac{\mathscr{H}'+u^2vk'^2}{\mathrm{D}'}\\ & -\frac{\alpha}{2\pi}\int du\frac{dv}{2}u\left[(1-u)(2\Pi-k')^\nu\cdot\frac{1}{\mathrm{D}'}-uvk'^\nu\frac{1}{\mathrm{D}'}\right]\\ & \times\left[m(1+u)-\frac{u}{2}(\gamma\Pi'+m)-\frac{uv}{2}\gamma k'\right]\\ & +\frac{\alpha}{2\pi}m\int du\frac{dv}{2}u(1-u^2)(2\Pi-k')^\nu\frac{1}{m^2u^2+\dfrac{1-v}{2}\mathscr{H}'}. \qquad (5\text{–}7.156)\end{aligned}$$

In working out this contribution to $k_\nu'\gamma_1^\nu(k')$, we encounter

$$u(1-u)k'(2\Pi-k')\to -u(1-u)\left[(\Pi-k')^2+m^2-eq\sigma F\right]=-\mathscr{H}', \quad (5\text{–}7.157)$$

and then find that

$$\begin{aligned}k_\nu'\gamma_1^\nu(k')\big|_e = {} & -\frac{\alpha}{2\pi}\int du\frac{dv}{2}u(1-u)eq\sigma F\left[m(1+u)+u\frac{1-v}{2}\gamma k'\right]\frac{1}{D'}\\ & +\frac{\alpha}{2\pi}\int_0^1 du\,u(1-u)eq\sigma F\cdot\left[m(1+u)+u\gamma k'\right]\frac{1}{D_1'}. \qquad (5\text{–}7.158)\end{aligned}$$

The result of combining (5–7.129), (5–7.136), and (5–7.158) [recalling the cancellation of (5–7.114) and (5–7.118)] is

$$k_\nu'\gamma_1^\nu(k')\big|_{a\cdots e} = -\frac{\alpha}{2\pi}\int_0^1 du\, u(1-u)(2-u)\,eq\sigma F\cdot\gamma k'\frac{1}{D_1'}$$

$$+\frac{\alpha}{2\pi}m\int_0^1 du\, u^2(1-u)\,eq\sigma F\left(\frac{1}{D_1'}-\frac{1}{m^2u^2}\right), \qquad (5\text{–}7.159)$$

which agrees precisely with the anticipated expression (5–7.101). This may occasion some surprise, since the calculation of $\gamma_1^\nu(k')$ is not yet complete. The clarifying observation is that all terms not yet considered involve the spin structure $k_\lambda'\sigma^{\lambda\nu}$, which vanishes identically on multiplication with $k_\nu'$.

To complete this first stage of our program, we return to Eqs. (5–7.74, 75), and isolate the spin term that has thus far been set aside:

$$-e^2\int ds\, s^2\, du\,\frac{dv}{2}\,u\gamma^\mu$$

$$\times\left\langle[m-\gamma(\Pi-k)]\exp\left[-is\frac{1+v}{2}\chi\right]ik_\lambda'\sigma^{\lambda\nu}\exp\left[-is\frac{1-v}{2}\chi'\right]\right\rangle\gamma_\mu. \qquad (5\text{–}7.160)$$

Let us write

$$\chi = \chi_1 - ueq\sigma F,$$
$$\chi_1 = (k-u\Pi)^2 + \mathscr{H} + m^2u^2, \qquad (5\text{–}7.161)$$

and begin by picking out the explicit $\sigma F$ terms. The associated matrix structure is

$$\gamma^\mu[m-\gamma(\Pi-k)]\left[isu\frac{1+v}{2}\sigma F\sigma^{\lambda\nu}+\sigma^{\lambda\nu}isu\frac{1-v}{2}\sigma F\right]\gamma_\mu$$

$$= isu\left[(1+v)\sigma F\sigma^{\lambda\nu}+(1-v)\sigma^{\lambda\nu}\sigma F\right]\gamma(\Pi-k)$$

$$+[m+\gamma(\Pi-k)]isu\left\{\frac{1+v}{2}\gamma^\mu\sigma F\sigma^{\lambda\nu}\gamma_\mu+\frac{1-v}{2}\gamma^\mu\sigma^{\lambda\nu}\sigma F\gamma_\mu\right\}. \qquad (5\text{–}7.162)$$

Now, the difference between $\sigma F\sigma^{\lambda\nu}$ and $\sigma^{\lambda\nu}\sigma F$, which is a commutator of $\sigma$-matrices, is itself a linear combination of $\sigma$-matrices, and is therefore annulled by the $\gamma^\mu\cdots\gamma_\mu$ operation. Accordingly,

$$\gamma^\mu\sigma F\sigma^{\lambda\nu}\gamma_\mu = \gamma^\mu\sigma^{\lambda\nu}\sigma F\gamma_\mu = 2i(\gamma^\lambda\sigma F\gamma^\nu - \gamma^\nu\sigma F\gamma^\lambda), \qquad (5\text{–}7.163)$$

and the combination found between braces in (5-7.162) reduces to the right-hand side of (5-7.163). Inasmuch as $\sigma F$ is everywhere in evidence, the exponential functions of $\chi_1$ and $\chi_1'$ can be directly combined and simplified as in (5-7.89-91). In particular, the vector $k$ that occurs linearly in (5-7.160) is effectively replaced by

$$k \to u\Pi - u\frac{1-v}{2}k' = u\Pi' + u\frac{1+v}{2}k'. \tag{5-7.164}$$

This gives

$$\gamma_1''(k')|_f = \frac{\alpha}{2\pi}\int du\,\frac{dv}{2}\,u^2 eq\left\{\left(\frac{1+v}{2}\sigma F ik_\lambda'\sigma^{\lambda\nu} + \frac{1-v}{2}ik_\lambda'\sigma^{\lambda\nu}\sigma F\right)\right.$$
$$\times\left[(1-u)\gamma\Pi' + \left(1 - u\frac{1+v}{2}\right)\gamma k'\right]$$
$$\left. + u\left(m + \frac{1-v}{2}\gamma k'\right)(\gamma^\nu\sigma F\gamma k' - \gamma k'\sigma F\gamma^\nu)\right\}\frac{1}{D'^2}. \tag{5-7.165}$$

After the $\sigma F$ terms are separated out, (5-7.160) becomes

$$-2e^2\int ds\, s^2\, du\,\frac{dv}{2}\,uik_\lambda'\sigma^{\lambda\nu}\left\langle\gamma(\Pi - k)\exp\left[-is\left(\frac{1+v}{2}\chi_1 + \frac{1-v}{2}\chi_1'\right)\right]\right.$$
$$\left.\times\left\{1 - s^2u^2\frac{1-v^2}{2}kieqFk'\right\}\right\rangle, \tag{5-7.166}$$

which makes use of the combined exponential form introduced in Eqs. (5-7.81, 86). According to (5-7.87), without the $\sigma F$ term, we have

$$\frac{1+v}{2}\chi_1 + \frac{1-v}{2}\chi_1' = \bar{k}^2 + \frac{1+v}{2}\mathscr{H} + \frac{1-v}{2}\mathscr{H}' + u^2\frac{1-v^2}{4}k'^2 + m^2u^2, \tag{5-7.167}$$

and we now exploit the possibility of effectively translating $\mathscr{H}$ to the extreme left, where it is replaced by $u(1-u)eq\sigma F$. This supplies another explicit $\sigma F$ term,

$$-2e^2\int ds\, s^2\, du\,\frac{dv}{2}\,u\left[-is\frac{1+v}{2}u(1-u)eq\sigma F\right]ik_\lambda'\sigma^{\lambda\nu}\left\langle\gamma(\Pi - k)e^{-is\bar{k}^2}\right\rangle e^{-isD'}, \tag{5-7.168}$$

or

$$\gamma_1''(k')\Big|_g = -\frac{\alpha}{2\pi}\int du\frac{dv}{2}\frac{1+v}{2}u^2(1-u)eq\sigma Fik_\lambda'\sigma^{\lambda\nu}$$
$$\times\left[(1-u)\gamma\Pi' + \left(1-u\frac{1+v}{2}\right)\gamma k'\right]\frac{1}{D'^2}. \qquad (5\text{–}7.169)$$

The explicit $F$-term in (5–7.166) contributes

$$2e^2\int ds\,s^2\,du\frac{dv}{2}us^2u^2\frac{1-v^2}{2}ik_\lambda'\sigma^{\lambda\nu}\left\langle\gamma(\Pi-k)kieqFk'e^{-is\bar{k}^2}\right\rangle e^{-isD'}, \quad (5\text{–}7.170)$$

which is

$$\gamma_1''(k')\Big|_h = -\frac{\alpha}{4\pi}\int du\frac{dv}{2}u^3\frac{1-v^2}{2}ik_\lambda'\sigma^{\lambda\nu}\gamma ieqFk'\frac{1}{D'^2}. \qquad (5\text{–}7.171)$$

The remainder of (5–7.166) is stated in

$$-2e^2\int ds\,s^2\,du\frac{dv}{2}uik_\lambda'\sigma^{\lambda\nu}\left\langle\gamma(\Pi-k)\exp\left[-is(\bar{k}^2+\mathbf{D}')\right]\right\rangle$$
$$= i\frac{\alpha}{2\pi}\int ds\,du\frac{dv}{2}uik_\lambda'\sigma^{\lambda\nu}\left[\gamma\Pi\,e^{-is\mathbf{D}'} - u\gamma\left(\Pi-\frac{1-v}{2}k'\right)\cdot e^{-is\mathbf{D}'}\right]. \quad (5\text{–}7.172)$$

Then, if one removes the indicated symmetrization with the aid of

$$\left[\gamma\Pi, e^{-is\mathbf{D}'}\right] \cong is(1-v)u(1-u)\gamma ieqFk'e^{-is\mathbf{D}'}, \qquad (5\text{–}7.173)$$

performs the $s$-integrals, and adds the $s$-integrated explicit spin term of (5–7.72), the final contribution to $\gamma_1''(k')$ is obtained:

$$\gamma_1''(k')\Big|_i = \frac{\alpha}{2\pi}\int du\frac{dv}{2}uik_\lambda'\sigma^{\lambda\nu}\left[(1-u)\gamma\Pi' + \left(1-u\frac{1+v}{2}\right)\gamma k'\right]\frac{1}{\mathbf{D}'}$$
$$+\frac{\alpha}{2\pi}m\int du\frac{dv}{2}u(1-u^2)ik_\lambda'\sigma^{\lambda\nu}\frac{1}{m^2u^2+\dfrac{1-v}{2}\mathscr{H}'}$$
$$+\frac{\alpha}{2\pi}\int du\frac{dv}{2}\frac{1-v}{2}u^3(1-u)ik_\lambda'\sigma^{\lambda\nu}\gamma ieqFk'\frac{1}{D'^2}. \qquad (5\text{–}7.174)$$

Of the nine sets of terms that constitute $\gamma_1''(k')$, only two, those labeled as $e$ and $i$, contain expressions that do not exhibit $F$ explicitly. These are subject to checks based on comparison with previous calculations. The simplest of those is the consideration of scattering, where the Dirac equation is applicable to simplify the right-hand as well as the left-hand side. Under such circumstances, which include setting $F = 0$, the distinction between $\mathscr{H}$ and $\boldsymbol{\mathscr{H}}$ disappears, symmetrized multiplication is unnecessary, and

$$\mathscr{H}' \to 0, \qquad \gamma\Pi' + m \to 0,$$

$$D' \to u^2\left(m^2 + \frac{1-v^2}{4}k'^2\right) + \mu^2, \tag{5-7.175}$$

which also incorporates the photon mass now required. After these and related rearrangements, such as

$$(\Pi + \Pi')^\nu \to 2m\gamma^\nu - ik_\lambda'\sigma^{\lambda\nu}, \tag{5-7.176}$$

we get

$$\gamma_1^\nu(k') \to [F_1(k') - 1]\gamma^\nu + \frac{\alpha}{2\pi}\frac{1}{2m}F_2(k')ik_\lambda'\sigma^{\lambda\nu}. \tag{5-7.177}$$

Here (omitting the prime on $k$)

$$F_2(k) = \int_0^1 dv \frac{1}{1 + \dfrac{1-v^2}{4}\dfrac{k^2}{m^2}}, \tag{5-7.178}$$

in agreement with (4-4.75), while

$$\begin{aligned} F_1(k) - 1 = &-\frac{\alpha}{2\pi}k^2\int du\,\frac{dv}{2} \\ &\times\left[u\frac{1-v^2}{2}\frac{\mu^2}{m^2u^2+\mu^2}\frac{1}{u^2\left(m^2 + \dfrac{1-v^2}{4}k^2\right)+\mu^2}\right. \\ &\left.+u\frac{1+v^2}{2}\frac{1}{u^2\left(m^2+\dfrac{1-v^2}{4}k^2\right)+\mu^2} - \frac{1-u+2uv^2}{2}\frac{1}{m^2+\dfrac{1-v^2}{4}k^2}\right] \\ = &-\frac{\alpha}{2\pi}\frac{k^2}{4m^2}\int_0^1 dv\,\frac{(1+v^2)\log\left(\dfrac{4m^2}{\mu^2}\dfrac{v^2}{1-v^2}\right) - 1 - 2v^2}{1+\dfrac{1-v^2}{4}\dfrac{k^2}{m^2}}, \end{aligned} \tag{5-7.179}$$

which uses the integrals of (4-14.67, 68) and the identity (4-12.42). This form-factor result also coincides with the known one displayed in Eqs. (4-4.68, 77).

After this lengthy interlude, it is well to recall what has to be calculated, namely [Eq. 5–7.63]

$$M_1^{(2)} = ie^2 \int \frac{(dk)}{(2\pi)^4} \gamma_1^{\nu}(k) \frac{1}{k^2} \frac{1}{\gamma(\Pi - k) + m} \gamma_\nu + \zeta'\left(\frac{\alpha}{2\pi}\frac{eq}{2m}\sigma F\right), \quad (5\text{–}7.180)$$

in which the prime on $k$ has been omitted, and we have made explicit the form of $M_1$ for a particle field obeying $(\gamma\Pi + m)\psi = 0$. Before engaging in further detailed operations, it is desirable to perform the exorcism that is implied by the contact term containing $\zeta'$.

This is assisted by rearranging the integrand of (5–7.180) according to

$$\frac{1}{\gamma(\Pi - k) + m}\gamma_\nu = \frac{1}{(\Pi - k)^2 + m^2 - eq\sigma F}\left[m - \gamma(\Pi - k)\right]\gamma_\nu$$

$$\rightarrow \frac{1}{(\Pi - k)^2 + m^2 - eq\sigma F}\left(2\Pi_\nu - k_\nu - ik^\lambda\sigma_{\lambda\nu}\right). \quad (5\text{–}7.181)$$

We then isolate the following piece of $M_1^{(2)}$:

$$ie^2 \int \frac{(dk)}{(2\pi)^4} \gamma_1^{\nu}(k) \frac{1}{k^2} \frac{1}{(\Pi - k)^2 + m^2 - eq\sigma F}(-k_\nu) + \zeta'\left(\frac{\alpha}{2\pi}\frac{eq}{2m}\sigma F\right). \quad (5\text{–}7.182)$$

The integrand here contains $k_\nu\gamma_1^{\nu}(k)$, which, as given in (5–7.101), already displays $\sigma F$, permitting us to discard that structure in the denominator of (5–7.182). We shall, furthermore, decompose $k_\nu\gamma_1^{\nu}(k)$ into two parts, the first of which is the contribution of the term $-1/m^2u^2$ in (5–7.101),

$$\frac{\alpha}{2\pi} m \int_0^1 du\, u^2(1 - u)\, eq\sigma F\left(-\frac{1}{m^2u^2}\right) = -\frac{\alpha}{2\pi}\frac{eq}{2m}\sigma F; \quad (5\text{–}7.183)$$

it states the asymptotic form of $k_\nu\gamma_1^{\nu}(k)$ for $|k^2| \gg m^2$. This initial contribution to $M_1^{(2)}$ is

$$M_1^{(2)}\Big|_\alpha = \frac{\alpha}{2\pi}\frac{eq}{2m}\sigma F\left[\zeta' + ie^2 \int \frac{(dk)}{(2\pi)^4} \frac{1}{k^2} \frac{1}{(\Pi - k)^2 + m^2}\right]. \quad (5\text{–}7.184)$$

We perform the momentum integral in our usual manner,

$$ie^2\left\langle \frac{1}{k^2}\frac{1}{(\Pi - k)^2 + m^2}\right\rangle = -ie^2 \int ds\, s\, du \left\langle \exp\{-is[(1-u)k^2 + u(k^2 - 2k\Pi)]\}\right\rangle$$

$$= -\frac{\alpha}{4\pi}\int \frac{ds}{s}\, du\, e^{-ism^2u^2}. \quad (5\text{–}7.185)$$

If we were now to combine this $s$-integral with the very similar expression for $\zeta'$ [Eq. (5–7.51)], the result would indeed be finite. There is, however, the danger that the purely mathematical procedure of stopping each $s$-integral at a common lower limit is not completely consistent under these circumstances, where the two terms have arisen in quite different ways. For this reason we proceed alternatively to introduce an effective lower limit to $s$, in a way that has an assured universal meaning, through a modification in the propagation function of the electromagnetically neutral photon:

$$\frac{1}{k^2} = \int_0^\infty i\,ds\,e^{-isk^2} \to \lim_{\delta\to+0} \int_0^\infty i\,ds\,\frac{(im^2s)^\delta}{\Gamma(1+\delta)} e^{-isk^2}. \tag{5–7.186}$$

(The presence of additional factors other than $s^\delta$ is merely for convenience.) When the calculation that produced $\zeta'$ is repeated with this modification (that of $\zeta''$ remains unchanged in the limit $\delta \to 0$), we find that

$$\begin{aligned} \zeta' &= \frac{\alpha}{2\pi} \int_0^1 du\,(1-u)^{1+\delta} \int_0^\infty \frac{ds}{s} \frac{(im^2s)^\delta}{\Gamma(1+\delta)} e^{-ism^2u^2} \\ &= \frac{\alpha}{2\pi}\frac{1}{\delta} \int_0^1 du\,(1-u)^{1+\delta} u^{-2\delta} \\ &= \frac{\alpha}{2\pi}\frac{1}{\delta} \frac{\Gamma(2+\delta)\Gamma(1-2\delta)}{\Gamma(3-\delta)} \cong \frac{\alpha}{2\pi}\left(\frac{1}{2\delta}+\frac{5}{4}\right), \end{aligned} \tag{5–7.187}$$

where the last form is a sufficient approximation for $\delta \ll 1$. The analogous modification in the calculation of (5–7.185) is

$$\begin{aligned} -\frac{\alpha}{4\pi} \int_0^1 du\,(1-u)^\delta \int_0^\infty \frac{ds}{s} \frac{(im^2s)^\delta}{\Gamma(1+\delta)} e^{-ism^2u^2} &= -\frac{\alpha}{4\pi}\frac{1}{\delta} \int_0^1 du\,(1-u)^\delta u^{-2\delta} \\ &= -\frac{\alpha}{4\pi}\frac{1}{\delta} \frac{\Gamma(1+\delta)\Gamma(1-2\delta)}{\Gamma(2-\delta)} \cong -\frac{\alpha}{2\pi}\left(\frac{1}{2\delta}+\frac{1}{2}\right), \end{aligned} \tag{5–7.188}$$

and the value thus assigned to the bracket in (5–7.184) is $\frac{3}{4}(\alpha/2\pi)$. Accordingly, the first contribution to $M_1^{(2)}$ is obtained as

$$M_1^{(2)}\Big|_\alpha = -\left(\frac{\alpha}{2\pi}\right)^2 (c_2)_\alpha \frac{eq}{2m}\sigma F, \tag{5–7.189}$$

with

$$(c_2)_\alpha = -\tfrac{3}{4}. \tag{5–7.190}$$

The remainder of (5–7.182) is

$$M_1^{(2)}|_\beta = ie^2\frac{\alpha}{2\pi}\int_0^1 du\, u(1-u)$$

$$\times\left\langle[(2-u)eq\sigma F\cdot\gamma k - mueq\sigma F]\frac{1}{D_1'}\frac{1}{k^2}\frac{1}{k^2-2k\Pi}\right\rangle, \quad (5\text{–}7.191)$$

where

$$D_1' = u(1-u)(\Pi-k)^2 + m^2u \to u(1-u)\left[k^2 - 2k\Pi + \frac{u}{1-u}m^2\right]. \quad (5\text{–}7.192)$$

The type of momentum integral met here has been dealt with in (5–7.20), from which we recognize that ($u$ in that formula is replaced by $y$)

$$\left\langle\frac{1}{k^2}\frac{1}{k^2-2k\Pi}\frac{1}{k^2-2k\Pi+\frac{u}{1-u}m^2}\right\rangle = \frac{i}{(4\pi)^2}\frac{1}{m^2}\int_0^1 dw\int_0^1 dy\frac{1}{y+w\frac{u}{1-u}}$$

$$= \frac{i}{(4\pi)^2}\frac{1}{m^2}\left[\frac{1-u}{u}\log\frac{1}{1-u} + \log\frac{1}{u}\right]$$

(5–7.193)

and

$$\left\langle k\frac{1}{k^2}\frac{1}{k^2-2k\Pi}\frac{1}{k^2-2k\Pi+\frac{u}{1-u}m^2}\right\rangle = \frac{i}{(4\pi)^2}\frac{\Pi}{m^2}\int_0^1 dw\int_0^1 dy\frac{y}{y+w\frac{u}{1-u}}$$

$$= \frac{i}{(4\pi)^2}\frac{\Pi}{m^2}\left[\frac{1-u}{2u}\log\frac{1}{1-u} + \frac{1}{2} - \frac{1}{2}\frac{u}{1-u}\log\frac{1}{u}\right]. \quad (5\text{–}7.194)$$

All the $u$-integrals then encountered in (5–7.191) are elementary, with the exception of [cf. (5–4.107)]

$$\int_0^1 du\frac{1}{u}\log\frac{1}{1-u} = \frac{\pi^2}{6}, \quad (5\text{–}7.195)$$

and we get

$$M_1^{(2)}|_\beta = -\left(\frac{\alpha}{2\pi}\right)^2(c_2)_\beta\frac{eq}{2m}\sigma F, \quad (5\text{–}7.196)$$

with

$$(c_2)_\beta = -\left(\frac{\pi^2}{12} + \frac{1}{2}\right). \tag{5-7.197}$$

Now we must evaluate

$$ie^2\left\langle \gamma_1^{\nu}(k)\frac{1}{k^2}\frac{1}{(\Pi - k)^2 + m^2 - eq\sigma F}(2\Pi_\nu - ik^\lambda\sigma_{\lambda\nu})\right\rangle, \tag{5-7.198}$$

where it will usually be convenient to employ one of the forms

$$ik^\lambda\sigma_{\lambda\nu} = \gamma_\nu\gamma k + k_\nu = -\gamma k\gamma_\nu - k_\nu. \tag{5-7.199}$$

There are some parts of (5-7.198) that are similar to the one just considered. They arise from the contributions to $\gamma_1^{\nu}(k)$ that are given by (5-7.128) and the first term of (5-7.156):

$$-\frac{\alpha}{2\pi}\int_0^1 du\left[u(1-u)^2\gamma^\nu eq\sigma F + ueq\sigma F\gamma^\nu\right]\frac{1}{D_1'}. \tag{5-7.200}$$

This piece of $c_2$ is found to be

$$(c_2)_\gamma = \tfrac{17}{12}\pi^2 - 2. \tag{5-7.201}$$

The momentum integral displayed in (5-7.193), for example, might have been computed more simply, in the sense that only a single parametric integral is required. This is a consequence of the similarity of two of the denominators, which differ by a constant. That makes the following partial fraction decomposition advantageous:

$$\frac{1}{k^2 - 2k\Pi}\,\frac{1}{k^2 - 2k\Pi + m^2\delta} = \frac{1}{m^2\delta}\left[\frac{1}{k^2 - 2k\Pi} - \frac{1}{k^2 - 2k\Pi + m^2\delta}\right], \tag{5-7.202}$$

where, here,

$$\delta = \frac{u}{1-u}. \tag{5-7.203}$$

Now the individual integrals are evaluated, as illustrated by

$$\left\langle \frac{1}{k^2}\,\frac{1}{k^2 - 2k\Pi + m^2\delta}\right\rangle = \frac{i}{(4\pi)^2}\int \frac{ds}{s}\,dy\exp[-ism^2(y^2 + y\delta)] \tag{5-7.204}$$

[the analogous result for $\delta = 0$ is already given in (5–7.185)], and they are combined in

$$\left\langle \frac{1}{k^2}\frac{1}{k^2-2k\Pi}\frac{1}{k^2-2k\Pi+m^2\delta}\right\rangle = \frac{i}{(4\pi)^2}\frac{1}{m^2\delta}\int_0^1 dy \int_0^\infty \frac{ds}{s}$$

$$\times\left\{\exp\left[-ism^2y^2\right] - \exp\left[-ism^2(y^2+y\delta)\right]\right\}$$

$$= \frac{i}{(4\pi)^2}\frac{1}{m^2\delta}\int_0^1 dy \log\frac{y+\delta}{y}. \qquad (5\text{–}7.205)$$

Here is the direct production of what is realized in (5–7.193) only after the $w$-integration is performed.

The point of this little lesson becomes clearer on examining the kinds of integrals encountered in the remainder of the calculation, an example of which is (the presence of $F$ as a factor is understood)

$$\left\langle \frac{1}{k^2}\frac{1}{k^2-2k\Pi}\frac{1}{\mathrm{D}'}\right\rangle,$$

$$\mathrm{D}' = \frac{1-v}{2}u(1-u)(k^2-2k\Pi) + u^2\left(m^2 + \frac{1-v^2}{4}k^2\right). \qquad (5\text{–}7.206)$$

A straightforward evaluation would yield a rather complicated double parametric integral. But here we remark that

$$\frac{1}{k^2-2k\Pi}\frac{1}{\mathrm{D}'} = \left[\frac{1}{k^2-2k\Pi} - \frac{\frac{1}{2}(1-v)u(1-u)}{\mathrm{D}'}\right]\frac{1}{u^2\left(m^2+\dfrac{1-v^2}{4}k^2\right)},$$

$$(5\text{–}7.207)$$

and, again, that

$$\frac{1}{k^2}\frac{1}{m^2+\dfrac{1-v^2}{4}k^2} = \left[\frac{1}{k^2} - \frac{1}{k^2+\dfrac{4m^2}{1-v^2}}\right]\frac{1}{m^2}. \qquad (5\text{–}7.208)$$

Now we evaluate the component integrals, of which the simpler one is

$$\left\langle \frac{1}{k^2 - 2k\Pi}\left[\frac{1}{k^2} - \frac{1}{k^2 + \dfrac{4m^2}{1-v^2}}\right]\right\rangle$$

$$= \frac{i}{(4\pi)^2}\int_0^1 dy \log \frac{y^2 + \dfrac{4}{1-v^2}(1-y)}{y^2}$$

$$= \frac{i}{(4\pi)^2}\left[\frac{2}{1+v}\log\frac{2}{1-v} + \frac{2}{1-v}\log\frac{2}{1+v}\right]. \qquad (5\text{–}7.209)$$

Somewhat more complicated is

$$\left\langle \frac{\frac{1}{2}(1-v)u(1-u)}{\mathbf{D}'}\left[\frac{1}{k^2} - \frac{1}{k^2 + \dfrac{4m^2}{1-v^2}}\right]\right\rangle$$

$$= \frac{i}{(4\pi)^2}\frac{2}{1-v^2}\frac{1}{1-u}\left\{(1-u)(1+v)\log\frac{2}{1+v} - u(1+v)\log\frac{1}{u}\right.$$

$$\left. +(1-v+2uv)\log\frac{2}{1-v+2uv} - (2-u+uv)\log\frac{2}{2-u+uv}\right\}, \qquad (5\text{–}7.210)$$

and the combination of the two gives

$$\left\langle \frac{1}{k^2}\frac{1}{k^2-2k\Pi}\frac{1}{\mathbf{D}'}\right\rangle = \frac{i}{(4\pi)^2}\frac{1}{(mu)^2}\left\{\frac{2}{1+v}\log\frac{2}{1-v} + \frac{2}{1-v}\frac{u}{1-u}\log\frac{1}{u}\right.$$

$$\left. + \frac{2}{1-v^2}\frac{1}{1-u}\left[(2-u+uv)\log\frac{2}{2-u+uv} - (1-v+2uv)\log\frac{2}{1-v+2uv}\right]\right\}.$$

$$(5\text{–}7.211)$$

We are not yet ready to proceed on to this final stage of integration, however, since some contributions must still be supplied with the contact terms that isolate the explicit $\sigma F$ dependence. They are found in the parts of $M_1^{(2)}$ that are implied by $\gamma_1''(k)|_{e+i}$, Eqs. (5–7.156, 174). An example of the kind of momentum integral encountered here is

$$\Pi^\nu\left\langle \frac{1}{\mathbf{D}'}\frac{1}{k^2}\frac{1}{(\Pi-k)^2+m^2-eq\sigma F}\right\rangle \Pi_\nu. \qquad (5\text{–}7.212)$$

The surrounding $\Pi$-factors have been retained in order to emphasize the following property. This object is dimensionless. If $m^2$ were everywhere accompanied by $-eq\sigma F$, as it is in the eventual substitution, $\Pi^2 \to -(m^2 - eq\sigma F)$, the outcome would be a pure number (it is multiplied by $m$ in the complete structure), which must be removed by a suitable contact term in order to maintain the normalization condition on $M$. But the required $\sigma F$ terms are lacking in

$$\mathbf{D}' = m^2u^2 + \frac{1-v}{2}u(1-u)\left[(\Pi-k)^2+m^2\right] + u^2\frac{1-v^2}{4}k^2. \quad (5\text{–}7.213)$$

Hence the effective value of (5–7.212) is the negative of that multiple of $\sigma F$ which is needed to repair the deficiency:

$$\Pi^\nu\left\langle \frac{1}{\mathbf{D}'}\frac{1}{k^2}\frac{1}{(\Pi-k)^2+m^2-eq\sigma F}\right\rangle \Pi_\nu$$

$$\to m^2 eq\sigma F\left[u^2+\frac{1-v}{2}u(1-u)\right]\left\langle \frac{1}{\mathbf{D}'^2}\frac{1}{k^2}\frac{1}{k^2-2k\Pi}\right\rangle; \quad (5\text{–}7.214)$$

it has been simplified by letting $\Pi^2 + m^2 \to 0$, as permitted by the explicit factor $\sigma F$.

While this argument is quite correct, it may not be convincing to some. And the procedure can fail if the presence of spin matrices interferes with the simple treatment of $m^2 - eq\sigma F$. For these reasons, we describe a more formal process. It begins by expanding the last denominator in (5–7.212), now designated as $I$,

$$I \cong \Pi^\nu\left\langle \frac{1}{\mathbf{D}'}\frac{1}{k^2}\frac{1}{(\Pi-k)^2+m^2}\right\rangle \Pi_\nu$$

$$-m^2 eq\sigma F\left\langle \frac{1}{\mathbf{D}'}\frac{1}{k^2}\frac{1}{(k^2-2k\Pi)^2}\right\rangle. \qquad (5\text{–}7.215)$$

The essential observation about the first integral is that both denominators involving $\Pi$ contain it in the same combination, $(\Pi - k)^2$. Hence they can be combined in our standard manner, without reference to $F$. And the $k$-integration then proceeds as though $F = 0$, since the implied error is $\sim F^2$: The result is a function of $\Pi^2$,

$$\left\langle \frac{1}{\mathbf{D}'}\frac{1}{k^2}\frac{1}{(\Pi - k)^2 + m^2}\right\rangle = f(\Pi^2) \cong f(-m^2) + (\Pi^2 + m^2)f'(-m^2), \quad (5\text{–}7.216)$$

which we have expanded with sufficient accuracy to deal with our situation, where $\Pi^2 + m^2 \to eq\sigma F$. The flanking factors of $\Pi$ do not interfere with the latter substitution, and thus

$$\Pi^\nu\left\langle \frac{1}{\mathbf{D}'}\frac{1}{k^2}\frac{1}{(\Pi - k)^2 + m^2}\right\rangle\Pi_\nu \cong (-m^2 + eq\sigma F)f - m^2 eq\sigma F f', \quad (5\text{–}7.217)$$

where we have omitted the arguments of $f$ and $f'$. Then, retaining only the explicit $\sigma F$ terms, we have

$$I \to eq\sigma F\left[f - m^2 f' - m^2\left\langle \frac{1}{\mathbf{D}'}\frac{1}{k^2}\frac{1}{(k^2 - 2k\Pi)^2}\right\rangle\right]. \quad (5\text{–}7.218)$$

To find an expression for $f'$, which is only required for $F = 0$, we differentiate (5–7.216) with respect to $\Pi^\nu$, and then multiply by $\frac{1}{2}\Pi^\nu$:

$$-m^2 f' = \left\langle \frac{1}{\mathbf{D}'}\frac{1}{k^2}\frac{k\Pi + m^2}{(k^2 - 2k\Pi)^2}\right\rangle + \frac{1 - v}{2}u(1 - u)\left\langle \frac{k\Pi + m^2}{\mathbf{D}'^2}\frac{1}{k^2}\frac{1}{k^2 - 2k\Pi}\right\rangle. \quad (5\text{–}7.219)$$

That leaves us with

$$I = eq\sigma F\left[f + \left\langle \frac{1}{\mathbf{D}'}\frac{1}{k^2}\frac{k\Pi}{(k^2 - 2k\Pi)^2}\right\rangle + \frac{1 - v}{2}u(1 - u)\left\langle \frac{k\Pi + m^2}{\mathbf{D}'^2}\frac{1}{k^2}\frac{1}{k^2 - 2k\Pi}\right\rangle\right]. \quad (5\text{–}7.220)$$

Now we return to $f$, which is (5–7.216), evaluated at $F = 0$, $\Pi^2 + m^2 = 0$, and consider its dependence upon $m$. Dimensional considerations show that it is a numerical multiple of $1/m^2$, or that

$$-\tfrac{1}{2}m\frac{\partial}{\partial m}f = f. \qquad (5\text{–}7.221)$$

On the other hand, direct differentiation of the integral, with $\Pi/m$ treated as invariable, gives

$$\tfrac{1}{2}m\frac{\partial}{\partial m}f = \left\langle \frac{1}{\mathbf{D}'}\frac{1}{k^2}\frac{k\Pi}{(k^2-2k\Pi)^2}\right\rangle$$

$$+\left\langle \frac{\frac{1-v}{2}u(1-u)k\Pi - m^2u^2}{\mathbf{D}'^2}\frac{1}{k^2}\frac{1}{k^2-2k\Pi}\right\rangle. \qquad (5\text{–}7.222)$$

The information supplied by adding (5–7.221) and (5–7.222) then simplifies (5–7.220) to

$$I = m^2eq\sigma F\left[u^2 + \frac{1-v}{2}u(1-u)\right]\left\langle \frac{1}{\mathbf{D}'^2}\frac{1}{k^2}\frac{1}{k^2-2k\Pi}\right\rangle, \qquad (5\text{–}7.223)$$

in agreement with (5–7.214).

To indicate the caution with which the elementary argument should be applied, consider the integral

$$J = \Pi^\nu\left\langle \frac{1}{\mathbf{D}'}\frac{1}{k^2}\frac{1}{(\Pi-k)^2+m^2-eq\sigma F}ik^\lambda\sigma_{\lambda\nu}\right\rangle, \qquad (5\text{–}7.224)$$

which is also dimensionless. If the situation were completely analogous to that of (5–7.212), the result would only refer to the effect of inserting $eq\sigma F$ into $\mathbf{D}'$, where the associated coefficient vanishes, since the vector integral containing $k^\lambda$ is proportional to $\Pi^\lambda$, which enters into an effectively null commutation relation with $\Pi^\nu$. But this implicitly assumes that the outcome obtained when only $m^2 - eq\sigma F$ occurs is just a constant to be removed by a contact term. In fact, it necessarily involves the combination

$$\Pi^\nu i\Pi^\lambda\sigma_{\lambda\nu} = eq\sigma F, \qquad (5\text{–}7.225)$$

owing to the antisymmetrical rather than symmetrical pairing of $\Pi$ components.

The presence of this factor indicates that the basic integral can be stated as though $F = 0$:

$$\left\langle \frac{1}{\mathrm{D}'} \frac{1}{k^2} \frac{1}{k^2 - 2k\Pi} k^\lambda \right\rangle = \Pi^\lambda \left\langle \left( -\frac{k\Pi}{m^2} \right) \frac{1}{\mathrm{D}'} \frac{1}{k^2} \frac{1}{k^2 - 2k\Pi} \right\rangle, \quad (5\text{–}7.226)$$

which yields the non-vanishing result

$$J = eq\sigma F \left\langle \left( -\frac{k\Pi}{m^2} \right) \frac{1}{\mathrm{D}'} \frac{1}{k^2} \frac{1}{k^2 - 2k\Pi} \right\rangle. \quad (5\text{–}7.227)$$

Among the more complicated of the integrals is

$$K = \left\langle \gamma k \frac{1}{\mathrm{D}'} \frac{1}{k^2} \frac{1}{(\Pi - k)^2 + m^2 - eq\sigma F} k\Pi \right\rangle$$

$$= \gamma^\mu \left\langle k_\mu k_\nu \frac{1}{\mathrm{D}'} \frac{1}{k^2} \frac{1}{(\Pi - k)^2 + m^2 - eq\sigma F} \right\rangle \Pi^\nu. \quad (5\text{–}7.228)$$

Since the integral in the latter form is dimensionless, it might appear that the elementary argument is applicable. This is not quite true, however. The integral is meaningless without its accompanying contact term, which results in an additional contribution. Let us concentrate on that effect by supposing that $m^2$ has everywhere been supplied with its partner $-eq\sigma F$. After removing a factor in $\mathrm{D}'$ to make the coefficient of $k^2$ unity, we get an expression of the form

$$-i\left[ u\frac{1-v}{2}\left(1 - u\frac{1-v}{2}\right)\right]^{-1} \int ds\, s^2\, dp\, \gamma^\mu$$

$$\times \left\langle k_\mu k_\nu \exp\left\{ -is\left[ (k - f(p)\Pi)^2 + g(p)(m^2 - eq\sigma F) \right] \right\} \right\rangle \Pi^\nu, \quad (5\text{–}7.229)$$

where the symbol $p$ stands for the two additional parameters, say $w$ and $y$, which are such that

$$\int dp = \int_0^1 dw\, w \int_0^1 dy = \frac{1}{2}. \quad (5\text{–}7.230)$$

Concerning the functions $f(p)$, $g(p)$, we need only remark that $g(p)$ is always positive. As in the result of (5–7.150), we have, effectively,

$$k_\mu k_\nu \to -\frac{i}{2s} g_{\mu\nu} + f(p)\Pi_\mu \cdot f(p)\Pi_\nu. \quad (5\text{–}7.231)$$

The elementary argument, leading to no $\sigma F$ contribution, does apply to the well-defined integral produced by the second of these terms, where the flanking factors of $\gamma$ and $\Pi$ finally combine as $\gamma\Pi \to -m$. What remains is

$$-\left[u\frac{1-v}{2}\left(1-u\frac{1-v}{2}\right)\right]^{-1}\frac{i}{(4\pi)^2}\frac{m}{2}$$

$$\times\int dp\int_0^\infty\frac{ds}{s}\{\exp[-isg(p)(m^2-eq\sigma F)]-\exp[-isg(p)m^2]\},\quad (5\text{–}7.232)$$

where the necessary contact term is now explicit. The $s$-integral that appears here is simply

$$\log\frac{m^2}{m^2-eq\sigma F}\cong\frac{eq\sigma F}{m^2},\quad (5\text{–}7.233)$$

and thus

$$K=-\frac{i}{(4\pi)^2}\frac{eq\sigma F}{4m}\frac{1}{u\dfrac{1-v}{2}\left(1-u\dfrac{1-v}{2}\right)}+K_F,\quad (5\text{–}7.234)$$

where

$$K_F=-\left[u^2+\frac{1-v}{2}u(1-u)\right]\left\langle\gamma k\frac{1}{\mathbf{D}'^2}\frac{1}{k^2}\frac{1}{k^2-2k\Pi}k\Pi\right\rangle eq\sigma F\quad (5\text{–}7.235)$$

cancels the $\sigma F$ terms that have been inserted in $\mathbf{D}'$.

Concerning the last integral, we note that the form

$$\left\langle k_\mu k_\nu\frac{1}{\mathbf{D}'^2}\frac{1}{k^2}\frac{1}{k^2-2k\Pi}\right\rangle=ag_{\mu\nu}+b\frac{\Pi_\mu\Pi_\nu}{m^2}\quad (5\text{–}7.236)$$

leads to

$$\left\langle\gamma k\frac{1}{\mathbf{D}'^2}\frac{1}{k^2}\frac{1}{k^2-2k\Pi}k\Pi\right\rangle=m(b-a),\quad (5\text{–}7.237)$$

in which

$$b-a=\left\langle\left(\frac{k\Pi}{m}\right)^2\frac{1}{\mathbf{D}'^2}\frac{1}{k^2}\frac{1}{k^2-2k\Pi}\right\rangle.\quad (5\text{–}7.238)$$

Accordingly, we get

$$K_F=-\left[u^2+\frac{1-v}{2}u(1-u)\right]\frac{eq\sigma F}{m}\left\langle(k\Pi)^2\frac{1}{\mathbf{D}'^2}\frac{1}{k^2}\frac{1}{k^2-2k\Pi}\right\rangle.\quad (5\text{–}7.239)$$

Another type of integral requiring some comment is

$$L = \left\langle k^2\gamma^\nu \frac{1}{\mathbf{D}'}\frac{1}{k^2}\frac{1}{(\Pi-k)^2+m^2-eq\sigma F} ik^\lambda \sigma_{\lambda\nu}\right\rangle$$

$$= -4\left\langle \gamma k\frac{1}{\mathbf{D}'}\frac{1}{(\Pi-k)^2+m^2}\right\rangle + \left\langle \gamma k\frac{1}{\mathbf{D}'}\frac{1}{(\Pi-k)^2+m^2-eq\sigma F}\right\rangle, \quad (5\text{–}7.240)$$

or, in effect,

$$-\tfrac{1}{3}L = \left\langle \gamma k\frac{1}{\mathbf{D}'}\frac{1}{(\Pi-k)^2+m^2+\frac{1}{3}eq\sigma F}\right\rangle. \quad (5\text{–}7.241)$$

We again consider the integral as it would be if only the combination $m^2 - eq\sigma F$ occurred. It has the form

$$-\left[u\frac{1-v}{2}\left(1-u\frac{1-v}{2}\right)\right]^{-1}$$

$$\times \int ds\, s\int_0^1 dy \left\langle \gamma k \exp\left\{-is\left[(k-f\Pi)^2+g\left(m^2-eq\sigma F\right)\right]\right\}\right\rangle, \quad (5\text{–}7.242)$$

where now it is necessary to know that

$$f(y) = 1 - \frac{u\dfrac{1+v}{2}}{1-u\dfrac{1-v}{2}}y. \quad (5\text{–}7.243)$$

After performing the $k$-integration and supplying the contact term, this reads

$$-\frac{i}{(4\pi)^2}\left[u\frac{1-v}{2}\left(1-u\frac{1-v}{2}\right)\right]^{-1} m$$

$$\times \int \frac{ds}{s}\, dy\, f(y)\left\{\exp\left[-isg\left(m^2-eq\sigma F\right)\right] - \exp\left[-isgm^2\right]\right\}$$

$$= -\frac{i}{(4\pi)^2}\frac{eq\sigma F}{m}\frac{1-\dfrac{1}{2}\dfrac{u\dfrac{1+v}{2}}{1-u\dfrac{1-v}{2}}}{u\dfrac{1-v}{2}\left(1-u\dfrac{1-v}{2}\right)}, \quad (5\text{–}7.244)$$

and

$$L = \frac{3i}{(4\pi)^2}\frac{eq\sigma F}{m}\frac{1 - \frac{1}{2}u - \frac{1}{2}u\frac{1-v}{2}}{u\frac{1-v}{2}\left(1 - u\frac{1-v}{2}\right)^2} + L_F. \qquad (5\text{–}7.245)$$

The latter object, which introduces the correct $\sigma F$ dependence, is

$$L_F = 3\left[u^2 + \frac{1-v}{2}u(1-u)\right]\left\langle \gamma k\frac{1}{\mathbf{D}'^2}\frac{1}{k^2 - 2k\Pi}\right\rangle eq\sigma F$$

$$+ 4\left\langle \gamma k\frac{1}{\mathbf{D}'}\frac{1}{(k^2 - 2k\Pi)^2}\right\rangle eq\sigma F. \qquad (5\text{–}7.246)$$

In both of the integrals that compose $L_F$ we can make the effective substitution

$$\gamma k \to \frac{k\Pi}{m}. \qquad (5\text{–}7.247)$$

We also remark that

$$\tfrac{1}{2}m\frac{\partial}{\partial m}\left\langle \frac{1}{\mathbf{D}'}\frac{1}{k^2 - 2k\Pi}\right\rangle = \left\langle \frac{1}{\mathbf{D}'}\frac{k\Pi}{(k^2 - 2k\Pi)^2}\right\rangle$$

$$-\left\langle \frac{m^2u^2 - \frac{1-v}{2}u(1-u)k\Pi}{\mathbf{D}'^2}\frac{1}{k^2 - 2k\Pi}\right\rangle, \qquad (5\text{–}7.248)$$

where it is important not to be misled by the dimensionless nature of the integral on the left. The form of this integral is indicated by

$$\left\langle \frac{1}{\mathbf{D}'}\frac{1}{k^2 - 2k\Pi}\right\rangle = -\left[u\frac{1-v}{2}\left(1 - u\frac{1-v}{2}\right)\right]^{-1}$$

$$\times \int ds\, s\, dy \left\langle \exp\left\{-is\left[(k - f\Pi)^2 + gm^2\right]\right\}\right\rangle$$

$$= \left[u\frac{1-v}{2}\left(1 - u\frac{1-v}{2}\right)\right]^{-1}\frac{i}{(4\pi)^2}\int \frac{ds}{s}\, dy\, e^{-isgm^2}, \qquad (5\text{–}7.249)$$

and

$$\tfrac{1}{2}m\frac{\partial}{\partial m}\int\frac{ds}{s}e^{-isgm^2} = -igm^2\int_0^\infty ds\, e^{-isgm^2} = -1. \tag{5-7.250}$$

As a result,

$$\begin{aligned} L_F = &-\frac{i}{(4\pi)^2}\frac{4}{u\frac{1-v}{2}\left(1-u\frac{1-v}{2}\right)}\frac{eq\sigma F}{m} \\ &+\left(3u^2-\frac{1-v}{2}u(1-u)\right)\left\langle k\Pi\frac{1}{\mathbf{D}'^2}\frac{1}{k^2-2k\Pi}\right\rangle\frac{eq\sigma F}{m} \\ &+4m^2u^2\left\langle\frac{1}{\mathbf{D}'^2}\frac{1}{k^2-2k\Pi}\right\rangle\frac{eq\sigma F}{m}. \end{aligned} \tag{5-7.251}$$

We have now illustrated the various labor-saving devices that can be used to reduce the remainder of $M_1^{(2)}$ to the final integration stage. Since the number of terms has become fairly large, we give this list as it appears after some algebraic combination has been effected. It is stated with the aid of the following set of $k$-integrals ($n = 1, 2$):

$$\begin{aligned} &\left\langle\frac{(m^2)^{n-1}}{\mathbf{D}'^n}\frac{1}{k^2}\frac{1}{k^2-2k\Pi}\left[1, \frac{k\Pi}{m^2}, \left(\frac{k\Pi}{m^2}\right)^2, \frac{k^2}{m^2}, \frac{k^2}{m^2}\frac{k\Pi}{m^2}\right]\right\rangle \\ &\equiv \frac{i}{(4\pi)^2}\frac{1}{m^2}[A_n, B_n, C_n, D_n, E_n], \end{aligned} \tag{5-7.252}$$

where only the $A$ and $B$ types appear for $n = 1$. There is also another set of integrals, which are produced by the substitution

$$\mathbf{D}' \to m^2u^2 + \frac{1-v}{2}u(1-u)(k^2-2k\Pi); \tag{5-7.253}$$

lowercase letters designate this type. This final contribution to $c_2$ is given, in terms of the double parametric integration

$$\int \equiv \int_0^1 du\int_{-1}^1 \tfrac{1}{2}\, dv, \tag{5-7.254}$$

by

$$
\begin{aligned}
(c_2)_8 = {} & \int u^2(1-u)(1-v)A_1 + \int u\left[-4+8u-3u^2-u(1-3u)(1-v)\right]B_1 \\
& + \int u^3\left[-\tfrac{20}{3}+\tfrac{8}{3}u+2u^2+\tfrac{7}{3}u(1-u)(1-v)\right]A_2 \\
& + \int \tfrac{4}{3}u^3[-1-2u+u(2+u)(1-v)]B_2 \\
& + \int u^2\Big[\tfrac{8}{3}(1-v)-u\left(\tfrac{8}{3}+2(1-v)+\tfrac{11}{6}(1-v)^2\right) \\
& + u^2(1-v)\left(-\tfrac{17}{3}+\tfrac{13}{3}(1-v)\right)+3u^3(1-v)\left(1-\tfrac{1}{2}(1-v)\right)\Big]C_2 \\
& + \int u^3\Big[-\tfrac{2}{3}+\tfrac{2}{3}(1-v)-\tfrac{2}{3}(1-v)^2+u(1-v) \\
& \times\left(\tfrac{4}{3}+\tfrac{2}{3}(1-v)+\tfrac{7}{12}(1-v)^2\right)-u^2(1-v)^2\left(\tfrac{3}{2}+\tfrac{7}{12}(1-v)\right)\Big]D_2 \\
& + \int u^2\frac{1-v}{2}\Big[-\tfrac{8}{3}+u\left(\tfrac{14}{3}+\tfrac{13}{6}(1-v)\right) \\
& + u^2\left(3-\tfrac{1}{6}(1-v)-\tfrac{1}{6}(1-v)^2\right)+u^3(1-v)\left(-3+\tfrac{1}{6}(1-v)\right)\Big]E_2 \\
& + \int u(1-u)(1+u)\left[4b_1+2u^2b_2-3u^2d_2\right] + \int\left[-2u+6\frac{1+u}{1-v}\right].
\end{aligned}
\tag{5-7.255}
$$

The form of the very last term emphasizes that the integrand, although here presented as the sum of a number of different contributions, should be considered as a unit. Indeed, it is a useful check of this rather involved algebra to verify that singularities at various end points of the integration, which occur in individual terms, are cancelled when the whole structure is examined.

Turning to the explicit forms of the functions $A_1, \ldots, E_2$, we recall that the first of these is already known [Eq. (5–7.211)]. We present it again, now written as

$$
A_1 = \frac{1}{u(1-u)}\frac{2}{1-v}L_1 + \frac{1}{u^2}\frac{2}{1+v}(L_2+L_3) + \frac{1}{u^2}\frac{2}{1-v}L_3, \tag{5-7.256}
$$

where

$$
L_1 = \log\frac{\dfrac{1-v}{2}+uv}{u\left(1-u\dfrac{1-v}{2}\right)}, \qquad L_2 = \log\frac{\dfrac{1-v}{2}+uv}{\dfrac{1-v}{2}}, \qquad L_3 = \log\frac{1}{1-u\dfrac{1-v}{2}}. \tag{5-7.257}
$$

The other functions of the capital class can now be found by combining direct integration with the use of identities such as

$$A_1 = u^2A_2 - u(1-u)(1-v)B_2 + u\frac{1-v}{2}\left(1 - u\frac{1-v}{2}\right)D_2 \quad (5\text{–}7.258)$$

and

$$\frac{\partial}{\partial u}\left(u^2A_1\right) = u^2\frac{1-v}{2}(D_2 - 2B_2),$$
$$2\frac{\partial}{\partial v}A_1 = u(1-u)(D_2 - 2B_2) + u^2vD_2. \quad (5\text{–}7.259)$$

The results are

$$B_1 = \frac{2}{(1-u)^2(1-v)^2}L_1 - \frac{2}{u^2(1+v)^2}(L_2 + L_3) - \frac{2}{u^2(1-v)^2}L_3,$$
$$A_2 = \frac{1}{u^4v}L_2, \quad (5\text{–}7.260)$$
$$B_2 = -\frac{1}{u^2}B_1 - \frac{2}{u^4v}L_2 + \frac{2}{u^4(1+v)}(L_2 + L_3) + \frac{2}{u^4(1-v)}L_3,$$

and

$$C_2 = -\frac{4}{u(1-u)^3(1-v)^3}L_1 + \frac{2}{u^2(1-u)(1-v)^2}\frac{1}{1 - u\frac{1-v}{2}} + \tfrac{1}{2}E_2,$$
$$D_2 = -\frac{4}{u^4v}L_2 + \frac{4}{u^4}\left(\frac{1}{(1+v)^2} + \frac{1}{1+v}\right)(L_2 + L_3)$$
$$+ \frac{4}{u^4}\left(\frac{1}{(1-v)^2} + \frac{1}{1-v}\right)L_3, \quad (5\text{–}7.261)$$
$$E_2 = \frac{8}{u^4v}L_2 - \frac{8}{u^4}\left(\frac{1}{(1+v)^3} + \frac{1}{(1+v)^2} + \frac{1}{1+v}\right)(L_2 + L_3)$$
$$- \frac{8}{u^4}\left(\frac{1}{(1-v)^3} + \frac{1}{(1-v)^2} + \frac{1}{1-v}\right)L_3$$
$$+ \frac{8}{u^3(1-v)^2(1+v)}\frac{1}{1 - u\frac{1-v}{2}}.$$

The lowercase functions appear in the single combination

$$4b_1 + 2u^2b_2 - 3u^2d_2 = \frac{\delta}{u^2}\left[\delta \log\frac{1+\delta}{\delta} - 4\right], \tag{5–7.262}$$

where, here,

$$\delta = \frac{u}{1-u}\frac{2}{1-v}. \tag{5–7.263}$$

The last two sets of terms in (5–7.255) combine into

$$\begin{aligned}(c_2)_{\delta}' &= \int\left[(1+u)\frac{2}{1-v}\left(\delta\log\frac{1+\delta}{\delta} - 1\right) - 2u\right]\\ &= \int_0^1 du\left\{-\frac{1+u}{1-u}\log\frac{1}{u} + 1 - u\right\}\\ &= -2\frac{\pi^2}{6} + \frac{3}{2}.\end{aligned} \tag{5–7.264}$$

Of the remaining types, perhaps the most accessible are the integrals of the form $[w = \frac{1}{2}(1-v)]$

$$\begin{aligned}I_{kl} &= \int_0^1 du\int_0^1 dw\, u^{k-1}w^{l-1}\log\frac{1}{1-uw}\\ &= \sum_{n=1}^{\infty}\frac{1}{n}\frac{1}{n+k}\frac{1}{n+l},\end{aligned} \tag{5–7.265}$$

where the last version is produced by expanding the logarithm. The specific examples encountered here for $k, l \geqslant 0$ are enumerated by

$$I_{00} = \sum_{n=1}^{\infty}\frac{1}{n^3} = \zeta(3), \tag{5–7.266}$$

which introduces the Riemann zeta function of argument three; then $(k > 0)$

$$\begin{aligned}I_{0k} &= \sum_{n=1}^{\infty}\frac{1}{n^2}\frac{1}{n+k} = \frac{1}{k}\sum_{n=1}^{\infty}\left[\frac{1}{n^2} - \frac{1}{k}\left(\frac{1}{n} - \frac{1}{n+k}\right)\right]\\ &= \frac{1}{k}\frac{\pi^2}{6} - \frac{1}{k^2}\sum_{n=1}^{k}\frac{1}{n},\end{aligned} \tag{5–7.267}$$

followed by

$$I_{kk} = \frac{1}{k}\sum_{n=1}^{\infty}\left[-\frac{1}{(n+k)^2} + \frac{1}{k}\left(\frac{1}{n} - \frac{1}{n+k}\right)\right]$$

$$= -\frac{1}{k}\left[\frac{\pi^2}{6} - \sum_{n=1}^{k}\frac{1}{n^2}\right] + \frac{1}{k^2}\sum_{n=1}^{k}\frac{1}{n}, \tag{5–7.268}$$

and finally ($k \neq l > 0$)

$$I_{kl} = \frac{1}{l-k}\sum_{n=1}^{\infty}\left[\frac{1}{k}\left(\frac{1}{n} - \frac{1}{n+k}\right) - \frac{1}{l}\left(\frac{1}{n} - \frac{1}{n+l}\right)\right]$$

$$= \frac{1}{l-k}\left[\frac{1}{k}\sum_{n=1}^{k}\frac{1}{n} - \frac{1}{l}\sum_{n=1}^{l}\frac{1}{n}\right]. \tag{5–7.269}$$

A related set of numbers is provided by the integrals

$$J_{kl} = \int_0^1 du \int_0^1 dw\, u^k w^l \frac{1}{1-uw} = \sum_{n=1}^{\infty}\frac{1}{n+k}\frac{1}{n+l}. \tag{5–7.270}$$

The various possibilities here, for $k, l \geqslant 0$, are set out as

$$J_{00} = \frac{\pi^2}{6}, \qquad J_{0k} = \frac{1}{k}\sum_{n=1}^{k}\frac{1}{n}, \qquad J_{kk} = \frac{\pi^2}{6} - \sum_{n=1}^{k}\frac{1}{n^2},$$

$$J_{kl} = \frac{1}{l-k}\left[\sum_{n=1}^{l}\frac{1}{n} - \sum_{n=1}^{k}\frac{1}{n}\right]. \tag{5–7.271}$$

One also meets forms of these integrals in which $l < 0$, specifically, $l = -1, -2$. To deal with them, we separate out the first, or the first two terms of the series in (5–7.265) and get

$$I_{kl} = \int du\, dw\, u^k w^l + I_{1\,k+1\,l+1}, \tag{5–7.272}$$

or

$$I_{kl} = \int du\, dw\left(u^k w^l + \tfrac{1}{2}u^{k+1}w^{l+1}\right) + I_{2\,k+2\,l+2}, \tag{5–7.273}$$

which introduces examples of the summation

$$I_{jkl} = \sum_{n=1}^{\infty}\frac{1}{n+j}\frac{1}{n+k}\frac{1}{n+l}. \tag{5–7.274}$$

The latter is specified by (again, the zero value and identity of the indices are handled separately)

$$I_{000} = \zeta(3), \qquad I_{00k} = I_{0k}, \qquad I_{0kk} = I_{kk}, \qquad I_{0kl} = I_{kl}, \qquad (5\text{–}7.275)$$

and by

$$I_{kkk} = \zeta(3) - \sum_{n=1}^{k} \frac{1}{n^3},$$

$$I_{kkl} = \frac{1}{l-k} \sum_{n=1}^{\infty} \left( \frac{1}{(n+k)^2} - \frac{1}{n+k} \frac{1}{n+l} \right) = \frac{1}{l-k} ( J_{kk} - J_{kl}),$$

$$I_{jkl} = \frac{1}{l-k} ( J_{jk} - J_{jl})$$

$$= \frac{1}{j-k} \frac{1}{l-j} \sum_{n=1}^{j} \frac{1}{n} + \frac{1}{k-l} \frac{1}{j-k} \sum_{n=1}^{k} \frac{1}{n} + \frac{1}{l-j} \frac{1}{k-l} \sum_{n=1}^{l} \frac{1}{n}. \qquad (5\text{–}7.276)$$

In the analogous situations for $J_{kl}$, we have the simpler relations

$$J_{kl} = \int du\, dw\, u^k w^l + J_{k+1\, l+1}, \qquad (5\text{–}7.277)$$

or

$$J_{kl} = \int du\, dw\, ( u^k w^l + u^{k+1} w^{l+1}) + J_{k+2\, l+2}. \qquad (5\text{–}7.278)$$

For the application of these results, we isolate all terms having $L_3$ as a factor, and also retain those pieces of

$$\frac{8}{u^3(1-v)^2(1+v)} \frac{1}{1 - u\dfrac{1-v}{2}}$$

$$= \frac{1}{u^3} \left( \frac{1}{w^2} + \frac{1}{w} \right) \frac{1}{1-uw} + \frac{2}{u^3} \frac{1}{1+v} \frac{1}{1 - u\dfrac{1-v}{2}} \qquad (5\text{–}7.279)$$

that have inverse $w$-factors. The net contribution obtained in this way is

$$(c_2)_{L_3} = -2\zeta(3) + \left(-23 - \tfrac{1}{3}\right)\frac{\pi^2}{6} + 37 + \tfrac{1}{3} + \tfrac{1}{4} + \tfrac{1}{8}, \qquad (5\text{–}7.280)$$

where we have set aside, for eventual cancellation elsewhere, the singular integral

$$-\frac{1}{2}\int \frac{dv}{1-v}. \qquad (5\text{–}7.281)$$

Next we consider the terms having $L_2$ as a factor. Rather untypically, only a relatively small number of them survive after heavy cancellation; the explicit structure is

$$(c_3)_{L_2} = \int \left[-\frac{16}{u} + 4\frac{v}{u} + 5v + 3v^2 + 8u - 3uv - 3uv^2\right] L_2. \qquad (5\text{–}7.282)$$

These integrals are uniformly evaluated by partial integration on $v$, combined with the introduction of the variable

$$z = 2u - 1, \qquad -1 < z < 1. \qquad (5\text{–}7.283)$$

As a simple example, consider

$$\int vL_2 = \int_0^1 du \int_{-1}^1 d\left(\frac{v^2-1}{4}\right) \log\left(1 + \frac{2uv}{1-v}\right)$$

$$= \int_{-1}^1 \frac{dv}{2} \int_{-1}^1 \frac{dz}{2} \frac{1}{2} \frac{(1+z)(1+v)}{1+vz} = \frac{1}{2}, \qquad (5\text{–}7.284)$$

which exploits the possibility of omitting powers of $v$ and $z$ that are odd under the reflection of both variables. Another example is

$$\int \frac{1}{u} L_2 = \int_0^1 \frac{du}{u} \int_{-1}^1 d\left(\frac{v-1}{2}\right) \log\left(1 + \frac{2uv}{1-v}\right)$$

$$= -\int_0^1 \frac{du}{u} \log\frac{1}{1-u} + 2\int_{-1}^1 \frac{dv}{2} \int_{-1}^1 \frac{dz}{2} \frac{1}{1+vz}. \qquad (5\text{–}7.285)$$

Here we meet the familiar integral

$$\int_0^1 \frac{du}{u} \log\frac{1}{1-u} = \sum_{n=1}^{\infty} \frac{1}{n^2} = \frac{\pi^2}{6}, \qquad (5\text{–}7.286)$$

and the related one,

$$2\int_{-1}^{1}\frac{dv}{2}\int_{-1}^{1}\frac{dz}{2}\frac{1}{1+vz}=\int_{-1}^{1}\frac{dv}{2}\frac{1}{v}\log\frac{1+v}{1-v}$$

$$=2\sum_{n=0}^{\infty}\frac{1}{(2n+1)^2}=\frac{3}{2}\frac{\pi^2}{6},\qquad(5\text{–}7.287)$$

which yields

$$\int\frac{1}{u}L_2=\frac{1}{2}\frac{\pi^2}{6}.\qquad(5\text{–}7.288)$$

The outcome obtained in this manner is

$$(c_2)_{L_2}=\left(-1-\tfrac{7}{16}\right)\frac{\pi^2}{6}-3-\frac{1}{8}.\qquad(5\text{–}7.289)$$

The terms containing $L_1$ that are of greatest difficulty occur in $C_2$, where the following combination appears:

$$\Gamma_1=-4\frac{u}{(1-u)^3(1-v)^3}L_1+\frac{2}{(1-u)(1-v)^2}\frac{1}{1-u\dfrac{1-v}{2}}.\qquad(5\text{–}7.290)$$

To appreciate its significance, one must know the limiting behaviors of

$$L_1=\log\frac{\dfrac{1-v}{2}+uv}{u\left(1-u\dfrac{1-v}{2}\right)}\qquad(5\text{–}7.291)$$

at the boundaries of the variables $u$ and $v$. These are

$$u\to1:\quad L_1\to(1-u)^2\frac{1-v}{1+v},\qquad u\to0:\quad L_1\to\log\frac{(1-v)/2}{u}$$

$$v\to1:\quad L_1\to\frac{(1-u)^2}{u}\frac{1-v}{2},\qquad v\to-1:\quad L_1\to\log\frac{1}{u}.\qquad(5\text{–}7.292)$$

As a consequence, $\Gamma_1$ has no singularity at $u=1$. But it behaves as $(1-v)^{-1}$ for $v\to1$; that singularity must be isolated. The technical problem in evaluating an integral like $\int u\Gamma_1$ is removing the large inverse powers of $1 - u$ and $1 - v$ without thereby introducing spurious singularities, specifically, at $v = -1$. Here is one procedure for that purpose.

We write

$$u\Gamma_1 = -\frac{4}{(1-v)^3}\frac{\partial}{\partial u}\left[\left(\frac{1}{2}\frac{u^2}{(1-u)^2} - \frac{u}{1-u} + \log\frac{1}{1-u}\right)L_1 + \frac{1}{2} - \frac{1}{2}\frac{1}{1 - u\dfrac{1-v}{2}}\right]$$
$$+ \frac{1}{(1-v)^2}\left[\frac{u^2}{(1-u)^2} + 2\log\frac{1}{1-u}\right]\frac{1}{1 - u\dfrac{1-v}{2}}$$
$$- \frac{1}{(1-v)^2}\frac{1}{\left(1 - u\dfrac{1-v}{2}\right)^2}$$
$$- \frac{2}{(1-v)^2}\left[\frac{1}{2}\frac{u}{(1-u)^2} - \frac{1}{1-u} + \frac{1}{u}\log\frac{1}{1-u}\right]\frac{1}{\dfrac{1-v}{2} + uv}, \tag{5–7.293}$$

where the quantity subject to differentiation has been contrived to vanish at both limits of $u$. We also decompose the other terms in partial fractions in order to exhibit the powers $(1 - v)^{-2}$ and $(1 - v)^{-1}$. This yields

$$\int u\Gamma_1 = \int\frac{dv}{(1-v)^2}\int_0^1 du\left[\frac{1}{u} - \frac{1}{u^2}\log\frac{1}{1-u} + \log\frac{1}{1-u}\right]$$
$$+ \int\frac{dv}{1-v}\frac{1}{2}\int_0^1 du\left\{\frac{1}{2}\frac{u^3}{(1-u)^2} + u\log\frac{1}{1-u} - \frac{u-\frac{1}{2}}{u^2}\left[\frac{u}{(1-u)^2} - \frac{2}{1-u} + \frac{2}{u}\log\frac{1}{1-u}\right] - u\right\}$$
$$+ \int_0^1 du\left\{\frac{u}{4}\left[\frac{u^2}{(1-u)^2} + 2\log\frac{1}{1-u}\right]\log\frac{1}{1-u} - \tfrac{1}{2}u\log\frac{1}{1-u} - \frac{1}{4}\frac{u^2}{1-u} - \frac{u-\frac{1}{2}}{2u^2}\left[\frac{u}{(1-u)^2} - \frac{2}{1-u} + \frac{2}{u}\log\frac{1}{1-u}\right]\log\frac{u}{1-u}\right\}, \tag{5–7.294}$$

where the last term, which lacks a $v$-singularity, has already been integrated over $v$. The $u$-integral that is the coefficient of $(1-v)^{-2}$ vanishes, as it should; that of $(1-v)^{-1}$ turns out to be $\frac{1}{4}$. The last term is decomposed into a number of integrals like

$$\int_0^1 du \frac{1}{u^3}\left[\log\frac{1}{1-u} - u - \tfrac{1}{2}u^2\right]\log\frac{1}{u} = \frac{1}{2}\frac{\pi^2}{6} - \frac{3}{8} \qquad (5\text{–}7.295)$$

and

$$\int_0^1 du\left[\frac{1}{(1-u)^2}\log\frac{1}{u} - \frac{u^2}{1-u}\right] = \frac{5}{2} \qquad (5\text{–}7.296)$$

to give the result,

$$-\frac{8}{3}\int u\Gamma_1 = -\frac{1}{3}\int\frac{dv}{1-v} + 2\frac{\pi^2}{6} - \frac{10}{3}. \qquad (5\text{–}7.297)$$

Another procedure that is sometimes useful is based on the relation

$$(1-v)\Gamma_1 = -\frac{4u}{(1-u)^3}\frac{\partial}{\partial v}\left(\frac{L_1}{1-v}\right) + \frac{(2u-1)^2}{(1-u)^3}\frac{1}{\dfrac{1-v}{2}+uv}$$

$$-\frac{u^3}{(1-u)^3}\frac{1}{1-u\dfrac{1-v}{2}} + \frac{u}{1-u}\frac{1}{1-u\dfrac{1-v}{2}}, \qquad (5\text{–}7.298)$$

with its consequent $v$-integral

$$\int_{-1}^1 \frac{dv}{2}(1-v)\Gamma_1 = \frac{d}{du}\left[\frac{1}{1-u}\log\frac{1}{u}\right] + \frac{1}{u}. \qquad (5\text{–}7.299)$$

An example of a component integral evaluated this way is

$$\int u^2(1-v)\Gamma_1 = \frac{7}{2} - 2\frac{\pi^2}{6}. \qquad (5\text{–}7.300)$$

Numbers of the latter kind, combinations of fractions and multiples of $\pi^2$, are the rule, with two exceptions. They arise from one term:

$$\int (1-v)^2 u \Gamma_1 = \int_0^1 du \left\{ \frac{1}{(1-u)^2} \log \frac{1}{u} - \frac{1}{1-u} - \frac{1}{1-u} \log \frac{1}{u} \right.$$
$$+ \frac{1}{u-\frac{1}{2}} \log \frac{u}{1-u} + \frac{4}{u}\left(\log \frac{1}{1-u}\right)^2$$
$$\left. - \frac{2}{u} \log \frac{1}{1-u} \log \frac{1}{u} - \frac{2}{u-\frac{1}{2}} \log \frac{1}{1-u} \log \frac{u}{1-u} \right\}. \tag{5–7.301}$$

The first of the novel structures occurs in the integrals

$$\int_0^1 \frac{du}{u}\left(\log \frac{1}{1-u}\right)^2 = 2\int_0^1 \frac{du}{u} \log \frac{1}{1-u} \log \frac{1}{u}, \tag{5–7.302}$$

which are connected by partial integration, combined with the substitution $u \to 1-u$. The introduction of a new variable,

$$u = 1 - e^{-x}, \tag{5–7.303}$$

converts the initial integral into

$$\int_0^\infty dx\, x^2 \frac{e^{-x}}{1-e^{-x}} = \sum_{n=1}^{\infty} \int_0^\infty dx\, x^2 e^{-nx} = 2 \sum_{n=1}^{\infty} \frac{1}{n^3}, \tag{5–7.304}$$

and we recognize the zeta function of argument three,

$$\int_0^1 \frac{du}{u}\left(\log \frac{1}{1-u}\right)^2 = 2\zeta(3),$$
$$\int_0^1 \frac{du}{u} \log \frac{1}{1-u} \log \frac{1}{u} = \zeta(3). \tag{5–7.305}$$

For the last term of (5–7.301), we make the substitution

$$u = \tfrac{1}{2}(1+v), \tag{5–7.306}$$

which, incidentally, also evaluates [cf. Eq. (5–7.287)]

$$\int_0^1 du\,\frac{1}{u-\frac{1}{2}}\log\frac{u}{1-u} = \int_{-1}^1 \frac{dv}{v}\log\frac{1+v}{1-v} = 3\frac{\pi^2}{6}, \tag{5–7.307}$$

and get

$$I \equiv \int_0^1 du\,\frac{1}{u-\frac{1}{2}}\log\frac{1}{1-u}\log\frac{u}{1-u} = \int_{-1}^1 \frac{dv}{v}\log\frac{2}{1-v}\log\frac{1+v}{1-v}$$

$$= \int_{-1}^1 \frac{dv}{v}\log\frac{1+v}{1-v}\log 2 + \int_0^1 \frac{dv}{v}\left[\left(\log\frac{1}{1-v}\right)^2 - \left(\log\frac{1}{1+v}\right)^2\right]. \tag{5–7.308}$$

The information needed to determine the only new integral appearing here is supplied by

$$\int_0^1 \frac{dv}{v}\left(\log\frac{1}{1-v} + \log\frac{1}{1+v}\right)^2 = \int_0^1 \frac{dv}{v}\left(\log\frac{1}{1-v^2}\right)^2$$

$$= \frac{1}{2}\int_0^1 \frac{du}{u}\left(\log\frac{1}{1-u}\right)^2 = \zeta(3), \tag{5–7.309}$$

and, employing the substitution

$$\frac{1+v}{1-v} = e^x, \tag{5–7.310}$$

by

$$\int_0^1 \frac{dv}{v}\left(\log\frac{1}{1-v} - \log\frac{1}{1+v}\right)^2 = 2\int_0^\infty dx\,x^2\frac{e^{-x}}{1-e^{-2x}}$$

$$= 4\sum_{n=0}^{\infty}\frac{1}{(2n+1)^3} = \tfrac{7}{2}\zeta(3). \tag{5–7.311}$$

Thus,

$$\int_0^1 \frac{dv}{v}\left(\log\frac{1}{1+v}\right)^2 = \tfrac{1}{4}\zeta(3), \tag{5–7.312}$$

and

$$I = \tfrac{1}{2}\pi^2\log 2 + \tfrac{7}{4}\zeta(3), \tag{5–7.313}$$

which then yields

$$\int (1 - v^2) u \Gamma_1 = \tfrac{5}{2}\zeta(3) - \pi^2 \log 2 + 2\frac{\pi^2}{6} + 1. \tag{5–7.314}$$

It should also be remarked that the core of the structure which produces $\zeta(3)$ and $\pi^2 \log 2$ is found in the integral

$$\int 2\frac{L_1}{(1-u)(1-v)} = \tfrac{1}{2}\pi^2 \log 2 - \tfrac{5}{4}\zeta(3); \tag{5–7.315}$$

it is a somewhat more complicated analogue of

$$\int 2\frac{L_3}{(1-u)(1-v)} = \zeta(3). \tag{5–7.316}$$

The complete list for this type of contribution is

$$(c_2)_{L_1} = \tfrac{5}{2}\zeta(3) - \pi^2 \log 2 + \left(7 - \tfrac{1}{3} - \tfrac{1}{4}\right)\frac{\pi^2}{6} - 5 - \tfrac{1}{3} - \tfrac{1}{8}, \tag{5–7.317}$$

together with a singular $v$-integral that precisely cancels (5–7.281).

Finally, we come to the terms involving

$$L_2 + L_3 = \log \frac{\dfrac{1-v}{2} + uv}{\dfrac{1-v}{2}\left(1 - u\dfrac{1-v}{2}\right)} \equiv L_{23}, \tag{5–7.318}$$

which also includes the contribution of the last part of (5–7.279). The first thing to observe is this. Under the substitution

$$u \leftrightarrow \frac{1-v}{2}, \tag{5–7.319}$$

we have

$$L_{23} \leftrightarrow L_1, \tag{5–7.320}$$

which would seem to indicate that the two sets of terms should have been united. Unfortunately, they are of sufficiently different structure that no great simplification would thereby result (which is the general characteristic of this calculation). Nevertheless a few integrals are conveniently evaluated by this transformation.

The most important of these is

$$\int 2\frac{L_{23}}{u(1+v)} \to \int 2\frac{L_1}{(1-u)(1-v)} = \tfrac{1}{2}\pi^2 \log 2 - \tfrac{5}{4}\zeta(3), \qquad (5\text{–}7.321)$$

which also constitutes the complete source of these terms in the $L_{23}$ structures. For the rest, we proceed somewhat analogously to the $L_1$ calculation, but without using the substitution (5–7.319).

We start with the following combination extracted from the expression for $E_2$ [Eq. (5–7.261)],

$$\Gamma_{23} = -\frac{1}{u^4 w^3} L_{23} + \frac{1}{u^3 w}\frac{1}{1-u(1-w)}, \qquad (5\text{–}7.322)$$

where

$$w = \tfrac{1}{2}(1+v). \qquad (5\text{–}7.323)$$

On remarking that

$$\Gamma_{23} = \frac{1}{2u^4}\frac{\partial}{\partial w}\left(\frac{1-w^2}{w^2}L_{23}\right) - \frac{1}{u^2}\frac{1}{1-u}\frac{1}{1-u(1-w)}$$
$$+\frac{2u-1}{2u^3}\left(\frac{2-3u}{(1-u)^2}\frac{1}{1-u-w+2uw} + \frac{1}{(1-u)^2}\frac{1.}{1-u(1-w)}\right), \qquad (5\text{–}7.324)$$

we evaluate the $w$-integral:

$$\int_0^1 dw\,\Gamma_{23} = -\frac{1}{2u^3}\frac{1}{1-u} + \frac{3u-2}{2u^3(1-u^2)}\log\frac{1}{u} - \frac{1}{2u^4}\log\frac{1}{1-u}, \qquad (5\text{–}7.325)$$

from which we deduce such integrals as

$$\int 2u^3\Gamma_{23} = 1 - 4\frac{\pi^2}{6}. \qquad (5\text{–}7.326)$$

Another set of integrals is inferred from the relation

$$w\Gamma_{23} = \frac{1}{u^4}\frac{\partial}{\partial w}\left[\frac{1-w}{w}L_{23}\right] + \frac{2u-1}{u^3}\frac{1}{1-u}\frac{1}{1-u-w+2uw}$$
$$-\frac{1}{u^2}\frac{1}{1-u}\frac{1}{1-u(1-w)}, \qquad (5\text{–}7.327)$$

with its consequence

$$\int_0^1 dw\, w\Gamma_{23} = -\frac{1}{u^3}\frac{1}{1-u}\log\frac{1}{u}; \qquad (5\text{–}7.328)$$

an example is

$$-\int 4u^3 w\Gamma_{23} = 4\frac{\pi^2}{6}. \qquad (5\text{–}7.329)$$

The complete contribution of the $\Gamma_{23}$ class is

$$-\tfrac{2}{3}\left(\tfrac{1}{2}\pi^2\log 2 - \tfrac{5}{4}\zeta(3)\right) - \left(3 + \tfrac{1}{3} + \tfrac{1}{4} + \tfrac{1}{8}\right)\frac{\pi^2}{6} + 7 + \tfrac{1}{2} - \tfrac{1}{3}. \qquad (5\text{–}7.330)$$

The remainder of the integrals contain $L_{23}$ and various powers of $u$ and $w$. Examples of these that can be inferred from known results are

$$\int \frac{L_{23}}{uw^2} = \int 2\frac{L_1}{(1-u)^2(1-v)} = 2\frac{\pi^2}{6},$$
$$\int \frac{L_{23}}{u} = \int 2\frac{L_1}{1-v} = \frac{3}{2}\frac{\pi^2}{6} - 1. \qquad (5\text{–}7.331)$$

As for the others, the following are cited as representative:

$$\int \frac{L_{23}}{w^2} = \frac{1}{2}\frac{\pi^2}{6},$$
$$\int uwL_{23} = \frac{19}{32}\frac{\pi^2}{6} - \frac{11}{16}. \qquad (5\text{–}7.332)$$

The final outcome, combining all terms related to $L_{23}$, is

$$(c_2)_{L_{23}} = \tfrac{5}{2}\zeta(3) - \pi^2\log 2 + \left(23 + \tfrac{1}{4} + \tfrac{1}{6} - \tfrac{1}{16}\right)\frac{\pi^2}{6} - 37 - \tfrac{1}{8}. \qquad (5\text{–}7.333)$$

The artificial nature of the separation of

$$(c_2)''_s = (c_2)_{L_3} + (c_2)_{L_2} + (c_2)_{L_1} + (c_2)_{L_{23}} \qquad (5\text{–}7.334)$$

into the four parts exhibited in Eqs. (5–7.280, 289, 317, 333) is evident from the relative simplicity of their total:

$$(c_2)''_s = 3\zeta(3) - 2\pi^2\log 2 + 5\frac{\pi^2}{6} - 8. \qquad (5\text{–}7.335)$$

The other pieces of $c_2$ listed in Eqs. (5–7.54, 190, 197, 201, 264) have the sum

$$(c_2)_{\bar{G}} + (c_2)_{\zeta''} + (c_2)_{\alpha} + (c_2)_{\beta} + (c_2)_{\gamma} + (c_2)'_{\delta} = 5\frac{\pi^2}{6} + \frac{1}{4}; \quad (5\text{–}7.336)$$

we note separately the vacuum-polarization contribution [Eq. (5–7.6)]

$$(c_2)_{\text{v.pol.}} = -8\frac{\pi^2}{6} + 13 + \tfrac{2}{9}. \quad (5\text{–}7.337)$$

The time has come to add (5–7.335, 336, 337) and, at long last, infer the numerical coefficient of $(\alpha/2\pi)^2$ in the additional electron magnetic moment. Here it is:

$$c_2 = 3\zeta(3) - 2\pi^2 \log 2 + \tfrac{1}{3}\pi^2 + 5 + \tfrac{17}{36} = -1.31392, \quad (5\text{–}7.338)$$

which uses the number

$$\zeta(3) = 1.20206. \quad (5\text{–}7.339)$$

The implied value of the electron magnetic moment, based upon the nominal fine structure constant

$$1/\alpha = 137.036, \quad (5\text{–}7.340)$$

is

$$\mu_e = 1 + 0.00116141 - 0.00000177 = 1.00115964, \quad (5\text{–}7.341)$$

which is in exceptional agreement with a recent measurement,

$$\mu_e = 1.0011596577 \pm 0.0000000035. \quad (5\text{–}7.342)$$

To take seriously the tiny residual discrepancy ($\sim 2 \times 10^{-8}$) would require a knowledge of the $(\alpha/2\pi)^3$ correction, and more accurate information about the value of $\alpha$.

The improved value of the electron magnetic moment will have implications for various subjects already extensively discussed, but it seems reasonable not to go further into these matters now. The length of calculation required for the evaluation of these small effects has begun to outweigh the instructional value provided by seeing all the details of source theory at work on varied problems. Indeed, as the just-concluded magnetic moment calculation made distressingly evident, it eventually becomes unfeasible to display the computational details completely (yet, hopefully, enough clues have been provided to permit a relatively painless repetition of the labor).

Harold looks appreciative.

H.: Let me congratulate you on joining the very small club of people who have successfully performed this magnetic moment calculation, first correctly carried out by C. Sommerfield [Harvard Ph.D. Thesis, 1957; Ann. Phys. (N.Y.) 5, 26 (1958)]. How do you feel at this moment?

S.: Exhausted. And somewhat disappointed. While the source calculation, which is rather similar in spirit to that of Sommerfield, is vastly simpler than his (it is quite staggering to find him, at one stage, manipulating as many as seven parameters), the anticipated reduction of the algebraic structure prior to the final integration never quite materialized. As we noted, the eventual numerical form is not particularly complex, in contrast with its component elements, which suggests that some other way of organizing the calculation might be even more effective. It would be pleasant to find it, particularly if one wanted to press on to the $(\alpha/2\pi)^3$ calculation without resorting, as others have in desperation, to computer assistance. But that is not, for us, an immediate prospect.

## 5–8 PHOTON PROPAGATION FUNCTION III

Although we have just forsworn further large-scale electrodynamic computations (despite some earlier tentative promises), it may still be worthwhile to explore additional aspects of the non-causal calculational methods. It would, for example, be interesting to see how these techniques fare in comparison with the rather extensive causal calculations set out in Section 5–4, where effects of relative order $\alpha^2$ in the photon propagation function were discussed. That is the purpose of this section.

The coupling between two component electromagnetic fields, $A_a$ and $A_b$, that is produced by the exchange of a single pair of non-interacting spin-$\frac{1}{2}$ particles is conveyed by [compare the action expression (4–3.9)]

$$W_{ab}^{(2)} = \tfrac{1}{2}i \operatorname{Tr} eq\gamma A_a G_+ eq\gamma A_b G_+ + \text{c.t.}; \tag{5–8.1}$$

the superscript records the power of $e$ in the formula. Recalling the expansion [Eq. (3–12.23), for example]

$$G_+{}^A = G_+ + G_+ eq\gamma A G_+ + \cdots, \tag{5–8.2}$$

we recognize that (5–8.1) is also part of

$$\tfrac{1}{2}i \operatorname{Tr} eq\gamma A_a G_+{}^{A_b} + \text{c.t.} \tag{5–8.3}$$

that is linear in $A_b$, as we might have inferred directly from (4–8.19).

The next dynamical level introduces the exchange of a single virtual photon. That is indicated in (5–8.3) by replacing the propagation function $G_+{}^A$ with the modified propagation function $\bar{G}_+{}^A$. Correct to the relative order $\alpha$, we have [Eqs. (5–6.2, 3)]

$$\bar{G}_+{}^A = G_+{}^A - G_+{}^A M^A G_+{}^A, \tag{5–8.4}$$

where, apart from the contact term,

$$M^A = ie^2 \int \frac{(dk)}{(2\pi)^4} \gamma^\nu \frac{1}{k^2} \frac{1}{\gamma(\Pi - k) + m} \gamma_\nu$$

$$\equiv ie^2 \left\langle \gamma^\nu \frac{1}{k^2} \frac{1}{\gamma(\Pi - k) + m} \gamma_\nu \right\rangle. \tag{5–8.5}$$

The desired effect of order $\alpha^2$ is then given by

$$W_{ab}^{(4)} = \tfrac{1}{2}i \operatorname{Tr} eq\gamma A_a(-G_+ M G_+)^{A_b} + \text{c.t.}, \tag{5–8.6}$$

where it is understood that only the term linear in $A_b$ is retained. This is made somewhat more explicit by writing (5–8.6) as

$$\begin{aligned} W_{ab}^{(4)} = {} & \tfrac{1}{2}i \operatorname{Tr} eq\gamma A_a G_+ eq\gamma A_b(-G_+ M_0 G_+) \\ & + \tfrac{1}{2}i \operatorname{Tr} eq\gamma A_a(-G_+ M_0 G_+) eq\gamma A_b G_+ \\ & - \tfrac{1}{2}i \operatorname{Tr} eq\gamma A_a G_+ M^{A_b} G_+ + \text{c.t.}, \end{aligned} \tag{5–8.7}$$

where $M_0$ refers to the form of $M$ in the absence of an electromagnetic potential.

We recognize that the first two terms of this formula state the effect of modifying the propagation functions of the two non-interacting particles. This was discussed in Section 5–4 as part of the three-particle causal exchange mechanism. Our major concern now is with the last part of (5–8.7), specifically, with the determination of the contact terms that are related to the internal single-photon exchange act. When $A_b$ is made explicit in the third term of (5–8.7), the latter can be written out as

$$-\tfrac{1}{2}i\,\mathrm{Tr}\left\langle G_+(\Pi)eq\gamma A_a G_+(\Pi)\gamma^\nu \frac{ie^2}{k^2} G_+(\Pi - k)eq\gamma A_b G_+(\Pi - k)\gamma_\nu\right\rangle. \quad (5\text{–}8.8)$$

The notation used here facilitates the consideration of an additional electromagnetic field, which, as in the preceding section, we exploit to help clarify the nature of the contact terms. An infinitesimal variation of that ambient field induces the following change in (5–8.8):

$$-\tfrac{1}{2}i\,\mathrm{Tr}\,\delta[G_+(\Pi)eq\gamma A_a G_+(\Pi)]\left\langle \gamma^\nu \frac{ie^2}{k^2} G_+(\Pi - k)eq\gamma A_b G_+(\Pi - k)\gamma_\nu\right\rangle$$

$$-\tfrac{1}{2}i\,\mathrm{Tr}\left\langle \gamma^\nu \frac{ie^2}{k^2} G(\Pi - k)eq\gamma A_a G_+(\Pi - k)\gamma_\nu\right\rangle \delta[G_+(\Pi)eq\gamma A_b G_+(\Pi)]. \quad (5\text{–}8.9)$$

The cyclic property of the trace, combined with the transformation $k \to -k$, has been used in producing the form of the second term. In each of the two contributions we see the infinitesimal action of an electromagnetic field modifying the process of single-photon exchange. The appropriate contact terms are known, and are illustrated in a related context in Eqs. (5–7.42, 43), for example. The alternative view of (5–8.9), where single-photon exchange is influenced by the differential action of two fields, requires no contact terms. Thus, the contact terms that supplement (5–8.9) are

$$\begin{aligned} &-\tfrac{1}{2}i\,\mathrm{Tr}\,\delta[G_+ eq\gamma A_a G_+]\zeta_c eq\gamma A_b \\ &-\tfrac{1}{2}i\,\mathrm{Tr}\,\zeta_c eq\gamma A_a \delta[G_+ eq\gamma A_b G_+] \\ &= 2\zeta_c\delta[-\tfrac{1}{2}i\,\mathrm{Tr}\,G_+ eq\gamma A_a G_+ eq\gamma A_b]. \end{aligned} \quad (5\text{–}8.10)$$

As in the preceding section, half of the contact term can be absorbed in producing the complete structure of $M$ in Eq. (5–8.6), which gives the following determination of the contact term in that formula:

$$\text{c.t.} = -\zeta_c \tfrac{1}{2}i\,\mathrm{Tr}\,eq\gamma A_a G_+ eq\gamma A_b G_+. \quad (5\text{–}8.11)$$

In addition to this "internal" contact term, there are, of course, "external" contact terms which exploit the non-overlapping arrangement of the fields $A_a$ and $A_b$ to satisfy the physical requirements of the theory.

A useful analysis of the structure of $W_{ab}^{(4)}$ follows from the known decomposition of $M$ into two parts [Eq. (5–7.58)]

$$M = M_0(\gamma\Pi) + M_1, \tag{5–8.12}$$

where $M_0$ is the gauge-covariant form of $M$ in the absence of an electromagnetic field, and $M_1$ depends explicitly on field strengths. We recognize in $-G_+M_0G_+$ the modification in the free-particle propagation function that is given by [cf. Eqs. (5–7.55, 56)]

$$-G_+M_0G_+ = \int dM' \left[\frac{A_+(M')}{\gamma\Pi + M'} + \frac{A_-(M')}{\gamma\Pi - M'}\right], \tag{5–8.13}$$

where

$$A_\pm(M') = \frac{\alpha}{4\pi}\frac{1}{M'}\,`\!\left(1 - \frac{m^2}{M'^2}\right)\!'\left(1 \mp \frac{2mM'}{(M' \mp m)^2}\right), \tag{5–8.14}$$

and we have indicated the necessity for an infra-red modification of the kinematical factor near the threshold at $M' = m$:

$$M' \gtrsim m + \mu: \qquad `\!\left(1 - \frac{m^2}{M'^2}\right)\!' \cong \frac{2}{m}\left[(M' - m)^2 - \mu^2\right]^{1/2}. \tag{5–8.15}$$

The use of the individual propagation functions $(\gamma\Pi \pm M')^{-1}$ in (5–8.6) reduces the problem to the known one of two-particle exchange without interaction, but with the mass substitution $m \to M'$ (the algebraic sign of $M'$ is without effect). Thus, we get a first contribution to the modified weight function of the photon propagation function, $a(M^2)$, by combining the elementary result [Eq. (5–4.186), for example, with $m \to M'$] and the spectral distribution of the mass $M'$ that is inferred from (5–8.14),

$$A_+(M') + A_-(M') = \frac{\alpha}{2\pi}\frac{1}{M'}\,`\!\left(1 - \frac{m^2}{M'^2}\right)\!'\left(1 - \frac{4m^2M'^2}{(M'^2 - m^2)^2}\right). \tag{5–8.16}$$

This gives

$$M^2\,\delta a(M^2)\big|_I = \frac{\alpha^2}{6\pi^2}\int_{m+\mu}^{M/2}\frac{dM'}{M'}\,`\!\left(1 - \frac{m^2}{M'^2}\right)\!'\left(1 - \frac{4m^2M'^2}{(M'^2 - m^2)^2}\right)\left(1 - \frac{4M'^2}{M^2}\right)^{1/2}$$
$$\times\left(1 + 2\frac{M'^2}{M^2}\right) + \frac{\alpha^2}{3\pi^2}\left(\log\frac{m}{\mu} - \frac{1}{2}\right)\left(1 - \frac{4m^2}{M^2}\right)^{1/2}\left(1 + 2\frac{m^2}{M^2}\right), \tag{5–8.17}$$

where we have also included the contribution associated with $\zeta''$ [cf. Eq. (5–7.52)], the infra-red-sensitive part of the contact term. That has the advantage of removing all reference to the fictitious mass $\mu$ at this initial stage of the calculation.

The latter remark is verified by isolating the part of the spectral integral in which $M'$ ranges from $m + \mu$ to $m + \delta M$, where

$$\mu \ll \delta M \ll m. \tag{5–8.18}$$

This is

$$-\frac{\alpha^2}{3\pi^2}\left(1 - \frac{4m^2}{M^2}\right)^{1/2}\left(1 + 2\frac{m^2}{M^2}\right)\int_\mu^{\delta M} d(M' - m)\frac{\left[(M' - m)^2 - \mu^2\right]^{1/2}}{(M' - m)^2}$$

$$= -\frac{\alpha^2}{3\pi^2}\left(\log\frac{2\,\delta M}{\mu} - 1\right)\left(1 - \frac{4m^2}{M^2}\right)^{1/2}\left(1 + 2\frac{m^2}{M^2}\right), \tag{5–8.19}$$

which applies a specialization of the integral (4–4.97), or (5–4.79). We thereby replace (5–8.17) with

$$M^2\,\delta a(M^2)|_I = \frac{\alpha^2}{6\pi^2}\int_{m+\delta M}^{M/2}\frac{dM'}{M'}\left[1 - \frac{m^2}{M'^2} - \frac{4m^2}{M'^2 - m^2}\right]$$

$$\times\left(1 - \frac{4M'^2}{M^2}\right)^{1/2}\left(1 + 2\frac{M'^2}{M^2}\right)$$

$$+\frac{\alpha^2}{3\pi^2}\left(\log\frac{m}{2\,\delta M} + \frac{1}{2}\right)\left(1 - \frac{4m^2}{M^2}\right)^{1/2}\left(1 + 2\frac{m^2}{M^2}\right). \tag{5–8.20}$$

The spectral integration is easily performed with the aid of the variable $v'$ defined by

$$M' = \tfrac{1}{2}M(1 - v'^2)^{1/2} = m\frac{(1 - v'^2)^{1/2}}{(1 - v^2)^{1/2}}; \tag{5–8.21}$$

it ranges from zero to $v - \delta v$, in which

$$\delta v = \frac{1 - v^2}{v}\frac{\delta M}{m}. \tag{5–8.22}$$

The outcome is

$$M^2\,\delta a(M^2)\big|_I = \frac{\alpha^2}{6\pi^2}\left[5v\chi(v) + \tfrac{5}{2}v - \tfrac{17}{6}v^3 - v(3 - v^2)\log\frac{4v^2}{1 - v^2}\right], \quad (5\text{–}8.23)$$

where, as in Eq. (5–4.67),

$$\chi(v) = \frac{1}{2v}\log\frac{1 + v}{1 - v}. \quad (5\text{–}8.24)$$

One can now compute a portion of the integral that is related to atomic energy displacements [Eq. (5–4.91)]

$$\int dM^2\,\frac{a(M^2)}{M^2} = \frac{1}{(2m)^2}\int dv^2\,M^2 a(M^2). \quad (5\text{–}8.25)$$

The use of (5–8.23), employing such integrals as

$$\int_0^1 dv\,v^{2n}\log\frac{4v^2}{1 - v^2} = \frac{2}{2n + 1}\sum_{k=0}^{n-1}\frac{1}{2k + 1}, \quad (5\text{–}8.26)$$

gives the result

$$\int dv^2\,M^2\,\delta a(M^2)\bigg|_I = \frac{4\alpha}{15\pi}\left[\frac{13}{2}\frac{\alpha}{4\pi}\right], \quad (5\text{–}8.27)$$

although it is possibly simpler to proceed directly from (5–8.20):

$$\int\frac{dM^2}{M^2}\,\delta a(M^2)\bigg|_I = \frac{\alpha}{2\pi}\int_{m+\delta M}^{M/2}\frac{dM'}{M'}\left[1 - \frac{m^2}{M'^2} - \frac{4m^2}{M'^2 - m^2}\right]\frac{\alpha}{15\pi}\frac{1}{M'^2}$$

$$+\frac{\alpha}{\pi}\left(\log\frac{m}{2\,\delta M} + \frac{1}{2}\right)\frac{\alpha}{15\pi}\frac{1}{m^2} = \frac{\alpha}{15\pi}\frac{1}{m^2}\left[\frac{13}{2}\frac{\alpha}{4\pi}\right], \quad (5\text{–}8.28)$$

where one utilizes the value of this integral that is applicable to the non-interacting system with particles of effective mass $M'$. The fractional correction exhibited in brackets in (5–8.27), for example, is less than half of the known result given in (5–4.197). In contrast, the high-mass behavior of (5–8.23) overshoots the mark:

$$M \gg m: \qquad M^2\,\delta a(M^2)\big|_I \sim \frac{\alpha^2}{6\pi^2}\left(\log\frac{M}{m} - \frac{1}{3}\right), \tag{5–8.29}$$

since the logarithmic factor is not present in the correct result of Eq. (5–4.201).

We now turn to the major problem posed by (5–8.6), which is the evaluation of

$$-\tfrac{1}{2}i\,\mathrm{Tr}\,G_+ eq\gamma A_a G_+\left(M_1^{A_b} + \zeta' eq\gamma A_b\right). \tag{5–8.30}$$

For that, we need an appropriate expression for the part of $M_1$ that is linear in the field strengths. The qualifier "appropriate" means that the simplification thus far employed in non-causal calculations, the rejection of the factor $\gamma\Pi + m$ on at least one side, is inapplicable here. We must first remove this lacuna in our knowledge (that information is available from the causal calculation of Section 4–6, but our purpose is to illustrate non-causal methods).

Let us begin with the form [equivalent to Eqs. (5–6.4, 5)]

$$M = -ie^2\int ds\,s\,du\,\gamma^\nu\left\langle\left[m - \gamma(\Pi - k)\right]e^{-is\chi}\right\rangle\gamma_\nu, \tag{5–8.31}$$

$$\chi = (k - u\Pi)^2 + u(1-u)(\Pi^2 + m^2) + m^2u^2 - ueq\sigma F,$$

where we have the intention of exhibiting $M_0$ explicitly in order to facilitate its removal (a different strategy from that followed in the last section, where $M_0$ was simply subtracted). Accordingly, we first commute $\gamma_\nu$ from the right-hand side to produce

$$M = 2ie^2\int ds\,s\,du\left\langle\left[2m + \gamma(\Pi - k)\right]e^{-is\chi}\right\rangle$$

$$+e^2\int ds\,s^2\,du\,u\frac{dv}{2}\gamma^\nu\left\langle\left[m - \gamma(\Pi - k)\right]e^{-is\chi\frac{1}{2}(1+v)}eq\left[\sigma F, \gamma_\nu\right]e^{-is\chi\frac{1}{2}(1-v)}\right\rangle. \tag{5–8.32}$$

The last term, being explicitly linear in the field strengths, is easily evaluated, and we set it aside. Next we decompose the first term of (5–8.32) through the rearrangement

$$2m + \gamma(\Pi - k) = 2m + (1-u)\gamma\Pi - \gamma(k - u\Pi), \tag{5–8.33}$$

and consider the integral containing the factor $k - u\Pi$. Here, we note that

$$0 = \frac{1}{2s}\left\langle\left[\xi, e^{-is\chi}\right]\right\rangle = \int_{-1}^{1} \frac{dv}{2}\left\langle e^{-is\chi\frac{1}{2}(1+v)}(k - u\Pi)e^{-is\chi\frac{1}{2}(1-v)}\right\rangle, \quad (5\text{–}8.34)$$

or

$$\langle ke^{-is\chi}\rangle = \int \frac{dv}{2} u\langle e^{-is\chi\frac{1}{2}(1+v)}\Pi e^{-is\chi\frac{1}{2}(1-v)}\rangle$$

$$= u\Pi\langle e^{-is\chi}\rangle + su\int \frac{dv}{2}\frac{1-v}{2}\left\langle e^{-is\chi\frac{1}{2}(1+v)}i[\Pi, \chi]e^{-is\chi\frac{1}{2}(1-v)}\right\rangle, \quad (5\text{–}8.35)$$

the last statement recording the result of a partial integration. Accordingly,

$$\left\langle(k - u\Pi)e^{-is\chi}\right\rangle = su^2\int \frac{dv}{2}\frac{1-v}{2}$$
$$\times\left\langle e^{-is\chi\frac{1}{2}(1+v)}eq\{-F\Pi + \Pi F + 2Fk - \partial\sigma F\}e^{-is\chi\frac{1}{2}(1-v)}\right\rangle, \quad (5\text{–}8.36)$$

which, again, is explicitly linear in the field strengths.

That focuses attention on the remaining structure:

$$2ie^2\int ds\, s\, du\left[2m + (1 - u)\gamma\Pi\right]\langle e^{-is\chi}\rangle. \quad (5\text{–}8.37)$$

Now we make use of the $\xi$-transformation device of Section 4–14, which is such that

$$\chi \to \hat{\chi} = \left[k + eq\int_0^u du'\, u'\, F(x - u'\xi)\xi\right]^2$$
$$+u(1 - u)\left[\Pi - eq\int_0^u du'\, F(x - u'\xi)\xi\right]^2$$
$$+m^2 u - ueq\sigma F(x - u\xi). \quad (5\text{–}8.38)$$

Since only linear field terms are of interest, we write this as

$$\hat{\chi} = k^2 + u(1 - u)\left[-(\gamma\Pi)^2 + m^2\right] + m^2u^2 + \int_0^u du'\, u'\, 2eqk \cdot F(x - u'\xi)\xi$$
$$-u(1 - u)\int_0^u du'\, 2eq\Pi \cdot F(x - u'\xi)\xi$$
$$-ueq\sigma F(x - u\xi) + u(1 - u)eq\sigma F(x). \quad (5\text{–}8.39)$$

When the explicit field-strength terms are omitted, and the $k$-integration performed by the usual formula

$$\langle e^{-isk^2}\rangle = \frac{1}{(4\pi)^2 is^2}, \tag{5-8.40}$$

the result obtained for (5-8.37) is precisely $M_0$, as presented in Eq. (5-6.54).

We therefore confine our attention to the explicit linear $F$-terms that appear in an expansion of $\langle e^{-is\chi}\rangle$. In doing that we need not include the $k\cdot F\xi$ term of (5-8.39), for, as detailed in (4-14.60), its rotational structure annuls the states symbolized by $\langle$ and $\rangle$. The remaining terms are

$$\begin{aligned}\langle e^{-is\chi}\rangle_F = {}& isu(1-u)\int_{-1}^{1}\frac{dv}{2}\int_0^u du' \\ &\times\left\langle e^{-is(k^2+\psi)\frac{1}{2}(1+v)}2eq\Pi\cdot F(x-u'\xi)\xi e^{-is(k^2+\psi)\frac{1}{2}(1-v)}\right\rangle \\ &+is\int_{-1}^{1}\frac{dv}{2}\left\langle e^{-is(k^2+\psi)\frac{1}{2}(1+v)}eq[u\sigma F(x-u\xi)-u(1-u)\sigma F(x)]\right. \\ &\left.\times e^{-is(k^2+\psi)\frac{1}{2}(1-v)}\right\rangle,\end{aligned} \tag{5-8.41}$$

where, as in (4-14.57), but without reference to a photon mass,

$$\psi = u(1-u)[\Pi^2+m^2]+m^2u^2. \tag{5-8.42}$$

It is convenient to consider a typical Fourier component of the fields, $e^{ipx}$, and thereby apply some of the results obtained in Section 4-14:

$$\langle e^{-isk^2\frac{1}{2}(1+v)}e^{-iup\xi}e^{-isk^2\frac{1}{2}(1-v)}\rangle = \exp\left[-isu^2\frac{1-v^2}{4}p^2\right]\langle e^{-isk^2}\rangle, \tag{5-8.43}$$

which is the content of Eq. (4-14.55), and [Eq. (4-14.61)]

$$\begin{aligned}&\langle e^{-isk^2\frac{1}{2}(1+v)}e^{-iu'p\xi}\xi e^{-isk^2\frac{1}{2}(1-v)}\rangle \\ &\quad = 2su'\frac{1-v^2}{4}p\exp\left[-isu'^2\frac{1-v^2}{4}p^2\right]\langle e^{-isk^2}\rangle.\end{aligned} \tag{5-8.44}$$

In presenting the outcome, it is also convenient to consider a typical particle matrix element in which $\Pi$, standing on the left or the right of $F$, is replaced by $p'$

or $p''$, respectively, where

$$p' - p'' = p. \tag{5-8.45}$$

With the aid of the symbols

$$D = u(1-u)\left[\frac{1+v}{2}(p'^2 + m^2) + \frac{1-v}{2}(p''^2 + m^2)\right] + m^2u^2 + u^2\frac{1-v}{4}p^2 \tag{5-8.46}$$

and

$$d = u(1-u)\left[\frac{1+v}{2}(p'^2 + m^2) + \frac{1-v}{2}(p''^2 + m^2)\right] + m^2u^2, \tag{5-8.47}$$

we find that

$$e^2\langle e^{-is\chi}\rangle_F = \frac{\alpha}{4\pi}\frac{u(1-u)}{s}\int\frac{dv}{2}eq(p'+p'')J\frac{1}{p^2}(e^{-isD} - e^{-isd})$$

$$+\frac{\alpha}{4\pi}\frac{u}{s}\int\frac{dv}{2}eq\sigma Fe^{-isD} - \frac{\alpha}{4\pi}\frac{u(1-u)}{s}\int\frac{dv}{2}eq\sigma Fe^{-isd}, \tag{5-8.48}$$

which has gone through an intermediate stage involving the integration

$$\frac{1-v^2}{4}\int_0^{u^2} du'^2\, e^{-isD'} = \frac{i}{s}\frac{1}{p^2}(e^{-isD} - e^{-isd}), \tag{5-8.49}$$

where the prime on $D$ signifies that $u'$ replaces $u$ in the $p^2$ term of $D$. Another way of stating this gauge-invariant form emerges on adopting a Lorentz gauge, so that

$$\frac{1}{p^2}J = A, \tag{5-8.50}$$

while recalling the relation

$$(p'+p'')A + \sigma F = 2m\gamma A - (\gamma p' + m)\gamma A - \gamma A(\gamma p'' + m). \tag{5-8.51}$$

The result of (5–8.48), then appears as

$$e^2\langle e^{-is\chi}\rangle_F = \frac{\alpha}{2\pi}m\frac{u(1-u)}{s}\int\frac{dv}{2}eq\gamma A(e^{-isD}-e^{-isd})$$

$$+\frac{\alpha}{4\pi}\frac{u^2}{s}\int\frac{dv}{2}eq\sigma F e^{-isD}$$

$$-(\gamma p'+m)\frac{\alpha}{4\pi}\frac{u(1-u)}{s}\int\frac{dv}{2}eq\gamma A(e^{-isD}-e^{-isd})$$

$$-\frac{\alpha}{4\pi}\frac{u(1-u)}{s}\int\frac{dv}{2}eq\gamma A(e^{-isD}-e^{-isd})(\gamma p''+m), \quad (5\text{–}8.52)$$

which puts into evidence examples of terms having the factors $\gamma p'+m$ and $\gamma p''+m$, terms which must be retained in this calculation.

It behooves us now to retrace our steps and evaluate the integrals that were set aside in Eqs. (5–8.32, 36). In doing this we meet $k$-integrals that are the specialization of (4–15.55) to $\lambda = 1$, as given by

$$\langle e^{-is\chi'\frac{1}{2}(1+v)}e^{-is\chi''\frac{1}{2}(1-v)}\rangle = \frac{1}{(4\pi)^2 is^2}e^{-isD}, \quad (5\text{–}8.53)$$

where corresponding primes on $\chi$ indicate the introduction of $p'$ or $p''$, and of (4–15.6), here written as

$$\langle e^{-is\chi'\frac{1}{2}(1+v)}k e^{-is\chi''\frac{1}{2}(1-v)}\rangle = u\left(\frac{1+v}{2}p'+\frac{1-v}{2}p''\right)\frac{1}{(4\pi)^2 is^2}e^{-isD}. \quad (5\text{–}8.54)$$

We find that

$$e^2\int ds\,s^2\,du\,u\frac{dv}{2}\gamma^\nu\langle[m-\gamma(\Pi-k)]e^{-is\chi\frac{1}{2}(1+v)}eq[\sigma F,\gamma_\nu]e^{-is\chi\frac{1}{2}(1-v)}\rangle$$

$$= -\frac{\alpha}{4\pi}\int du\,u\frac{dv}{2}[4umeq\sigma F+2(1-u)(\gamma p'+m)eq\sigma F$$

$$+2(1-u)eq\sigma F(\gamma p''+m)-2(1-u)eq\gamma J]\frac{1}{D} \quad (5\text{–}8.55)$$

and

$$-2ie^2\int ds\, s\, du\, \gamma\langle(k-u\Pi)e^{-isx}\rangle$$

$$=\frac{\alpha}{2\pi}\int du\, u^2\frac{dv}{2}\frac{1-v}{2}$$

$$\times[-(1-u)(\gamma p'+m)eq\sigma F+(1-u)eq\sigma F(\gamma p''+m)+(1+uv)eq\gamma J]\frac{1}{D}. \tag{5-8.56}$$

In combining these results with the evaluation of (5–8.37) that is provided by (5–8.48), the following identity is valuable:

$$\int_{-1}^{1}dv\frac{\partial}{\partial v}\log\frac{D}{d}$$

$$=\int_{-1}^{1}\frac{dv}{2}\left[\frac{u(1-u)(p'^2-p''^2)-u^2vp^2}{D}-\frac{u(1-u)(p'^2-p''^2)}{d}\right]=0. \tag{5-8.57}$$

The resulting expression for $M_1$ is usefully presented as the sum of three parts,

$$M_1=M_\alpha+M_\beta+M_\gamma, \tag{5-8.58}$$

where [we again resort to the integration sign simplification of (5–7.254)]

$$M_\alpha=-\frac{\alpha}{2\pi}m\int u^2(1-u)\frac{eq\sigma F}{D}+\frac{\alpha}{\pi}m^2\int u(1-u)(1+u)eq\gamma A\left(\frac{1}{D}-\frac{1}{d}\right)$$

$$+\frac{\alpha}{2\pi}\int u\left(1-\tfrac{1}{2}u-\tfrac{1}{2}u^2v^2\right)\frac{eq\gamma J}{D}, \tag{5-8.59}$$

while

$$M_\beta=-\frac{\alpha}{2\pi}\int u(1-u)\left[\left(1-u\frac{1+v}{2}\right)(\gamma p'+m)eq\sigma F\right.$$

$$\left.+\left(1-u\frac{1-v}{2}\right)eq\sigma F(\gamma p''+m)\right]\frac{1}{D}$$

$$-\frac{\alpha}{2\pi}m\int u(1-u)(1+u)[(\gamma p'+m)eq\gamma A+eq\gamma A(\gamma p''+m)]\left(\frac{1}{D}-\frac{1}{d}\right)$$

$$+\frac{\alpha}{2\pi}\int u(1-u)^2\tfrac{1}{2}(p'^2+m^2+p''^2+m^2)eq\gamma A\left(\frac{1}{D}-\frac{1}{d}\right) \tag{5-8.60}$$

and

$$M_\gamma=-\frac{\alpha}{2\pi}\int u(1-u)^2(\gamma p'+m)eq\gamma A(\gamma p''+m)\left(\frac{1}{D}-\frac{1}{d}\right). \tag{5-8.61}$$

This decomposition evidently catalogues the presence of $\gamma p' + m$ and $\gamma p' + m$ factors, which are also implicit in the last term of $M_\beta$. Only $M_\alpha$ survives when both factors are rejected, and there are several (successful) tests to which it can be subjected. One is provided by the form factors displayed in (5–7.178, 179). Another refers to the situation considered in Section 4–5, as described by the coupling stated in Eq. (4–5.20). Here we deal with a photon field ($p^2 = 0$), and restrict one of the particle momenta to be on the mass shell ($p'^2 + m^2 = 0$). The resulting pure spin coupling inferred from (5–8.59) is represented by the form factor

$$G_2(p'') = 2m^2 \int \frac{u^2(1-u)}{u(1-u)\dfrac{1-v}{2}(p''^2 + m^2) + m^2u^2}$$

$$= \int dM^2 \, \frac{g_2(M^2)}{p''^2 + M^2}, \tag{5–8.62}$$

where

$$g_2(M^2) = \int_0^1 du \int_{-1}^1 \frac{dv}{2} \frac{4u}{1-v} \delta\left(\frac{M^2}{m^2} - 1 - \frac{u}{1-u}\frac{2}{1-v}\right)$$

$$= \frac{m^2}{M^2}\left(1 - \frac{m^2}{M^2}\right), \tag{5–8.63}$$

in agreement with the weight function shown in (4–5.20). The calculation of the preceding section provides a test of $M_\alpha + M_\beta$. When the homogeneous field considered there vanishes, the contributions to $\gamma_1{}''(k')$ that are labelled $e$ and $i$ survive to give the functional derivative of $M_1$ with respect to $-eqA_\nu$, with $\gamma p' + m$ and $p'^2 + m^2$ set equal to zero. After some rearrangement, which involves the use of the identity (5–8.57), the equivalence of the two versions is indeed realized. This comparison also leaves one with the impression that the forms used in the calculation of Section 5–7 are somewhat lacking in felicity.

We are now better prepared for the calculation of (5–8.30). But it might be well to also have before us the analogous non-causal evaluation of the elementary $W_{ab}^{(2)}$ compling, which has thus far been discussed only by the causal method of Section 4–3. The action expression of (5–8.1), without contact terms, is written out as

$$W_{ab}^{(2)} = ie^2 \int (dx)(dx') \operatorname{tr}_{(4)} \gamma A_a(x) G_+(x-x') \gamma A_b(x') G_+(x'-x), \tag{5–8.64}$$

where the trace in the charge space has already been performed. Let us comment here that a gauge transformation on $A_a(x)$, for example, when combined with the identity

$$\gamma\partial\lambda(x) = i\left[\gamma\frac{1}{i}\partial + m, \lambda(x)\right], \tag{5–8.65}$$

results in the following change of (5–8.64):

$$e^2\int(dx)(dx')\,\mathrm{tr}_{(4)}[\lambda(x)\delta(x-x')\gamma A_b(x')G_+(x'-x)$$

$$-\lambda(x)G_+(x-x')\gamma A_b(x')\delta(x'-x)] = 0, \tag{5–8.66}$$

which we infer from the non-causal circumstance of non-overlap between $A_b(x)$ and $A_a(x)$ [therefore also $\lambda(x)$]. Thus, if the coupling of (5–8.64) were applied without this space-time restriction, contact terms would generally be required to maintain gauge invariance. The preferred procedure here is to rearrange the coupling for the non-overlapping arrangement so that gauge invariance is maintained after the space-time extrapolation.

The introduction of the non-causal forms of the particle propagation functions converts (5–8.64) into

$$W_{ab}^{(2)} = ie^2\int\frac{(dp')}{(2\pi)^4}\frac{(dp'')}{(2\pi)^4}A_a^{\mu}(p''-p')A_b^{\nu}(p'-p'')$$

$$\times\mathrm{tr}_{(4)}[\gamma_\mu(m-\gamma p')\gamma_\nu(m-\gamma p'')]\frac{1}{p'^2+m^2}\frac{1}{p''^2+m^2}, \tag{5–8.67}$$

where Fourier transforms of the fields now appear. We change the momentum integration variables in two stages, first, by writing

$$p' = p + \tfrac{1}{2}k, \qquad p'' = p - \tfrac{1}{2}k, \tag{5–8.68}$$

and then, through the standard observation that

$$\frac{1}{p'^2+m^2}\frac{1}{p''^2+m^2}$$

$$= -\int_0^\infty ds\,s\int_{-1}^{1}\frac{dv}{2}\exp\left[-is\left(\frac{1+v}{2}p'^2 + \frac{1-v}{2}p''^2 + m^2\right)\right]$$

$$= -\int ds\,s\frac{dv}{2}\exp\left\{-is\left[\left(p+\tfrac{1}{2}vk\right)^2 + m^2 + \frac{1-v^2}{4}k^2\right]\right\}, \tag{5–8.69}$$

by the translation

$$p \to p - \tfrac{1}{2}vk. \tag{5–8.70}$$

The net transformation,

$$p' = p + \frac{1-v}{2}k, \qquad p'' = p - \frac{1+v}{2}k, \tag{5–8.71}$$

is combined with the trace evaluation

$$\tfrac{1}{4}\,\mathrm{tr}_{(4)}\left[\gamma_\mu(m-\gamma p')\gamma_\nu(m-\gamma p'')\right] = p'_\mu p''_\nu + p''_\mu p'_\nu - g_{\mu\nu}(p'p''+m^2), \tag{5–8.72}$$

and the evenness of the exponential in the final $p$-variable, to give the coupling

$$W_{ab}^{(2)} = -\int \frac{(dk)}{(2\pi)^4} A_a^{\mu}(-k) P_{\mu\nu}^{(2)}(k) A_b^{\nu}(k), \tag{5–8.73}$$

with

$$\begin{aligned} P_{\mu\nu}^{(2)}(k) = 4ie^2 \int ds\, s\frac{dv}{2}\int\frac{(dp)}{(2\pi)^4} \\ \times\left[2p_\mu p_\nu - g_{\mu\nu}\left(p^2+m^2+\frac{1-v^2}{4}k^2\right) - \frac{1-v^2}{2}(k_\mu k_\nu - g_{\mu\nu}k^2)\right] \\ \times\exp\left[-is\left(p^2+m^2+\frac{1-v^2}{4}k^2\right)\right]. \end{aligned} \tag{5–8.74}$$

If we now use the integral evaluation

$$\int\frac{(dp)}{(2\pi)^4}\,2p_\mu p_\nu e^{-isp^2} = -\frac{i}{s}g_{\mu\nu}\int\frac{(dp)}{(2\pi)^4}e^{-isp^2} \tag{5–8.75}$$

to provide an effective replacement for $2p_\mu p_\nu$, the first ($g_{\mu\nu}$) terms of (5–8.74) involve the $s$-integral

$$\begin{aligned} \int_0^\infty ds\, s\left[-\frac{i}{s} - p^2 - m^2 - \frac{1-v^2}{4}k^2\right]\exp\left[-is\left(p^2+m^2+\frac{1-v^2}{4}k^2\right)\right] \\ = -i\int_0^\infty ds\frac{\partial}{\partial s}\left\{s\exp\left[-is\left(p^2+m^2+\frac{1-v^2}{4}k^2\right)\right]\right\} = 0, \end{aligned} \tag{5–8.76}$$

which leaves an explicitly gauge-invariant coupling. But we must not fail to notice that if the $p$-integration were performed first, in accordance with the familiar result

$$\int \frac{(dp)}{(2\pi)^4} e^{-isp^2} = \frac{1}{(4\pi)^2} \frac{1}{is^2}, \tag{5-8.77}$$

the $s$ factor in the brace of (5-8.76) would become $s^{-1}$, and the indicated null outcome would no longer follow. It is here that the context of non-overlap intervenes, since any polynomial in $k^2$, specifically that implied by the first two terms of the exponential expression in (5-8.76), can be removed from $P_{\mu\nu}^{(2)}(k)$ without changing the coupling (5-8.73). Thus, we are still able to make $P_{\mu\nu}^{(2)}(k)$ assume the necessary gauge-covariant form without altering the coupling in the non-overlap situation.

At this stage, we have

$$P_{\mu\nu}^{(2)}(k) = -\left(k_\mu k_\nu - g_{\mu\nu}k^2\right)\frac{\alpha}{2\pi}$$
$$\times \int \frac{ds}{s} \int_0^1 dv\,(1 - v^2)\exp\left[-is\left(m^2 + \frac{1 - v^2}{4}k^2\right)\right], \tag{5-8.78}$$

which we proceed to rearrange by a partial integration on $v$:

$$\int_0^1 d\left(v - \frac{v^3}{3}\right)\exp\left[-is\left(m^2 + \frac{1 - v^2}{4}k^2\right)\right]$$
$$= \tfrac{2}{3}e^{-ism^2} - \tfrac{1}{2}isk^2\int_0^1 dv\,v^2\left(1 - \tfrac{1}{3}v^2\right)\exp\left[-is\left(m^2 + \frac{1 - v^2}{4}k^2\right)\right]. \tag{5-8.79}$$

The first term on the right side does not contribute to the non-overlapping situation. If it were nevertheless retained in performing the space-time extrapolation, it would constitute a local coupling of the field strengths. It is therefore excluded by the normalization condition that accompanies the initial description of the photon. In this way, we come to the final statement that accords with all the physical requirements,

$$P_{\mu\nu}^{(2)}(k) = \left(k_\mu k_\nu - g_{\mu\nu}k^2\right)k^2\frac{\alpha}{4\pi}\int_0^1 dv\,\frac{v^2\left(1 - \frac{1}{3}v^2\right)}{m^2 + \frac{1 - v^2}{4}k^2}; \tag{5-8.80}$$

the presence of $-i\varepsilon$ in the denominator is understood. The addition to the action then implied by (5-8.73) is exactly that given in (4-3.70), with the spectral weight function appearing in the form

$$M^2 a^{(2)}(M^2) = \frac{\alpha}{2\pi} v\left(1 - \tfrac{1}{3}v^2\right), \tag{5-8.81}$$

as presented in Eq. (5-4.186).

The first priority in evaluating (5-8.30) is, clearly, the identification of the structure that the $\zeta'$ contact term is intended to excise. A clue is provided by the observation that the denominators $D$ and $d$, which are exhibited in Eqs. (5-8.46, 47), change their character in the limit $u \to 1$, there becoming independent of the particle momenta $p'$ and $p''$. If we inspect the structure of $M_1$, as detailed in $M_\alpha$, $M_\beta$, $M_\gamma$, we see that all terms have a compensating factor $1 - u$ in the numerator, with one exception. That is the last contribution to $M_\alpha$ [Eq. (5-8.59)], which has in its numerator the factor

$$1 - \tfrac{1}{2}u - \tfrac{1}{2}u^2 v^2 = \tfrac{1}{2}(1 - v^2) + \tfrac{1}{2}(1 - u)\left[1 + (1 + u)v^2\right]. \tag{5-8.82}$$

Thus, attention is directed to this piece of $M_\alpha$,

$$M_\alpha' = \frac{\alpha}{4\pi} \int u(1 - v^2) \frac{eq\gamma J}{D}. \tag{5-8.83}$$

Note, incidentally, that we have not focused on the factor of $u$ that is common to the whole of $D$ and $d$, since every term of $M_1$ has a compensating factor of $u$ in its numerator.

A small digression is now in order, however. As in the preceding section, the comparison of singular integrals requires that a limiting process of universal significance be employed; an illustration is the modified photon propagation function of (5-7.186). We proceed analogously here, but find it convenient to alter the convergence factor adopted in the latter equation:

$$\frac{(im^2 s)^\delta}{\Gamma(1 + \delta)} \to 1 - e^{-i\lambda^2 s}; \tag{5-8.84}$$

the equivalent of the limit $\delta \to 0$ is $\lambda \to \infty$. Thus, the construction of $\zeta'$ given in (5-7.187) now reads

$$\begin{aligned} \zeta' &= \frac{\alpha}{2\pi} \int_0^1 du\,(1 - u) \int_0^\infty \frac{ds}{s}\left(1 - e^{-is\lambda^2(1-u)}\right) e^{-ism^2u^2} \\ &= \frac{\alpha}{2\pi} \int_0^1 du\,(1 - u) \log \frac{m^2u^2 + \lambda^2(1 - u)}{m^2u^2} \cong \frac{\alpha}{2\pi}\left(\log \frac{\lambda}{m} + \frac{5}{4}\right), \end{aligned} \tag{5-8.85}$$

where the last form refers to the situation $\lambda \gg m$. The advantage of the substitution (5–8.84) lies in the new form of the photon propagation function,

$$\frac{1}{k^2} \to \frac{1}{k^2} - \frac{1}{k^2 + \lambda^2}. \qquad (5\text{–}8{,}86)$$

in which the additional term mimics, but with the wrong sign, the propagation function of a particle with the mass $\lambda$. Then, in discussing (5–8.83), one has only to make the substitution

$$\frac{1}{D} \to \frac{1}{D} - \frac{1}{D_\lambda}, \qquad (5\text{–}8.87)$$

with

$$D_\lambda = D + (1 - u)\lambda^2. \qquad (5\text{–}8.88)$$

Harold seems disturbed.

H.: Your use of the alternative convergence factors exhibited in (5–8.84) reminds one of some variants of the device known as "regularization", which is widely employed nowadays to attribute mathematical meaning to physically ambiguous theories that give divergent integral answers to physical questions. Are you incorporating regularization into the principles of source theory?

S.: Certainly not. The contrary should be evident in the present context from the fact that the problem under discussion has already been solved by causal methods without reference to such concepts. The non-causal techniques lead to finite, well-defined expressions which, for convenience of evaluation, are decomposed in ways that require convergence factors to give meaning to the separate parts. Since there is no question about the existence of the whole structure, any reasonable convergence factor may be employed. To apply the term "regularization" to this procedure is to encumber it with a heavy load of misleading associations, which is always the danger when current terminology is transferred to a logically different situation. As K'ung Fu-tzu pointed out some time ago, the necessary prelude to the solution of any problem is "the rectification of names".

We now write out that part of the action expression (5–8.30) which is contributed by $M_\alpha'$ [Eq. (5–8.83)], as modified by (5–8.87). It is

$$-ie^2\frac{\alpha}{4\pi}\int u(1 - v^2)\int \frac{(dp')}{(2\pi)^4}\frac{(dp'')}{(2\pi)^4} k^2 A_a{}^{\mu}(-k)A_b{}^{\nu}(k)$$

$$\times \operatorname{tr}_{(4)}\left[\gamma_\mu(m - \gamma p')\gamma_\nu(m - \gamma p'')\right] \frac{1}{p'^2 + m^2}\frac{1}{p''^2 + m^2}\left(\frac{1}{D} - \frac{1}{D_\lambda}\right), \qquad (5\text{–}8.89)$$

where the Lorentz gauge has been made explicit, and we have continued the use of the symbol $k$, rather than $p$, to denote the photon momentum,

$$k = p' - p''. \qquad (5\text{–}8.90)$$

The trace appearing here is already known from the elementary calculation [cf. Eq. (5–8.72)]. The principal new feature is the presence of the additional denominator, $D$ or $D_\lambda$. For our present purposes, the simplest procedure for handling the three denominators may be the use of the representation that is illustrated by

$$\frac{1}{p'^2+m^2}\,\frac{1}{p''^2+m^2}\,\frac{u(1-u)}{D} = -i\int_0^\infty ds\, s^2\int_0^1 dw\, w\int_{-1}^1 \frac{dv'}{2}e^{-is\Delta}, \qquad (5\text{–}8.91)$$

where

$$\begin{aligned}\Delta &= w\frac{1+v'}{2}(p'^2+m^2) + w\frac{1-v'}{2}(p''^2+m^2) + (1-w)\frac{D}{u(1-u)}\\ &= \left[w\frac{1+v'}{2} + (1-w)\frac{1+v}{2}\right](p'^2+m^2)\\ &\quad + \left[w\frac{1-v'}{2} + (1-w)\frac{1-v}{2}\right](p''^2+m^2) + \frac{u}{1-u}(1-w)\left(m^2+\frac{1-v^2}{4}k^2\right),\end{aligned} \qquad (5\text{–}8.92)$$

and

$$\Delta_\lambda = \Delta + \frac{1-w}{u}\lambda^2 \qquad (5\text{–}8.93)$$

is the appropriate replacement when $D_\lambda$ occurs. The introduction of the variable $p$, according to

$$p' = p + \tfrac{1}{2}k, \qquad p'' = p - \tfrac{1}{2}k, \qquad (5\text{–}8.94)$$

converts $\Delta$ into

$$\begin{aligned}\Delta &= p^2 + m^2 + \tfrac{1}{4}k^2 + \bar{v}pk + \frac{u}{1-u}(1-w)\left(m^2+\frac{1-v^2}{4}k^2\right)\\ &= \left(p+\tfrac{1}{2}\bar{v}k\right)^2 + m^2 + \tfrac{1}{4}(1-\bar{v}^2)k^2\\ &\quad + \frac{u}{1-u}(1-w)\left(m^2+\frac{1-v^2}{4}k^2\right),\end{aligned} \qquad (5\text{–}8.95)$$

in which we have employed the symbol

$$\bar{v} = wv' + (1-w)v. \qquad (5\text{–}8.96)$$

The translation

$$p \to p - \tfrac{1}{2}\bar{v}k, \qquad (5\text{–}8.97)$$

which is analogous to (5–8.70), similarly converts the trace into [cf. Eqs. (5–8.72, 74, 75)]

$$\tfrac{1}{4}\,\mathrm{tr}_{(4)}[\ ] \to \left[\frac{i}{s} - m^2 + \tfrac{1}{4}(1-\bar{v}^2)k^2\right] g_{\mu\nu}, \tag{5–8.98}$$

which refers specifically to the Lorentz-gauge simplification of the gauge-invariant coupling. Thus, the action expression of Eq. (5–8.89) has become

$$-\int \frac{(dk)}{(2\pi)^4} A_a^{\mu}(-k) g_{\mu\nu} P^{(4)}(k)\big|_{\alpha'} A_b^{\nu}(k), \tag{5–8.99}$$

with

$$\begin{aligned} P^{(4)}(k)\big|_{\alpha'} &= 4e^2 \frac{\alpha}{4\pi} k^2 \int \frac{1-v^2}{1-u} \int \frac{(dp)}{(2\pi)^4} \\ &\quad \times \int ds\, s^2\, dw\, w \frac{dv'}{2} \left[\frac{i}{s} - m^2 + \frac{1-\bar{v}^2}{4} k^2\right] \left[e^{-is\Delta} - e^{-is\Delta_\lambda}\right] \\ &= -i\left(\frac{\alpha}{2\pi}\right)^2 k^2 \int \frac{1-v^2}{1-u} \int ds\, dw\, w \frac{dv'}{2} \\ &\quad \times \left[\frac{i}{s} - m^2 + \frac{1-\bar{v}^2}{4} k^2\right] \left[e^{-is\delta} - e^{-is\delta_\lambda}\right]. \end{aligned} \tag{5–8.100}$$

In the latter form, the following symbols are used:

$$\begin{aligned} \delta &= m^2 + \tfrac{1}{4}(1-\bar{v}^2)k^2 + \frac{u}{1-u}(1-w)\left(m^2 + \frac{1-v^2}{4}k^2\right) \\ &= \frac{1-uw}{1-u} m^2 + \tfrac{1}{4}(1-\bar{v}^2)k^2 + \frac{u}{1-u}(1-w)\frac{1-v^2}{4}k^2 \end{aligned} \tag{5–8.101}$$

and

$$\delta_\lambda = \delta + \frac{1-w}{u}\lambda^2. \tag{5–8.102}$$

The $s$-integration is now performed to give

$$\begin{aligned} P^{(4)}(k)\big|_{\alpha'} &= \left(\frac{\alpha}{2\pi}\right)^2 k^2 \int \frac{1-v^2}{1-u} \int dw\, w \frac{dv'}{2} \\ &\quad \times \left[\log\frac{\delta_\lambda}{\delta} + \frac{m^2 - \tfrac{1}{4}(1-\bar{v}^2)k^2}{\delta}\right], \end{aligned} \tag{5–8.103}$$

which anticipates that $\delta_\lambda^{-1}$ does not contribute in the limit $\lambda \to \infty$. Indeed, the only role of $\lambda$ is to appear in the integral

$$\int_0^1 \frac{du}{1-u} \log \frac{\delta_\lambda}{\delta} \cong \int_0^1 \frac{du}{1-u} \log\left[(1-w)\frac{(1-u)\lambda^2 + m^2 + \dfrac{1-v^2}{4}k^2}{u(1-u)\delta}\right], \quad (5\text{–}8.104)$$

where the form of the right-hand side exploits the fact that $\delta_\lambda$ is dominated by $\lambda^2(1-w)/u$, except for values of $u$ that are very close to unity, when it is necessary to retain the term having the factor $(1-u)^{-1}$. We shall separate this expression into two parts:

$$\int_0^1 \frac{du}{1-u} \log \frac{(1-u)\lambda^2 + m^2 + \dfrac{1-v^2}{4}k^2}{m^2 + \dfrac{1-v^2}{4}k^2}$$
$$+ \int_0^1 \frac{du}{1-u} \log\left[(1-w)\frac{m^2 + \dfrac{1-v^2}{4}k^2}{u(1-u)\delta}\right]. \quad (5\text{–}8.105)$$

For the first of these, which is independent of $w$ and $v'$, we perform a partial integration on $v$, taking into account the factor $1 - v^2$ in (5–8.103):

$$\int_0^1 d\left(v - \tfrac{1}{3}v^3\right) \int_0^1 \frac{du}{1-u} \log \frac{(1-u)\lambda^2 + m^2 + \dfrac{1-v^2}{4}k^2}{m^2 + \dfrac{1-v^2}{4}k^2}$$
$$= \tfrac{2}{3}\int_0^1 \frac{du}{1-u} \log\left[1 + (1-u)\frac{\lambda^2}{m^2}\right]$$
$$- \tfrac{1}{2}k^2 \int_0^1 dv\, v^2\left(1 - \tfrac{1}{3}v^2\right) \frac{1}{m^2 + \dfrac{1-v^2}{4}k^2}$$
$$\times \int_0^1 du \frac{\lambda^2}{(1-u)\lambda^2 + m^2 + \dfrac{1-v^2}{4}k^2}. \quad (5\text{–}8.106)$$

The first term on the right-hand side here produces a local coupling in (5–8.99), and is therefore without effect in the non-overlapping arrangement. The second one is evaluated as

$$-\tfrac{1}{2}k^2\int_0^1 dv\, v^2\left(1-\tfrac{1}{3}v^2\right)\frac{1}{m^2+\dfrac{1-v^2}{4}k^2}$$

$$\times\left[\log\frac{\lambda^2}{m^2}+\frac{5}{2}-\log\left(1+\frac{1-v^2}{4}\frac{k^2}{m^2}\right)-\frac{5}{2}\right], \qquad (5\text{–}8.107)$$

which is so written that one recognizes the structure of $\zeta'$ as given in (5–8.85). And indeed, according to the prediction of (5–8.30), this contribution to $P^{(4)}(k)$ is cancelled by $-\zeta' P^{(2)}(k)$, where $P^{(2)}(k)$ is the coefficient of $g_{\mu\nu}$ in (5–8.80). The remaining part of (5–8.107) can be rewritten with the aid of an identity, which is analogous to and inferrable from the identity of (4–12.42):

$$\int_0^1 dv\, v^2\left(1-\tfrac{1}{3}v^2\right)\frac{\log\left(1+\dfrac{1-v^2}{4}\dfrac{k^2}{m^2}\right)+\dfrac{5}{2}}{m^2+\dfrac{1-v^2}{4}k^2}$$

$$=\int_0^1 dv\, v^2\frac{\left(1-\tfrac{1}{3}v^2\right)\log\dfrac{4v^2}{1-v^2}+\dfrac{1}{2}+\dfrac{1}{18}v^2}{m^2+\dfrac{1-v^2}{4}k^2}. \qquad (5\text{–}8.108)$$

The resulting form of $P^{(4)}(k)|_{\alpha'}$ which now includes the $\zeta'$ contact term, is

$$P^{(4)}(k)\big|_{\alpha'}=(k^2)^2\left(\frac{\alpha}{4\pi}\right)^2\int_0^1 dv\, v^2$$

$$\times\frac{\left(1-\tfrac{1}{3}v^2\right)\log\dfrac{4v^2}{1-v^2}+\dfrac{1}{2}+\dfrac{1}{18}v^2}{m^2+\dfrac{1-v^2}{4}k^2}$$

$$+k^2\left(\frac{\alpha}{2\pi}\right)^2\int\frac{1-v^2}{1-u}\int dw\, w\frac{dv'}{2}\left[\frac{m^2-\frac{1}{4}(1-\bar{v}^2)k^2}{\delta}-\frac{1-u}{1-uw}\right], \qquad (5\text{–}8.109)$$

where the required contact term has been incorporated into the last contribution.

The comparison between the meaning of $P(k)$ in the Lorentz gauge, as illustrated by (5–8.99), and the Lorentz-gauge, momentum form of the action expression of (4–3.70), shows that

$$P(k) = -(k^2)^2 \int dM^2 \frac{a(M^2)}{k^2 + M^2} = -(k^2)^2 \int \frac{dM^2}{(M^2)^2} \frac{M^2 a(M^2)}{1 + (k^2/M^2)}. \quad (5\text{–}8.110)$$

The first term on the right side of (5–8.109) thus provides an immediately identifiable piece of $M^2 \delta a(M^2)$, expressed in the parametrization

$$M^2 = \frac{4m^2}{1 - v^2}; \quad (5\text{–}8.111)$$

it is

$$M^2 \delta a(M^2)\big|_{\alpha' 1} = -\left(\frac{\alpha}{4\pi}\right)^2 2v\left[\left(1 - \tfrac{1}{3}v^2\right)\log\frac{4v^2}{1 - v^2} + \frac{1}{2} + \frac{1}{18}v^2\right]. \quad (5\text{–}8.112)$$

We draw attention to the $M^2 \gg m^2$ limit of this part,

$$M^2 \delta a(M^2)\big|_{\alpha' 1} \sim -\frac{\alpha^2}{6\pi^2}\left(\log\frac{M}{m} + \frac{5}{12}\right), \quad (5\text{–}8.113)$$

for it removes the incorrect logarithmic $M$-dependence of (5–8.29). Here is further reassurance that all is going well.

In the following, we shall be primarily concerned with the simpler properties of the weight function $a(M^2)$. They comprise the computation of the integral of (5–8.25); the evaluation of the constant that, to the present accuracy, states the high-energy limit

$$\lim_{M \to \infty} M^2 a(M^2) = C_\infty; \quad (5\text{–}8.114)$$

and the evaluation of the constant that gives the altered threshold behavior,

$$M^2 a(M^2)\big|_{M^2 = 4m^2} = C_0. \quad (5\text{–}8.115)$$

All of these are obtainable from limiting aspects of $P(k)$. Thus,

$$-\lim_{k^2 \to 0} \frac{P(k)}{(k^2)^2} = \frac{1}{(2m)^2} \int dv^2\, M^2 a(M^2), \quad (5\text{–}8.116)$$

and

$$-\frac{P(k)}{k^2}\bigg|_{k^2 \gg m^2} \sim C_\infty\left(\log\frac{k^2}{m^2} + \text{const.}\right), \tag{5–8.117}$$

while

$$-\frac{P(k)}{k^2}\bigg|_{-k^2 \approx 4m^2} \sim C_0\left(\log\frac{4m^2}{k^2 + 4m^2} + \text{const.}\right). \tag{5–8.118}$$

Accordingly, we now proceed to set out the various contributions to $P^{(4)}(k)$.

The simplest term in $M_1$ [Eqs. (5–8.59–61)] is $M_\gamma$. This remark refers to the evaluation of (5–8.30), where the two free-particle Green's functions are cancelled by the corresponding factors in $M_\gamma$. The definition of (5–8.99), *mutatis mutandis*, gives

$$P^{(4)}(k)\big|_\gamma = ie^2\frac{\alpha}{2\pi}4\int\frac{(dp)}{(2\pi)^4}\int du\,\frac{dv}{2}u(1-u)^2\left(\frac{1}{D} - \frac{1}{d}\right), \tag{5–8.119}$$

where

$$p = p' - \tfrac{1}{2}k = p'' + \tfrac{1}{2}k. \tag{5–8.120}$$

The resulting form of the denominators is

$$D = u(1-u)\left(p + \tfrac{1}{2}vk\right)^2 + u\left(m^2 + \frac{1-v^2}{4}k^2\right),$$

$$d = u(1-u)\left(p + \tfrac{1}{2}vk\right)^2 + u\left(m^2 + (1-u)\frac{1-v^2}{4}k^2\right), \tag{5–8.121}$$

and the $p$-integral is parametrized as

$$u(1-u)^2\int\frac{(dp)}{(2\pi)^4}\left(\frac{1}{D} - \frac{1}{d}\right)$$

$$= \frac{1}{(4\pi)^2}\frac{1}{u}\int\frac{ds}{s^2}\left\{\exp\left[-isu\left(m^2 + \frac{1-v^2}{4}k^2\right)\right]\right.$$

$$\left. - \exp\left[-isu\left(m^2 + (1-u)\frac{1-v^2}{4}k^2\right)\right]\right\}. \tag{5–8.122}$$

Two partial integrations on $v$, supplemented by appropriate contact terms, then yield

$$\int \frac{dv}{2} u(1-u)^2 \int \frac{(dp)}{(2\pi)^4}\left(\frac{1}{D}-\frac{1}{d}\right)$$

$$\to \frac{i}{(4\pi)^2}(k^2)^2 \int_0^1 dv\, \frac{v^4}{12}\left[\frac{1}{m^2+\dfrac{1-v^2}{4}k^2} - \frac{(1-u)^2}{m^2+(1-u)\dfrac{1-v^2}{4}k^2}\right] \tag{5–8.123}$$

and

$$P^{(4)}(k)\big|_\gamma = -(k^2)^2 \frac{\alpha^2}{24\pi^2}\int_0^1 du \int_0^1 dv\, v^4$$
$$\times\left[\frac{1}{m^2+\dfrac{1-v^2}{4}k^2} - \frac{(1-u)^2}{m^2+(1-u)\dfrac{1-v^2}{4}k^2}\right]. \tag{5–8.124}$$

Utilizing the $k^2 \to 0$ limit, in accordance with (5–8.116), we get

$$\int dv^2\, M^2 \delta a(M^2)\bigg|_\gamma = \frac{\alpha^2}{6\pi^2}\int_0^1 dv\, v^4 \int_0^1 du\left[1-(1-u)^2\right]$$
$$= \frac{4}{15}\frac{\alpha}{\pi}\left[\frac{\alpha}{12\pi}\right], \tag{5–8.125}$$

while, for large $k^2$, with only $\log(k^2/m^2)$ retained,

$$P^{(4)}(k^2)\big|_\gamma \sim -k^2 \frac{\alpha^2}{12\pi^2}\int_0^1 du\left[\log\frac{k^2}{m^2} - (1-u)\log\frac{k^2}{m^2}\right], \tag{5–8.126}$$

or [Eq. (5–8.117)]

$$C_\infty\big|_\gamma = \left(\frac{\alpha}{2\pi}\right)^2\left[\frac{1}{6}\right]. \tag{5–8.127}$$

There is no contribution here to $C_0$.

The calculations associated with $M_\alpha - M_{\alpha'}$ and $M_\beta$ make use of all the same devices and auxiliary computations that we have already amply illustrated.

Accordingly, we now directly state these contributions to $P^{(4)}(k)$:

$$P^{(4)}(k)\big|_{\beta} = \frac{\alpha^2}{2\pi^2}k^2\int du\,\frac{dv}{2}\,dw\,(1-w)\left(1-u\frac{1-v}{2}\right)(1-v)\log\left[\frac{m^2}{\delta_+}\frac{1-uw}{1-u}\right]$$

$$-\frac{\alpha^2}{\pi^2}m^2\int du\,\frac{dv}{2}\,dw\,(1+u)\left[\log\frac{\partial_+}{\delta_+}+\frac{1-v^2}{4}\frac{k^2}{m^2}\frac{u(1-w)}{1-uw}\right]$$

$$-\frac{\alpha^2}{2\pi^2}\int du\,\frac{dv}{2}\,dw\,(1-u)$$

$$\times\left\{k^2\frac{1-\bar{v}_+{}^2}{2}\log\frac{\partial_+}{\delta_+}+\frac{1-v^2}{4}k^2\frac{u(1-w)}{1-u}\log\left[\frac{m^2}{\delta_+}\frac{1-uw}{1-u}\right]\right.$$

$$\left.+m^2\frac{u(1-w)}{1-u}\left[\log\frac{\partial_+}{\delta_+}+\frac{1-v^2}{4}\frac{k^2}{m^2}\frac{u(1-w)}{1-uw}\right]\right\}, \quad (5\text{–}8.128)$$

where the three terms are associated with the respective three terms of (5–8.60), and, similarly,

$$P^{(4)}(k)\big|_{\alpha-\alpha'} = \frac{\alpha^2}{2\pi^2}k^2\int du\,\frac{dv}{2}\,dw\,\frac{dv'}{2}\,uw\left(\frac{m^2}{\delta}-\frac{1-u}{1-uw}\right)$$

$$+\frac{\alpha^2}{\pi^2}m^2\int du\,\frac{dv}{2}\,dw\,\frac{dv}{2}(1+u)w$$

$$\times\left[\log\frac{\partial}{\delta}+\frac{1-v^2}{4}\frac{k^2}{m^2}\frac{u(1-w)}{1-uw}+\left(m^2-\frac{1-\bar{v}^2}{4}k^2\right)\left(\frac{1}{\delta}-\frac{1}{\partial}\right)\right.$$

$$\left.+\frac{1-v^2}{4}\frac{k^2}{m^2}\frac{u(1-u)(1-w)}{(1-uw)^2}\right]$$

$$+\frac{\alpha^2}{4\pi^2}\int du\,\frac{dv}{2}\,dw\,\frac{dv'}{2}\left[1+(1+u)v^2\right]w$$

$$\times\left\{k^2\log\left[\frac{m^2}{\delta}\frac{1-uw}{1-u}\right]+k^2\left(\frac{m^2}{\delta}-\frac{1-u}{1-uw}\right)-\frac{1-\bar{v}^2}{4}(k^2)^2\frac{1}{\delta}\right\}.$$

(5–8.129)

In writing these out we have used the symbol $\delta$ of (5-8.101), and introduced the related structure

$$\partial = \delta - \frac{1-v^2}{4}k^2\frac{u(1-w)}{1-u}. \tag{5-8.130}$$

The quantities with + subscripts in (5-8.128) are obtained from these and related symbols by placing $v' = +1$, as illustrated by

$$\bar{v}_+ = w + (1-w)v. \tag{5-8.131}$$

The asymptotic behavior of the weight function that is expressed by the constant $C_\infty$ [Eqs. (5-8.114, 117)] is easily inferred from the various pieces that we have exhibited. The contributions with labels $\alpha - \alpha'$, $\beta$, $\gamma$, are, respectively,

$$C_\infty\Big|_{\alpha-\alpha',\beta,\gamma} = \left(\frac{\alpha}{2\pi}\right)^2\left(\tfrac{3}{4}, \tfrac{7}{12}, \tfrac{1}{6}\right). \tag{5-8.132}$$

To these we add the result of combining (5-8.113) with (5-8.29), which gives

$$C_\infty = \left(\frac{\alpha}{2\pi}\right)^2\left(\tfrac{3}{4} + \tfrac{7}{12} + \tfrac{1}{6} - \tfrac{1}{2}\right) = \left(\frac{\alpha}{2\pi}\right)^2, \tag{5-8.133}$$

coinciding with what is exhibited in (5-4.201). As for the constant $C_0$ that gives the threshold behavior [Eqs. (5-8.115, 118)], we find that its only source is the second term of the piece labeled $\alpha'$, which is exhibited in Eq. (5-8.109). Its evaluation for $-k^2 = 4m^2$ has the following singular part, arising from $\delta^{-1}$:

$$\frac{P^{(4)}(k)}{k^2}\Bigg|_{\alpha'2} \sim \frac{\alpha^2}{2\pi^2}\int du\,\frac{dv}{2}\,dw\,\frac{dv'}{2}\,\frac{w}{1-u}\,\frac{1}{\bar{v}^2 + \dfrac{u}{1-u}(1-w)v^2}$$

$$= \frac{\alpha^2}{8\pi^2}\int\frac{du\,dw}{[u(1-u)(1-w)]^{1/2}}\int\frac{d\bar{v}\,d\bar{\bar{v}}}{\bar{v}^2+\bar{\bar{v}}^2}, \tag{5-8.134}$$

where

$$\bar{\bar{v}} = \frac{u}{1-u}(1-w)v^2. \tag{5-8.135}$$

When modified to incorporate a small non-zero value of $k^2 + 4m^2$, the singular integral factor becomes

$$\int\frac{d\bar{v}\,d\bar{\bar{v}}}{\bar{v}^2+\bar{\bar{v}}^2} \sim \pi\log\frac{4m^2}{k^2+4m^2}, \tag{5-8.136}$$

while

$$\int_0^1 \frac{du}{[u(1-u)]^{1/2}} \int_0^1 \frac{dw}{(1-w)^{1/2}} = 2\pi, \tag{5-8.137}$$

and we get

$$C_0 = \tfrac{1}{4}\alpha^2, \tag{5-8.138}$$

in agreement with (5–4.202). The third computation we have mentioned [Eq. (5–8.116)], which is the extraction of the coefficient of $(k^2)^2$ in an expansion of $-P^{(4)}(k)$, requires the evaluation of a number of elementary integrals of known types, and we refrain from supplying the details [cf. Eq. (5–4.197) for the result].

Harold has a question.

H.: Now that you have done the calculation in both a causal and noncausal manner, which method do you consider to be simpler?

S.: That is not easy to answer categorically, since each method appears to have characteristic advantages and disadvantages. The non-causal method seems to come more quickly to a form from which the asymptotic behavior of the weight function can be inferred, but the causal method might be preferable if one wants the detailed structure of the weight function (note that we did not complete this aspect of the non-causal version). I think that the emphasis should be placed on the flexibility of the source approach, which permits the use of whatever computational method is most effective for the purpose, as seen in the light of expanding experience with that type of problem.

H.: I have another question. The photon spectral weight function, as evaluated thus far, seems to have a very simple asymptotic behavior. And yet, in each calculational method, that emerges only after detailed cancellations between different contributions. Is there, perhaps, yet another way of viewing things which would make this behavior plausible, at least, without detailed calculations?

S.: I'm glad you asked that, since I intend to close this section with a discussion of just such an investigation of the asymptotic behavior of the photon propagation function, as it is described by the spectral weight function.

We begin with the general form of the modified photon propagation function as given in Eq. (5–4.109), for example, but multiplied by $e^2$ to produce the combination that would occur in any dynamical application of this function:

$$e^2\overline{D}_+(k) = \frac{1}{k^2} \frac{e^2}{1 - k^2 \int dM^2 \, \dfrac{a(M^2)}{k^2 + M^2}}. \tag{5-8.139}$$

It is more convenient to replace the weight function $a(M^2)$ by the dimensionless combination $M^2 a(M^2)$. And we make evident the dynamical nature of this function by exhibiting a factor of $e^2$, without prejudicing the otherwise arbitrary $e^2$ dependence:

$$M^2 a(M^2) = e^2 s(M^2/m^2, e^2). \tag{5-8.140}$$

The mass $m$ appearing here either belongs to a specific charged particle, as in the calculation just performed, or is representative of various particles of not too disparate masses. The spectral region now of interest to us refers to magnitudes of $k$ and $M$ that are large in comparison with $m$. Its consideration is facilitated by partitioning the spectral integral in

$$e^2\overline{D}_+(k) = \frac{1}{k^2} \frac{e^2}{1 - k^2 e^2 \int_0^\infty \frac{dM^2}{M^2} \frac{s(M^2/m^2, e^2)}{k^2 + M^2}} \tag{5-8.141}$$

at a value of

$$M^2 = \lambda^2 \gg m^2, \tag{5-8.142}$$

where the last denominator of (5-8.141), under the circumstances

$$|k^2| \gg \lambda^2, \tag{5-8.143}$$

can be approximated as

$$1 - e^2 \int_0^{\lambda^2} \frac{dM^2}{M^2} s - k^2 e^2 \int_{\lambda^2}^\infty \frac{dM^2}{M^2} \frac{s}{k^2 + M^2}. \tag{5-8.144}$$

The introduction of the symbol

$$e_\lambda^2 = \frac{e^2}{1 - e^2 \int_0^{\lambda^2} \frac{dM^2}{M^2} s} \tag{5-8.145}$$

then permits the asymptotic form of (5-8.141) to be written as

$$e^2\overline{D}_+(k) \sim \frac{1}{k^2} \frac{{e_\lambda}^2}{1 - k^2 {e_\lambda}^2 \int_{\lambda^2}^\infty \frac{dM^2}{M^2} \frac{s}{k^2 + M^2}}, \tag{5-8.146}$$

in which $e_\lambda$ evidently plays the role of an effective charge for a description in which all masses less than $\lambda$ are explicitly removed from the spectral integral.

It is now natural to rewrite the dimensionless weight function $s(M^2/m^2, e^2)$ in a similar manner, by introducing $e_\lambda{}^2$ in place of $e^2$, and referring $M^2$ to $\lambda^2$ instead of $m^2$:

$$s\left(\frac{M^2}{m^2}, e^2\right) = \sigma\left(\frac{M^2}{\lambda^2}, \frac{m^2}{\lambda^2}, e_\lambda{}^2\right). \tag{5–8.147}$$

This is an identity, in which a new functional form accompanies the appearance of a new dimensionless variable, $m^2/\lambda^2$. It is precisely here that a plausible simplification can be introduced. In a description that operates at a high level of momentum and mass, all explicit reference to small masses can presumably be ignored, thus permitting the neglect of $m^2/\lambda^2 \ll 1$. Indeed, one could raise this to the status of a principle of self-consistency. The actual dependence on $m$ then becomes implicit in the structure of $e_\lambda{}^2$. Accepting this, we have

$$s\left(\frac{M^2}{m^2}, e^2\right) \cong \sigma\left(\frac{M^2}{\lambda^2}, 0, e_\lambda{}^2\right) = \sigma\left(1, 0, e_M{}^2\right), \tag{5–8.148}$$

where the last form recognizes that the arbitrary parameter $\lambda$ can, in particular, be set equal to $M \gg m$. Thus, the spectral weight function is asymptotically dependent on the single variable

$$e_M{}^2 = \frac{e^2}{1 - e^2 \displaystyle\int_0^{M^2} \frac{dM'^2}{M'^2} s\left(\frac{M'^2}{m^2}, e^2\right)}, \tag{5–8.149}$$

which relation,

$$s\left(\frac{M^2}{m^2}, e^2\right) = \sigma\left(e_M{}^2\right), \tag{5–8.150}$$

is a functional equation for the weight function.

Let us suppose, reasonably enough, that the function $\sigma(e_M{}^2)$ can be expanded in a power series when $e_M{}^2 \ll 1$,

$$\sigma\left(e_M{}^2\right) = \sigma_0 + e_M{}^2\sigma_1 + \left(e_M{}^2\right)^2\sigma_2 + \cdots. \tag{5–8.151}$$

One can also expand ${e_M}^2$ in an $e^2$ power series:

$$ {e_M}^2 = e^2 + (e^2)^2 \int_0^{M^2} \frac{dM'^2}{M'^2} s\left(\frac{M'^2}{m^2}, 0\right) + \cdots $$

$$ \cong e^2 + (e^2)^2 \sigma_0 \log \frac{M^2}{\sim m^2} + \cdots, \tag{5–8.152} $$

where the asymptotic form of $s$ has been used to infer the logarithmic dependence on $M^2$, but does not fix the additive constant. The combination of (5–8.151) and (5–8.152) then yields

$$ \frac{M^2}{m^2} \gg 1: \qquad s\left(\frac{M^2}{m^2}, e^2\right) \cong \sigma_0 + e^2\sigma_1 + (e^2)^2\left(\sigma_0\sigma_1 \log \frac{M^2}{\sim m^2} + \sigma_2\right) + \cdots, \tag{5–8.153} $$

which, in its two leading terms, exhibits just the simple asymptotic behavior that we set out to understand. The explicit result of (5–4.201), for spin-$\frac{1}{2}$ charged particles, is conveyed by the parameters

$$ e^2\sigma_0 = \frac{\alpha}{3\pi}, \qquad (e^2)^2\sigma_1 = \frac{\alpha^2}{4\pi^2}. \tag{5–8.154} $$

In addition, we have learned the logarithmic $M^2$ dependence of the next power of $\alpha$, as exhibited in

$$ M^2\alpha(M^2) \sim \frac{\alpha}{3\pi} + \frac{\alpha^2}{4\pi^2} + \frac{\alpha^3}{12\pi^3}\left(\log \frac{M^2}{m^2} + \text{const.}\right). \tag{5–8.155} $$

The functional equation of (5–8.149, 150) merits some further discussion. Let us introduce the variable

$$ x = \frac{1}{{e_M}^2} = \frac{1}{e^2} - \int_0^{M^2} \frac{dM'^2}{M'^2} s\left(\frac{M'^2}{m^2}, e^2\right), \tag{5–8.156} $$

which is such that

$$ -M^2 \frac{dx}{dM^2} = s\left(\frac{M^2}{m^2}, e^2\right). \tag{5–8.157} $$

Then the functional equation reads

$$-M^2\frac{dx}{dM^2} = \sigma\left(\frac{1}{x}\right), \tag{5–8.158}$$

which is also conveyed by

$$\frac{df(x)}{d\log M^2} = 1, \tag{5–8.159}$$

where

$$\frac{df(x)}{dx} = -\frac{1}{\sigma(1/x)}. \tag{5–8.160}$$

If one extrapolates these definitions down to $M^2 \sim m^2$, where $x$ essentially reduces to $1/e^2$, the following approximate integrated form is obtained:

$$f(x) = f\left(\frac{1}{e^2}\right) + \log\frac{M^2}{m^2}. \tag{5–8.161}$$

The weight function thus depends only on the above combination of coupling constant and spectral mass, as given explicitly by

$$s\left(\frac{M^2}{m^2}, e^2\right) = -\frac{1}{f'\left[f^{-1}\left(f(1/e^2) + \log\frac{M^2}{m^2}\right)\right]}; \tag{5–8.162}$$

here $f'$ and $f^{-1}$ signify the derivative function and the inverse function, respectively. The form of $f(x)$ that corresponds to the expansion of the preceding paragraph, which refers to $x \gg 1$, is

$$f(x) = -\frac{x}{\sigma_0} + \frac{\sigma_1}{\sigma_0^{\,2}}\log x + \left(\frac{\sigma_1^{\,2}}{\sigma_0^{\,3}} - \frac{\sigma_2}{\sigma_0^{\,2}}\right)\frac{1}{x} + \cdots. \tag{5–8.163}$$

With the solution of the functional equation given by (5–8.161), one can verify that (5–8.148) is indeed independent of the arbitrary parameter $\lambda$. For this, we have only to observe that the replacement of $M^2$ by $\lambda^2$ in the variable $x$ produces $1/e_\lambda^{\,2}$, according to (5–8.145), and then

$$f\left(\frac{1}{e_\lambda^{\,2}}\right) = f\left(\frac{1}{e^2}\right) + \log\frac{\lambda^2}{m^2}, \tag{5–8.164}$$

or

$$f\left(\frac{1}{e^2}\right) + \log\frac{M^2}{m^2} = f\left(\frac{1}{e_\lambda{}^2}\right) + \log\frac{M^2}{\lambda^2}. \qquad (5\text{–}8.165)$$

It is thus apparent that the replacement of $e$ and $m$ by $e_\lambda$ and $\lambda$, for arbitrary $\lambda$, also maintains the asymptotic functional form.

In Section 4–3 it has been seen that the use of the asymptotic weight function

$$M^2 a(M^2) \sim \frac{\alpha}{3\pi} \qquad (5\text{–}8.166)$$

leads to a photon propagation function with an inadmissible spacelike singularity (of unphysically large magnitude). This situation would persist if the $\alpha^2$ term were added to (5–8.166). Now, let us turn matters about and ask what is implied by the demand that no such singularity occur. We first note that the second denominator of (5–8.141) is a monotonically decreasing function of $k^2$ ($> 0$),

$$\frac{d}{dk^2}\left[1 - k^2 e^2 \int_0^\infty \frac{dM^2}{M^2}\frac{s}{k^2 + M^2}\right] = -e^2 \int_0^\infty dM^2 \frac{s}{(k^2 + M^2)^2} < 0. \quad (5\text{–}8.167)$$

Accordingly, if this denominator, which equals unity for $k^2 = 0$, is not to have a zero for some finite $k^2 > 0$, it must remain non-negative as $k^2 \to \infty$, or

$$e^2 \int_0^\infty \frac{dM^2}{M^2} s\left(\frac{M^2}{m^2}, e^2\right) \leqslant 1. \qquad (5\text{–}8.168)$$

The first consequence, which is sparked by the necessary existence of the integral, is that $s$ must vanish as $M^2 \to \infty$, at least slightly more rapidly than $1/\log M^2$. It is this requirement that is violated by the simple form (5–8.166).

Suppose the equality sign in (5–8.168) is not realized. Then the function $k^2 e^2 \overline{D}_+(k)$ approaches a definite limit as $k^2 \to \infty$:

$$\lim_{k^2 \to \infty} k^2 e^2 \overline{D}_+(k) = e_\infty{}^2, \qquad (5\text{–}8.169)$$

where, indeed, $e_\infty{}^2$ is the limit of $e_\lambda{}^2$ as $\lambda \to \infty$. In this situation, which is expressed by

$$k^2 \to \infty: \quad e^2 \overline{D}_+(k) \to \frac{e_\infty{}^2}{k^2}, \qquad (5\text{–}8.170)$$

one could interpret $e_\infty$ as the charge appropriate to a more fundamental description whose basic concepts refer to an instantaneous characterization of certain irreducible entities. None of the practical arguments for source theory would be diminished if this should eventually turn out to be the true situation, and it is only for definiteness that we adopt the contrary position (hard-core sourcery), as expressed by the equality sign in (5–6.168):

$$\frac{1}{e^2} = \int_0^\infty \frac{dM^2}{M^2} s\left(\frac{M^2}{m^2}, e^2\right). \tag{5–8.171}$$

The variable $x$ of (5–8.156) could then be written

$$x = \int_{M^2}^\infty \frac{dM'^2}{M'^2} s\left(\frac{M'^2}{m^2}, e^2\right), \tag{5–8.172}$$

and thus approaches zero as $M^2 \to \infty$. The weight function must vanish in the latter limit. The way that it does this is restricted by the implication of (5–8.161) that

$$f(0) = \infty, \tag{5–8.173}$$

which, according to (5–8.160), requires that $\sigma(1/x)$ approach zero, as $x \to 0$, at least as rapidly as $x$. If this behavior is, in fact, linear in $x$,

$$x \to 0: \qquad \sigma(1/x) \sim \kappa x, \quad \kappa > 0, \tag{5–8.174}$$

we have

$$f(x) \sim \frac{1}{\kappa} \log \frac{1}{x} + \text{const.} \tag{5–8.175}$$

and

$$s\left(\frac{M^2}{m^2}, e^2\right) \sim \frac{A}{(M^2)^\kappa}. \tag{5–8.176}$$

Another possibility,

$$s\left(\frac{M^2}{m^2}, e^2\right) \sim \frac{B}{(\log M^2)^2}, \tag{5–8.177}$$

corresponds to $s$ vanishing as $x^2$. These examples only illustrate that, without further physical information, there are endless mathematical possibilities of extrapolating from the initial asymptotic form of the weight function, given in (5–8.153), to the ultimate asymptotic form as $M^2 \to \infty$. And, whereas the initial behavior, with the coefficients (5–8.154), refers to pure electrodynamics, the final asymptotic limit involves the totality of physics. It is for the latter reason that the statement of (5–8.171), which has the form of an eigenvalue equation for $\alpha = e^2/4\pi$, cannot be exploited for that purpose in the present or foreseeable state of physical knowledge.

## 5-9 PHOTON DECAY OF THE PION. A CONFRONTATION

We have already made some reference to the physical process $\pi^0 \to 2\gamma$ in Section 4–3. A phenomenological description of that coupling is given in Eq. (4–3.125), and the associated decay rate appears in Eq. (4–3.139). This topic is taken up again in order to discuss a dynamical model of the mechanism. Neither the model nor our handling of it is realistic; we are not yet ready to conclude the strong-interaction aspects of the process. Rather, the emphasis still remains on electrodynamics. We have a twofold purpose in discussing this problem. The first is to provide another illustration of non-causal computational methods; the second is to confront views of this situation that have gained widespread credence and popularity in the recent literature. The nature of this confrontation will be indicated as the development proceeds.

The analogy between the pion and an electron-positron combination of equivalent quantum numbers has been commented on in Section 4–3, and is implicit in the more extended discussion of Section 5–5. We base a dynamical model on that analogy, in which the pseudoscalar pion field, $\phi(x)$, is locally coupled to the appropriate bilinear combination of fields, $\psi(x)$, that are associated with charged, spin-$\frac{1}{2}$ particles. Those particles can be thought of as protons, but the only explicit characterization of them that enters our model is a restriction to large mass, relative to that of the pion,

$$m \gg m_\pi . \tag{5–9.1}$$

The primitive interaction we have described is exhibited as the Lagrange function term

$$g\phi(x)\,\tfrac{1}{2}\psi(x)\gamma^0\gamma_5\psi(x), \tag{5–9.2}$$

and the initial action expression that refers to the charged particles is therefore

$$W = \int (dx)[\psi\gamma^0\eta + \mathcal{L}],$$

$$\mathcal{L} = -\tfrac{1}{2}\psi\gamma^0(\gamma^\mu \Pi_\mu - g\gamma_5\phi + m)\psi, \tag{5–9.3}$$

where, as always,

$$\Pi_\mu = \frac{1}{i}\partial_\mu - eqA_\mu. \tag{5-9.4}$$

Suppose, now, that field and source are redefined by the local transformation

$$\psi \to \exp\left(\frac{g}{2m}\gamma_5\phi\right)\psi, \qquad \eta \to \exp\left(-\frac{g}{2m}\gamma_5\phi\right)\eta, \tag{5-9.5}$$

although, despite the appearance of the exponential function, only terms at most linear in $\phi$ are of interest in the present discussion. The antisymmetry of $\gamma_5$ and its anticommutativity with $\gamma^0$ combine to maintain the form of the source term, while

$$\begin{aligned}\tfrac{1}{2}\psi\gamma^0(m - g\gamma_5\phi)\psi &\to \tfrac{1}{2}\psi\gamma^0(m - g\gamma_5\phi)\exp\left(\frac{g}{m}\gamma_5\phi\right)\psi \\ &\to \tfrac{1}{2}\psi\gamma^0 m\psi,\end{aligned} \tag{5-9.6}$$

and

$$\begin{aligned}\tfrac{1}{2}\psi\gamma^0\gamma^\mu\Pi_\mu\psi &\to \tfrac{1}{2}\psi\gamma^0\gamma^\mu\exp\left(-\frac{g}{2m}\gamma_5\phi\right)\Pi_\mu\exp\left(\frac{g}{2m}\gamma_5\phi\right)\psi \\ &= \tfrac{1}{2}\psi\gamma^0\left(\gamma^\mu\Pi_\mu - \frac{g}{2m}i\gamma^\mu\gamma_5\partial_\mu\phi\right)\psi.\end{aligned} \tag{5-9.7}$$

Note that only the first of these transformations has actually involved the restriction to no more than linear $\phi$-terms. We recognize in this way that, correct to terms linear in the field $\phi$, there is an equivalence between pseudoscalar and pseudovector coupling:

$$g\gamma_5\phi \leftrightarrow \frac{g}{2m}i\gamma^\mu\gamma_5\partial_\mu\phi. \tag{5-9.8}$$

The question whether this equivalence is indeed realized in explicit calculations, specifically of the radiative decay, is at the heart of the somewhat controversial problems to be studied.

The elementary coupling between the field $\phi$ and the fields of two photons is produced by considering the exchange of a pair of charged particles. Such considerations are entirely analogous to those given in Section 4-8 for the discussion of light by light scattering, and we have only to apply the formula of Eq. (4-8.19), through the substitution

$$eq\gamma\delta A \to g\gamma_5\phi, \tag{5-9.9}$$

to get the following interaction expression for pseudoscalar coupling:

$$\begin{aligned} W &= \tfrac{1}{2}i \operatorname{Tr} g\gamma_5\phi G_+^{\ A} \\ &= \tfrac{1}{2}i \operatorname{Tr} g\gamma_5\phi \frac{m - \gamma\Pi}{\Pi^2 + m^2 - eq\sigma F}. \end{aligned} \tag{5–9.10}$$

Owing to the presence of the matrix $\gamma_5$, the $\gamma\Pi$ term disappears from the trace, and the desired interaction is produced entirely through the $\sigma F$ term in the denominator. Thus, we have

$$W_{\pi^0 \to 2\gamma} = \tfrac{1}{2}im \operatorname{Tr} g\gamma_5\phi \frac{1}{\Pi^2 + m^2} eq\sigma F \frac{1}{\Pi^2 + m^2} eq\sigma F \frac{1}{\Pi^2 + m^2}. \tag{5–9.11}$$

We shall evaluate this expression in an approximation that is based on the mass inequality (5–9.1). Since the photons share equally the total energy $m_\pi$ in the pion rest frame, they carry small momenta on the scale set by $m$. Accordingly, it suffices to regard their fields as slowly varying. Then, recalling that [Eq. (4–8.76)]

$$\operatorname{tr}_{(4)} \gamma_5(\sigma F)^2 = -8\mathbf{E} \cdot \mathbf{H}, \tag{5–9.12}$$

and employing the kind of Tr evaluation exhibited in (4–8.56), we get

$$\begin{aligned} W_{\pi^0 \to 2\gamma} &= -8imge^2 \int \frac{(dx)(dp)}{(2\pi)^4} \phi(x)\mathbf{E} \cdot \mathbf{H}(x) \frac{1}{(p^2 + m^2)^3} \\ &= \frac{\alpha}{\pi} \frac{g}{m} \int (dx)\phi\mathbf{E} \cdot \mathbf{H}. \end{aligned} \tag{5–9.13}$$

In stating the final form, we have used the integral

$$\int \frac{(dp)}{(2\pi)^4} \frac{1}{(p^2 + m^2)^3} = i\frac{1}{32\pi^2} \frac{1}{m^2}, \tag{5–9.14}$$

which is analogous to, and derivable from, (4–10.40). As in the discussion of the latter, we also remark on the alternative evaluation through transformation to a Euclidean metric ($p_0 \to ip_4$) and the use of the surface area of a unit sphere in four dimensions, $2\pi^2$:

$$\int \frac{(dp)}{(2\pi)^4} \frac{1}{(p^2 + m^2)^3} = i\frac{2\pi^2}{(2\pi)^4} \int_0^\infty \frac{\tfrac{1}{2}p^2\, dp^2}{(p^2 + m^2)^3} = 1\frac{1}{32\pi^2} \frac{1}{m^2}. \tag{5–9.15}$$

Then, applying the results of Section 4–3 that were cited in the first paragraph of this section, we infer the pion decay rate as

$$\gamma_{\pi^0 \to 2\gamma} = \left(\frac{\alpha}{4\pi}\right)^2 \frac{g^2}{4\pi} \left(\frac{m_\pi}{m}\right)^2 m_\pi . \tag{5–9.16}$$

Now let us see what happens when the equivalent pseudovector coupling of (5–9.8) is used in (5–9.10):

$$\begin{aligned} W &= \tfrac{1}{2} i \operatorname{Tr} \frac{g}{2m} i\gamma^\mu \gamma_5 \partial_\mu \phi G_+{}^A \\ &= \tfrac{1}{2} i \operatorname{Tr} \frac{g}{2m} i\gamma^\mu \gamma_5 \partial_\mu \phi \frac{m - \gamma\Pi}{\Pi^2 + m^2 - eq\sigma F} . \end{aligned} \tag{5–9.17}$$

In this situation, the presence of the additional matrix $\gamma^\mu$ singles out the $\gamma\Pi$ term in the trace and, as we shall verify in a moment, it suffices to consider one power of $\sigma F$ in the expansion of the denominator,

$$W_{\pi^0 \to 2\gamma} = -\tfrac{1}{2} i \operatorname{Tr} \frac{g}{2m} i\gamma^\mu \gamma_5 \partial_\mu \phi \gamma\Pi \frac{1}{\Pi^2 + m^2} eq\sigma F \frac{1}{\Pi^2 + m^2} . \tag{5–9.18}$$

Contact with the pseudoscalar form of the coupling should appear on transferring the derivative from the pion field. In doing this, it is important to maintain the appearance of gauge invariance, which is accomplished by writing

$$i\partial_\mu \phi = \left[\phi, \Pi_\mu\right], \tag{5–9.19}$$

and then

$$W = -\tfrac{1}{2} i \frac{g}{2m} \operatorname{Tr} \gamma^\mu \gamma_5 \phi \left[\Pi_\mu, \gamma\Pi \frac{1}{\Pi^2 + m^2} eq\sigma F \frac{1}{\Pi^2 + m^2}\right] . \tag{5–9.20}$$

The approximation of slowly varying fields ($m_\pi \ll m$) is used again to rewrite this as

$$W = \tfrac{1}{2} i \frac{g}{2m} \operatorname{Tr} \gamma_5 \gamma^\mu \gamma^\nu \phi eq\sigma F \left[\Pi_\mu, \Pi_\nu \frac{1}{(\Pi^2 + m^2)^2}\right] . \tag{5–9.21}$$

It is the presence of the commutator that introduces an additional field-strength factor, thus justifying the omission of $(\sigma F)^2$. The use of the basic commutator

$$\left[\Pi_\mu, \Pi_\nu\right] = ieqF_{\mu\nu} , \tag{5–9.22}$$

in the sufficiently accurate approximate form

$$\left[\Pi_\mu, f(\Pi)\right] = ieqF_{\mu\lambda}\frac{\partial}{\partial \Pi_\lambda}f(\Pi), \tag{5–9.23}$$

enables (5–9.21) to be presented as

$$W_{\pi^0\to 2\gamma} = e^2\frac{g}{2m}\int (dx)\,\mathrm{tr}_{(4)}\,\gamma_5\sigma^{\mu\nu}F_{\mu\lambda}\sigma F\phi i\int\frac{(dp)}{(2\pi)^4}\frac{\partial}{\partial p_\lambda}\frac{p_\nu}{\left(p^2+m^2\right)^2}. \tag{5–9.24}$$

The introduction of the Euclidean metric, and an evident symmetry, permits the latter momentum integral to be evaluated as a surface integral extended over a momentum sphere of large radius:

$$i\int\frac{(dp)}{(2\pi)^4}\frac{\partial}{\partial p_\lambda}\frac{p_\nu}{\left(p^2+m^2\right)^2} = -\tfrac{1}{4}\delta_\nu{}^\lambda\int\frac{(dp)_E}{(2\pi)^4}\frac{\partial}{\partial p}\frac{p}{\left(p^2+m^2\right)^2}$$

$$= -\tfrac{1}{4}\delta_\nu{}^\lambda\frac{2\pi^2}{(2\pi)^4}\frac{p^4}{\left(p^2+m^2\right)^2}\Bigg|_{p^2\gg m^2} = -\frac{1}{32\pi^2}\delta_\nu{}^\lambda. \tag{5–9.25}$$

Its consequence for (5–9.24) is

$$W_{\pi^0\to 2\gamma} = \frac{\alpha}{\pi}\frac{g}{m}\int (dx)\phi\mathbf{E}\cdot\mathbf{H}, \tag{5–9.26}$$

in agreement with (5–9.13).

Harold lifts an eyebrow.

H.: It is very interesting to see the emergence of a surface integral in momentum space as the instrument for maintaining the equivalence theorem of (5–9.8). But why have you abandoned your usual exponential representation method for handling such problems? Could it be that it has no counterpart to the comparative subtlety of a surface integral?

S.: Shall we find out?

Let us return to the pseudoscalar expression (5–9.10) and proceed to run rapidly through the evaluation

$$W = \tfrac{1}{2}i\,\mathrm{Tr}\,g\gamma_5\phi(m-\gamma\Pi)i\int_0^\infty ds\,\exp\left[-is\left(\Pi^2+m^2-eq\sigma F\right)\right]$$

$$\to \tfrac{1}{4}e^2m\,\mathrm{Tr}\,g\gamma_5\phi(\sigma F)^2\int_0^\infty ds\,s^2\exp\left[-is\left(\Pi^2+m^2\right)\right], \tag{5–9.27}$$

or

$$W_{\pi^0\to 2\gamma} = -4e^2gm\int\frac{(dx)(dp)}{(2\pi)^4}\phi\mathbf{E}\cdot\mathbf{H}\int_0^\infty ds\, s^2\exp[-is(p^2+m^2)]$$

$$= \frac{\alpha}{\pi}gm\int(dx)\phi\mathbf{E}\cdot\mathbf{H}i\int ds\, e^{-ism^2}$$

$$= \frac{\alpha}{\pi}\frac{g}{m}\int(dx)\phi\mathbf{E}\cdot\mathbf{H}, \qquad (5\text{–}9.28)$$

as expected. It is the pseudovector coupling that needs our attention. The latter [Eq. (5–9.17)] is now written as

$$W = \tfrac{1}{2}i\,\mathrm{Tr}\,\frac{g}{2m}i\gamma^\mu\gamma_5\partial_\mu\phi(m-\gamma\Pi)i\int_0^\infty ds\exp[-is(\Pi^2+m^2-eq\sigma F)]$$

$$\to \tfrac{1}{2}i\frac{g}{2m}\,\mathrm{Tr}\,eq\sigma Fi\gamma^\mu\gamma_5\partial_\mu\phi\gamma\Pi\int_0^\infty ds\, s\exp[-is(\Pi^2+m^2)], \qquad (5\text{–}9.29)$$

which again anticipates that one $\sigma F$ factor will suffice. The use of the commutator form (5–9.19) gives

$$W_{\pi^0\to 2\gamma} = -\tfrac{1}{2}i\frac{g}{2m}\,\mathrm{Tr}\,eq\sigma F\gamma^\mu\gamma^\nu\gamma_5\phi\left[\Pi_\mu, \Pi_\nu\int_0^\infty ds\, s\exp[-is(\Pi^2+m^2)]\right], \qquad (5\text{–}9.30)$$

but this time we evaluate the commutator directly, in the sufficient approximation of Eq. (5–9.23):

$$\left[\Pi_\mu, \Pi_\nu e^{-is\Pi^2}\right] = ieqF_{\mu\nu}e^{-is\Pi^2} - 2isieqF_{\mu\lambda}\Pi^\lambda\Pi_\nu e^{-is\Pi^2}. \qquad (5\text{–}9.31)$$

The Tr evaluation presents the interaction as

$$W_{\pi^0\to 2\gamma} = -ie^2\frac{g}{2m}\int(dx)\,\mathrm{tr}_{(4)}\,\sigma F\sigma^{\mu\nu}\gamma_5\phi$$

$$\times\int_0^\infty ds\, s\int\frac{(dp)}{(2\pi)^4}\left(F_{\mu\nu} - 2isF_{\mu\lambda}p^\lambda p_\nu\right)e^{-is(p^2+m^2)}$$

$$= 8e^2\frac{g}{m}\int(dx)\phi\mathbf{E}\cdot\mathbf{H}\int_0^\infty ds\, s\int\frac{(dp)_E}{(2\pi)^4}\left(1-\tfrac{1}{2}sp^2\right)e^{-s(p^2+m^2)}, \qquad (5\text{–}9.32)$$

where the latter form introduces the Euclidean metric, with the related transformation $s \to -is$, and has employed an obvious four-dimensional symmetry. The question of equivalence thus reduces to whether the final integrations will yield

$$\int_0^\infty ds\, s \int \frac{(dp)_E}{(2\pi)^4}\left(1 - \tfrac{1}{2}sp^2\right)e^{-s(p^2+m^2)} = \frac{1}{32\pi^2}. \qquad (5\text{–}9.33)$$

That the answer requires some care becomes apparent on inserting the infinite-momentum-space Euclidean integrals

$$\int \frac{(dp)_E}{(2\pi)^4} e^{-sp^2} = \frac{1}{(4\pi)^2}\frac{1}{s^2},$$

$$\int \frac{(dp)_E}{(2\pi)^4} p^2 e^{-sp^2} = -\frac{\partial}{\partial s}\int \frac{(dp)_E}{(2\pi)^4} e^{-sp^2} = \frac{2}{s}\frac{1}{(4\pi)^2}\frac{1}{s^2}. \qquad (5\text{–}9.34)$$

The left side of (5–9.33) vanishes!

This is where we must stop, and think about physics. Implicit in any non-causal calculation is the requirement of initial non-overlap between the fields of the emitting and absorbing sources—here, the pion field $\phi$ and the two photon fields, respectively. The complement to this insistence on a finite space-time interval between the two classes of fields is a limitation on the magnitude of the momentum that is exchanged between them. To convey this restriction in a calculation that has not made it explicit, which is our present situation, we must allow the domain of momentum integration to become infinite only at the end of the calculation, corresponding to the final extrapolation to the circumstance of overlapping fields. (A similar remark occurs in Section 4–8, in the context of light-by-light scattering.) With this comment in mind, we return to the test of (5–9.33), and first compute the momentum integrals for a Euclidean sphere of finite radius $P$:

$$\int \frac{(dp)_E}{(2\pi)^4} e^{-sp^2} = \int_0^{P^2} \frac{\pi^2 p^2 dp^2}{(2\pi)^4} e^{-sp^2} = \frac{1}{(4\pi)^2}\left(-\frac{d}{ds}\right)\frac{1 - e^{-sP^2}}{s}$$

$$= \frac{1}{(4\pi)^2}\left[\frac{1}{s^2} + \frac{d}{ds}\frac{e^{-sP^2}}{s}\right], \qquad (5\text{–}9.35)$$

and

$$\int \frac{(dp)_E}{(2\pi)^4} p^2 e^{-sp^2} = \frac{1}{(4\pi)^2}\left[\frac{2}{s^3} - \frac{d^2}{ds^2}\frac{e^{-sP^2}}{s}\right]. \qquad (5\text{–}9.36)$$

The left side of (5–9.33) then reads

$$\frac{1}{(4\pi)^2}\int_0^\infty ds\, se^{-sm^2}\left(\frac{d}{ds}+\tfrac{1}{2}s\frac{d^2}{ds^2}\right)\frac{e^{-sP^2}}{s}$$

$$=\frac{1}{32\pi^2}\int_0^\infty ds\, se^{-sm^2}\frac{d^2}{ds^2}e^{-sP^2}, \tag{5–9.37}$$

where, indeed, if $P^2 \to \infty$ inside the integration sign, for any finite $s$, the result is zero. But, if we retain a finite but large $P^2 \gg m^2$ until the $s$-integration is performed, the final integral of (5–9.37) becomes, effectively,

$$P^4\int_0^\infty ds\, se^{-sP^2} = 1, \tag{5–9.38}$$

and (5–9.33) is verified. In retrospect, it is evident that the surface-integral evaluation also refers to a large but finite momentum domain, and that the two computations are equivalent. To the latter remark we add the specific observations that first performing the $s$-integration in (5–9.33) gives

$$\int\frac{(dp)_E}{(2\pi)^4}\left[\frac{1}{(p^2+m^2)^2}-\frac{p^2}{(p^2+m^2)^3}\right]=\int\frac{(dp)_E}{(2\pi)^4}\frac{1}{4}\frac{\partial}{\partial p}\frac{p}{(p^2+m^2)^2} \tag{5–9.39}$$

and that the four-dimensional momentum integral on the left of (5–9.39) is also the one encountered in the pseudoscalar calculation [Eq. (5–9.14)].

Now that we have brought to the surface the circumstance of initial non-overlap and final space-time extrapolation, it is natural to ask how these matters go in a causal calculation, where the extrapolation procedure is quite explicit. Consider, then, a causal arrangement in which an extended pion source emits a pair of charged particles that eventually annihilate to produce a pair of photons. [The causal diagram can be drawn as an isosceles triangle standing upon its apex, to which the virtual pion line is attached. The opposing, horizontally drawn side refers to a virtual charged particle.] The primitive pseudoscalar coupling defines an effective two-particle emission source:

$$i\eta_2(x)\eta_2(x')\gamma^0\big|_{\text{eff.}} = \delta(x-x')g\gamma_5\phi(x), \tag{5–9.40}$$

while the effective two-particle detection source associated with the two-photon emission process is [cf. Eq. (4–8.3)]

$$i\eta_1(x)\eta_1(x')\gamma^0\big|_{\text{eff.}} = eq\gamma A(x)G_+(x-x')eq\gamma A(x'). \tag{5–9.41}$$

The vacuum amplitude describing the two-particle exchange, as inferred from (4–8.4), is then

$$-\frac{1}{2}\int(dx)(dx')(dx'')$$

$$\times\operatorname{tr}[eq\gamma A(x)G_+(x-x')eq\gamma A(x')G_+(x'-x'')g\gamma_5\phi(x'')G_+(x''-x)], \quad (5\text{–}9.42)$$

which also follows from (5–9.10) on inserting the expansion of $G_+^{\,A}$.

The causal situation under consideration is conveyed by the propagation function forms

$$G_+(x'-x'') = i\int d\omega_p\, e^{ip(x'-x'')}(m-\gamma p),$$

$$G_+(x''-x) = i\int d\omega_{p'}\, e^{ip'(x-x'')}(m+\gamma p'), \quad (5\text{–}9.43)$$

and by writing [cf. Eq. (4–9.3)]

$$A^\mu(x) = A_a^{\,\mu}(x) + A_b^{\,\mu}, \quad (5\text{–}9.44)$$

where $a$ and $b$ designate the two photons, with, for example,

$$A_a^{\,\mu}(x) = iJ_a^{\,*}(d\omega_a)^{1/2}e_a^{\,\mu *}e^{-ik_a x}. \quad (5\text{–}9.45)$$

The resulting expression for the vacuum amplitude is

$$iJ_a^{\,*}iJ_b^{\,*}(d\omega_a\, d\omega_b)^{1/2}e^2 g\int(dK)\delta(k_a+k_b-K)I\phi(K), \quad (5\text{–}9.46)$$

in which

$$I = \int d\omega_p\, d\omega_{p'}\,(2\pi)^4\delta(p+p'-K)$$

$$\times\operatorname{tr}_{(4)}\left[\left\{\gamma e_a^{\,*}\frac{1}{\gamma(k_a-p')+m}\gamma e_b^{\,*} + \gamma e_b^{\,*}\frac{1}{\gamma(k_b-p')+m}\gamma e_a^{\,*}\right\}\right.$$

$$\left.\times(m-\gamma p)\gamma_5(m+\gamma p')\right]. \quad (5\text{–}9.47)$$

Owing to the presence of matrix $\gamma_5$, the significant structure of the trace, which is illustrated by

$$\operatorname{tr}_{(4)}\left[\gamma e_a{}^*(m-\gamma(k_a-p'))\gamma e_b{}^*(m-\gamma p)\gamma_5(m+\gamma p')\right], \tag{5-9.48}$$

reduces to the products of four different components of $\gamma^\mu$, as given in

$$\tfrac{1}{4}\operatorname{tr}_{(4)}\gamma^\kappa\gamma^\lambda\gamma^\mu\gamma^\nu\gamma_5=-\varepsilon^{\kappa\lambda\mu\nu}. \tag{5-9.49}$$

These contributions are associated, in three different ways, with the presence of a single factor of $m$ in (5–9.48):

$$\begin{aligned} &m\left[\operatorname{tr}_{(4)}\gamma e_a{}^*\gamma e_b{}^*\gamma p\gamma p'\gamma_5-\operatorname{tr}_{(4)}\gamma e_a{}^*\gamma e_b{}^*\gamma k_a\gamma p'\gamma_5-\operatorname{tr}_{(4)}\gamma e_a{}^*\gamma e_b{}^*\gamma p\gamma k_b\gamma_5\right]\\ &\quad=-m\operatorname{tr}_{(4)}\gamma e_a{}^*\gamma e_b{}^*\gamma k_a\gamma k_b\gamma_5\\ &\quad=-4m\varepsilon_{\kappa\lambda\mu\nu}k_a{}^\kappa e_a{}^{\lambda *}k_b{}^\mu e_b{}^{\nu *}; \end{aligned} \tag{5-9.50}$$

the intermediate step in this reduction exploits the equality of the vectors $p-k_a$ and $-(p'-k_b)$.

The consequent form of $I$ is the product of (5–9.50) with the invariant momentum-space integral

$$\int d\omega_p\,d\omega_{p'}\,(2\pi)^4\delta(p+p'-K)\left[\frac{1}{-2k_ap'}+\frac{1}{-2k_bp'}\right]. \tag{5-9.51}$$

This integral is easily evaluated in the rest frame of $K$ ($\mathbf{K}=0,\ K^0=M$), where all particles and photons have the energy $\frac{1}{2}M$, and the integration reduces to one over a scattering angle of cosine $z$:

$$\begin{aligned} &\frac{1}{4\pi}\frac{1}{M^2}\left[\left(\tfrac{1}{2}M\right)^2-m^2\right]^{1/2}\int_{-1}^{1}dz\,\frac{1}{\frac{1}{2}M-\left[\left(\frac{1}{2}M\right)^2-m^2\right]^{1/2}z}\\ &\quad=\frac{1}{4\pi}\frac{1}{M^2}\log\frac{1+v}{1-v}, \end{aligned} \tag{5-9.52}$$

in which

$$v=\left(1-\frac{4m^2}{M^2}\right)^{1/2}. \tag{5-9.53}$$

Having devised the vacuum amplitude (5–9.46) for causal circumstances, we now make that space-time arrangement explicit by writing

$$(2\pi)^4\delta(k_a + k_b - K) = \int (dx) e^{-ik_a x} e^{-ik_b x} e^{iKx},$$

$$\phi(K) = \int (dx') e^{-iKx'}\phi(x'), \tag{5-9.54}$$

and then proceed to reconstruct the electromagnetic field, which is already expressed in terms of gauge-invariant field strengths. After inserting the relation

$$\frac{(dK)}{(2\pi)^3} = d\omega_K\, dM^2, \tag{5-9.55}$$

the vacuum amplitude becomes

$$i\int (dx)(dx')\, dM^2 \frac{\alpha}{\pi}\frac{g}{m}(-\tfrac{1}{4})\tfrac{1}{2}\varepsilon_{\kappa\lambda\mu\nu}F^{\kappa\lambda}(x)F^{\mu\nu}(x)$$

$$\times\left[i\int d\omega_K\, e^{iK(x-x')}\right]\phi(x')\frac{2m^2}{M^2}\log\frac{1+v}{1-v}, \tag{5-9.56}$$

and the space-time extrapolation is performed by the substitution

$$i\int d\omega_K\, e^{iK(x-x')} \to \Delta_+(x - x', M^2). \tag{5-9.57}$$

The resulting contribution to the action can be presented as

$$W_{\pi^0\to 2\gamma} = \frac{\alpha}{\pi}\frac{g}{m}\int (dx)(dx')\mathbf{E}\cdot\mathbf{H}(x)F(x - x')\phi(x'), \tag{5-9.58}$$

where the form factor has the momentum version

$$F(k^2) = \int_{(2m)^2}^{\infty} dM^2 \frac{2m^2}{M^2}\log\frac{1+v}{1-v}\frac{1}{k^2 + M^2}. \tag{5-9.59}$$

It is normalized at $k^2 = 0$:

$$F(0) = \int_{(2m)^2}^{\infty} dM^2 \frac{2m^2}{(M^2)^2}\log\frac{1+v}{1-v} = \int_0^1 dv\, v \log\frac{1+v}{1-v} = 1. \tag{5-9.60}$$

Under circumstances of small momentum transfer, $|k^2| \ll 4m^2$, $F(x - x')$ effectively becomes $\delta(x - x')$ and (5–9.58) reduces to the local coupling of (5–9.13), as expected. In the actual situation, where $-k^2 = m_\pi^2$, there is a small correction factor which, according to (5–9.59), is

$$F\left(-m_\pi^2\right) \cong 1 + m_\pi^2 \int_{(2m)^2}^{\infty} dM^2 \frac{2m^2}{\left(M^2\right)^2} \log\frac{1+v}{1-v}$$

$$= 1 + \frac{1}{12}\left(\frac{m_\pi}{m}\right)^2. \tag{5–9.61}$$

If $m$ is taken to be the proton mass, so that $m_\pi/m \cong 1/6.7$, the correction is about 0.2%.

The replacement of pseudoscalar by pseudovector coupling is expressed, in the causal vacuum amplitude of (5–9.46, 47), by the substitution

$$\gamma_5 \to \frac{1}{2m}\gamma_5\gamma K, \tag{5–9.62}$$

or

$$\gamma_5 \to \gamma_5 - \frac{1}{2m}(m + \gamma p)\gamma_5 - \frac{1}{2m}\gamma_5(m - \gamma p'). \tag{5–9.63}$$

Then, since $p$ and $p'$ are real particle momenta, $p^2 + m^2 = p'^2 + m^2 = 0$, the effect in (5–9.47) is such as to immediately recover the pseudoscalar coupling. Why, then, is there any question about the equivalence of the two coupling forms in this process? To provoke that problem one must retain the pseudovector form until after the space-time extrapolation is performed.

The illustrative trace of (5–9.48) now becomes

$$\mathrm{tr}_{(4)}\left[\gamma e_a^*(m - \gamma(k_a - p'))\gamma e_b^*(m - \gamma p)\gamma_5\gamma^\mu(m + \gamma p')\right]. \tag{5–9.64}$$

It can be decomposed into two distinct contributions, of which the first is the following three terms with the factor $m^2$,

$$m^2\left\{-2\,\mathrm{tr}_{(4)}\,\gamma e_a^*\gamma(k_a - p')\gamma e_b^*\gamma_5\gamma^\mu - \mathrm{tr}_{(4)}\,\gamma e_a^*\gamma e_b^*\gamma p\gamma_5\gamma^\mu + \mathrm{tr}_{(4)}\,\gamma e_a^*\gamma e_b^*\gamma_5\gamma^\mu\gamma p'\right\}$$

$$= -4m^2\varepsilon^{\kappa\lambda\mu\nu}e_{a\kappa}^* e_{b\lambda}^*(k_a - k_b)_\nu, \tag{5–9.65}$$

while the remainder is

$$\mathrm{tr}_{(4)}\left[\gamma e_a^*\gamma(k_a - p')\gamma e_b^*\left(\gamma p\gamma_5\gamma^\mu\gamma p' + m^2\gamma_5\gamma^\mu\right)\right]. \tag{5–9.66}$$

The evaluation of the latter trace is assuredly feasible, but we shall not trouble to do it, since

$$\left(\gamma p \gamma_5 \gamma^\mu \gamma p' + m^2 \gamma_5 \gamma^\mu\right) K_\mu = 0. \tag{5–9.67}$$

This means that the term in question does not contribute to the causal process and can only enter as a contact term. It may therefore be put aside until the end of the calculation, where any necessary contact term is to be inserted on the basis of the physical requirements that attend the calculation. In the present situation, the only requirement is that gauge invariance be maintained after the space-time extrapolation.

With the alterations we have indicated, the causal vacuum amplitude of (5–9.56) will be replaced by

$$i\int (dx)(dx')\, dM^2 \frac{\alpha}{\pi}\frac{g}{m}\left(-\frac{1}{4}\right)\left(-2\,{}^*F^{\mu\nu}(x)A_\nu(x)\right)$$

$$\times\left[i\int d\omega_K\, e^{iK(x-x')}\right]\partial'_\mu \phi(x')\frac{2m^2}{M^2}\log\frac{1+v}{1-v}, \tag{5–9.68}$$

where, as in Eq. (3–8.14),

$${}^*F^{\mu\nu} = \tfrac{1}{2}\varepsilon^{\mu\nu\kappa\lambda}F_{\kappa\lambda}. \tag{5–9.69}$$

The issue can now be squarely drawn. Does the pseudovector structure ${}^*F^{\mu\nu}A_\nu$ have an independent existence that demands for it an explicitly gauge-invariant form, or is it merely one factor in a complete expression which should be gauge-invariant only in its entirety? The proponents of current algebra take the first view; we champion the second one. With the latter attitude, one has only to remark that the transfer of the derivative in (5–9.68) produces

$$\partial_\mu\left[2\,{}^*F^{\mu\nu}A_\nu\right] = {}^*F^{\mu\nu}F_{\mu\nu} = \tfrac{1}{2}\varepsilon_{\kappa\lambda\mu\nu}F^{\kappa\lambda}F^{\mu\nu}, \tag{5–9.70}$$

and we immediately recover (5–9.56), which, with no need for a contact term, directly yields the action contribution of (5–9.58). But, if ${}^*F^{\mu\nu}A_\nu$ must be made explicitly gauge-invariant, while maintaining the pseudovector field, it is necessary to make the following substitution

$$\partial'_\mu \phi(x') \to \frac{1}{M^2}\partial'_\mu \partial^{\lambda\prime}\partial'_\lambda \phi(x'), \tag{5–9.71}$$

and then the transfer of the first two derivatives produces the vacuum amplitude

$$i\int(dx)(dx')\,dM^2\frac{\alpha}{\pi}\frac{g}{m}\left[\partial_\lambda\tfrac{1}{4}{}^*F^{\mu\nu}F_{\mu\nu}(x)\right]$$

$$\times\left[i\int d\omega_K\,e^{iK(x-x')}\right]\partial'_\lambda\phi(x')\frac{2m^2}{(M^2)^2}\log\frac{1+v}{1-v}. \qquad (5\text{–}9.72)$$

If the resulting coupling is applied to the decay process where, effectively, $\partial'^2 \to m_\pi^{\ 2}$, the additional factor of $m_\pi^{\ 2}/M^2 < m_\pi^{\ 2}/4m^2$ yields an essentially null result for the pion decay constant. And, indeed, this was the conclusion drawn from the initial application of current algebra to the process. [D. Sutherland, M. Veltman, 1967. These and other references, as well as a careful discussion of the current-algebra viewpoint, can be found in the contribution of R. Jackiw to *Lectures on Current Algebra and Its Applications*, Princeton University Press, N.J., 1972].

The breakdown of current algebra thus revealed could be traced to the neglect of 'anomalous' equal-time commutators of certain current components. This customary language is unfortunate, for although these additional terms do not appear when formal operator manipulations are employed, their presence is demanded by general physical requirements. Here is the reason that no 'anomaly' occurs in the source-theory discussion—we have utilized the physical requirements directly, without reference to operators. Incidentally, we have also seen the possibility of obtaining a null result by purely mathematical manipulations, and then recognized that it originated in insufficient attention to the physical context of the calculation. It may appear to be a trivial semantic point to deplore the use of the term 'anomaly', since the final current-algebraic description of the pion-photon coupling, at the level of dynamics now under consideration, is the expected one. Yet, like all inappropriate usages of language, it can and has led to error.

The point at issue refers to higher dynamical levels, where the internal exchange of additional photons is taken into account. To the current algebraist, the coupling of (5–9.13),

$$W_{\pi^0\to 2\gamma} = \frac{\alpha}{\pi}\frac{g}{m}\int(dx)\phi\mathbf{E}\cdot\mathbf{H}, \qquad (5\text{–}9.73)$$

is an anomaly associated with the peculiarities of the 'triangle diagram', and no further contributions are expected from more elaborate mechanisms [S. Adler, 1969]. Thus the coupling (5–9.73) is alleged to be valid 'to all orders in $\alpha$'. And detailed calculations have been carried out by various authors with results that are interpreted as supporting this dictum. But independent source-theoretic calculations of the causal type have also been performed by members of the UCLA Sourcery Group [Lester L. De Raad, Jr., Kimball A. Milton, Wu-Yang Tsai, Phys. Rev. D 6, 1766 (1972); Kimball A. Milton, Wu-Yang Tsai, Lester L. De Raad, Jr., Phys. Rev. D 6, 3491 (1972)] which show that, at the level of one internally exchanged photon, the coupling of (5–9.73) *is* modified, by the factor

$$1+\frac{\alpha}{2\pi}. \tag{5–9.74}$$

These calculations have met the source-theory requirement of internal consistency by being performed for two different causal arrangements, with concordant results. But, since they are rather elaborate, one might wish for a more transparent attack on this conceptually important question. We shall fill this need by using the noncausal approach, and therefore refer the interested reader to the above cited papers for the alternative causal calculations. That we present a simpler method detracts in no way from the significant achievement of the three Sourcerers in pushing their calculations through to a conclusion, and defending it against the firm Establishment ruling that no such effect could exist.

The discussion to follow is quite similar to that of Section 5–8, with one of the photon interactions replaced by the pion coupling, in accordance with (5–9.9) or the alternative of (5–9.8). However, we lack some of the information that was available in the purely electrodynamic discussion, namely, the dynamical modification (to order $\alpha$) of the primitive interaction (5–9.2) and of its pseudovector equivalent. To that end, let us consider the appropriate modification of the Eqs. (5–6.2, 3) as expressed, for pseudoscalar coupling, by

$$(\gamma\Pi - g\gamma_5\phi + m + M)\overline{G}_+ = 1,$$

$$M = ie^2\int\frac{(dk)}{(2\pi)^4}\gamma^\nu\frac{1}{k^2}\frac{1}{\gamma(\Pi-k)-g\gamma_5\phi+m}\gamma_\nu + \text{c.t.} \tag{5–9.75}$$

Attention now focuses on the part of $M$ that is linear in $\phi$:

$$M_{\text{ps.}} = ie^2\int\frac{(dk)}{(2\pi)^4}\gamma^\nu\frac{1}{k^2}\frac{1}{\gamma(\Pi-k)+m}g\gamma_5\phi\frac{1}{\gamma(\Pi-k)+m}\gamma_\nu + \text{c.t.}, \tag{5–9.76}$$

which we write out as a typical matrix element involving the particle momenta $p'$, $p''$, with the pion field supplying the momentum

$$p = p' - p'', \tag{5–9.77}$$

namely

$$M_{\text{ps.}} = ie^2 \int \frac{(dk)}{(2\pi)^4} \gamma^\nu \frac{1}{k^2} \frac{m - \gamma(p' - k)}{(p' - k)^2 + m^2} g\gamma_5\phi \frac{m - \gamma(p'' - k)}{(p'' - k)^2 + m^2} \gamma_\nu + \text{c.t.} \quad (5\text{–}9.78)$$

Here, we have

$$\begin{aligned}\gamma^\nu[m - \gamma(p' - k)]\gamma_5[m - \gamma(p'' - k)]\gamma_\nu \\ = 2\gamma_5\left[(p' - k)^2 + m^2 + (p'' - k)^2 + m^2 - p^2 - 2m^2\right] \\ +2m[m + \gamma(p' - k)]\gamma_5 + 2m\gamma_5[m + \gamma(p'' - k)]\end{aligned} \quad (5\text{–}9.79)$$

and

$$\frac{1}{(p' - k)^2 + m^2} \frac{1}{(p'' - k)^2 + m^2} \frac{1}{k^2} = -i\int_0^\infty ds\, s^2 \int_0^1 du\, u \int_{-1}^1 \tfrac{1}{2}\, dv\, e^{-is\chi}, \quad (5\text{–}9.80)$$

with

$$\chi = \left[k - u\left(p'\frac{1+v}{2} + p''\frac{1-v}{2}\right)\right]^2 + D, \quad (5\text{–}9.81)$$

$$\begin{aligned}D = u(1 - u)\left[\frac{1+v}{2}(p'^2 + m^2) + \frac{1-v}{2}(p''^2 + m^2)\right] \\ + u^2\left(m^2 + \frac{1 - v^2}{4}p^2\right) + (1 - u)\mu^2,\end{aligned}$$

which also incorporates a finite photon mass. In addition, owing to the appearance of the factors $(p' - k)^2 + m^2$ and $(p'' - k)^2 + m^2$ in (5–9.79), we need the simpler combination

$$\frac{1}{(p' - k)^2 + m^2} \frac{1}{k^2} = -\int ds\, s\, du\, e^{-is\chi_1'}, \quad (5\text{–}9.82)$$

$$\chi_1' = (k - up')^2 + D_1',$$

$$D_1' = u(1 - u)(p'^2 + m^2) + u^2m^2 + (1 - u)\mu^2$$

and its analogue with $p', \chi_1', D_1' \to p'', \chi_1'', D_1''$.

After making substitutions such as

$$k \to k + u\left(p'\frac{1+v}{2} + p''\frac{1-v}{2}\right), \tag{5–9.83}$$

and performing the $k$-integration, we find that

$$\begin{aligned} M_{\text{ps.}} = & -\frac{\alpha}{2\pi}i\int ds\,du\,u\frac{dv}{2}e^{-isD} \\ & \times\left[m(m+\gamma p')g\gamma_5\phi + mg\gamma_5\phi(m+\gamma p'') - g\gamma_5\phi(2m^2+p^2)\right] \\ & -\frac{\alpha}{2\pi}\int\frac{ds}{s}\,du\,g\gamma_5\phi(e^{-isD_1'} + e^{-isD_1''}) + \text{c.t.} \end{aligned} \tag{5–9.84}$$

The contact term is now inferred by imposing the physical normalization condition that, in the situation of real particle propagation ($\gamma p' + m, \gamma p'' + m \to 0$) and small momentum transfer ($p^2 \to 0$), the presence of $M_{\text{ps.}}$ shall imply no modification in the initial coupling. Hence,

$$\text{c.t.} = \zeta_{\text{ps.}}g\gamma_5\phi, \tag{5–9.85}$$

with

$$\zeta_{\text{ps.}} = \zeta'_{\text{ps.}} + \zeta''_{\text{ps.}} \tag{5–9.86}$$

and

$$\begin{aligned} \zeta'_{\text{ps.}} &= \frac{\alpha}{\pi}\int\frac{ds}{s}\,du\,e^{-ism^2u^2}, \\ \zeta''_{\text{ps.}} &= -\frac{\alpha}{\pi}\int_0^1 du\,u\frac{1}{u^2+(\mu/m)^2} = -\frac{\alpha}{\pi}\log\frac{m}{\mu}. \end{aligned} \tag{5–9.87}$$

We illustrate the effect of the final combination, for the situation of real particles and arbitrary $p^2$, by exhibiting the form factor

$$\begin{aligned} F(p^2) &= 1 - \frac{\alpha}{2\pi}\int du\,u\frac{dv}{2}\left[\frac{2m^2+p^2}{u^2\left(m^2+\dfrac{1-v^2}{4}p^2\right)+\mu^2} - \frac{2}{u^2+(\mu/m)^2}\right] \\ &= 1 - \frac{\alpha}{2\pi}p^2\int_0^1 dv\,(1+v^2)\frac{\log\left(\dfrac{4m^2}{\mu^2}\dfrac{v^2}{1-v^2}\right)}{4m^2+(1-v^2)p^2}, \end{aligned} \tag{5–9.88}$$

where it has been recognized that the structure of the integral is almost identical with that of (4–14.65), lacking only the factor of $1 - u$. Note that the normalization condition, in the form

$$F(0) = 1, \tag{5–9.89}$$

effectively applies to any momentum transfer such that $|p^2| \ll m^2$, which includes $-p^2 = m_\pi^2$, according to the simplifying restriction of (5–9.1).

Since it is convenient to work with the separate parts of the mass operator structure, we made them well defined by introducing the convergence factor

$$1 - e^{-is\lambda^2(1-u)}, \tag{5–9.90}$$

as in (5–8.85), which is only required in the second of the terms in (5–9.84), and in $\zeta'_{\text{ps.}}$. The latter becomes ($\lambda \gg m$)

$$\zeta'_{\text{ps.}} = \frac{\alpha}{\pi}\int_0^1 du \int_0^\infty \frac{ds}{s} e^{-ism^2u^2}\left(1 - e^{-is\lambda^2(1-u)}\right)$$

$$= \frac{\alpha}{\pi}\int_0^1 du \log\frac{\lambda^2(1-u)}{m^2u^2} = \frac{\alpha}{\pi}\left(2\log\frac{\lambda}{m} + 1\right). \tag{5–9.91}$$

The replacement of pseudoscalar by pseudovector coupling in (5–9.78) is conveyed by [this is analogous to (5–9.63)]

$$\gamma_5 \to \gamma_5 - \frac{1}{2m}[m + \gamma(p' - k)]\gamma_5 - \frac{1}{2m}\gamma_5[m + \gamma(p'' - k)], \tag{5–9.92}$$

and therefore

$$M_{\text{pv.}} = M_{\text{ps.}} + \frac{1}{2m} g\gamma_5\phi M_0(p'') + \frac{1}{2m} M_0(p')g\gamma_5\phi, \tag{5–9.93}$$

where

$$M_0(p') = ie^2\int\frac{(dk)}{(2\pi)^4}\gamma^\nu\frac{1}{k^2}\frac{1}{\gamma(p'-k)+m}\gamma_\nu, \tag{5–9.94}$$

for example, is the familiar mass operator (without contact term) for a free particle. Accordingly, the contact term associated with $M_{\text{pv.}}$,

$$\zeta_{\text{pv.}}\frac{g}{2m} i\gamma^\mu\gamma_5\partial_\mu\phi, \tag{5–9.95}$$

$$\zeta_{\text{pv.}} = \zeta_{\text{ps.}} - \frac{m_c}{m},$$

$$\cdots \frac{\phantom{\alpha}}{2\pi}\int \frac{ds}{s}\, du\,(1+u)\,e^{-ism^2u^2}\left(1 - e^{-is\lambda^2(1-u)}\right)$$

$$= \frac{\alpha}{2\pi}\int_0^1 du\,(1+u)\log\frac{\lambda^2(1-u)}{m^2u^2} = \frac{3\alpha}{2\pi}\left(\log\frac{\lambda}{m} + \frac{1}{4}\right). \qquad (5\text{–}9.97)$$

we also decompose $\zeta_{\text{pv.}}$ into high- and low-mass-sensitive parts,

$$\zeta_{\text{pv.}} = \zeta'_{\text{pv.}} + \zeta''_{\text{pv.}}, \qquad (5\text{–}9.98)$$

we have

$$\zeta'_{\text{pv.}} = \zeta'_{\text{ps.}} - \frac{m_c}{m} = \frac{\alpha}{2\pi}\left(\log\frac{\lambda}{m} + \frac{5}{4}\right) \qquad (5\text{–}9.99)$$

and

$$\zeta''_{\text{pv.}} = \zeta''_{\text{ps.}} = -\frac{\alpha}{\pi}\log\frac{m}{\mu}. \qquad (5\text{–}9.100)$$

Let us also remark on the relationship with the electromagnetic contact-term parameters [Eqs. (5–7.52), (5–8.85)], namely

$$\zeta'_{\text{pv.}} = \zeta',$$

$$\zeta''_{\text{pv.}} = \zeta'' - \frac{\alpha}{2\pi}, \qquad (5\text{–}9.101)$$

whence

$$\zeta_{\text{pv.}} = \zeta_c - \frac{\alpha}{2\pi}. \qquad (5\text{–}9.102)$$

Armed with the required information, we begin the discussion of pseudovector coupling by simply following the path that starts at Eq. (5–8.6), where now

$$eq\gamma A_a \rightarrow \frac{g}{2m} i\gamma\gamma_5\partial\phi. \qquad (5\text{–}9.103)$$

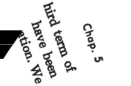

reached Eq. (5–8.9), which refers specifically to the
wo photons that are of interest in the present problem
, one explicitly as $A_b$ and the other implicitly in the field var
slate the statement of contact terms given in (5–8.10) as

$$-\tfrac{1}{2}i\,\mathrm{Tr}\,\delta\left[G_+\frac{g}{2m}i\gamma\gamma_5\partial\phi G_+\right]\zeta_c eq\gamma A_b$$

$$-\tfrac{1}{2}i\,\mathrm{Tr}\,\zeta_{\mathrm{pv.}}\frac{g}{2m}i\gamma\gamma_5\partial\phi\delta[G_+eq\gamma A_bG_+]$$

$$=(\zeta_c+\zeta_{\mathrm{pv.}})\delta\left[-\tfrac{1}{2}i\,\mathrm{Tr}\frac{g}{2m}i\gamma\gamma_5\partial\phi G_+eq\gamma A_bG_+\right], \qquad (5\text{–}9.104)$$

or, reuniting all photon fields into the field $A$,

$$-(\zeta_c+\zeta_{\mathrm{pv.}})\,\tfrac{1}{2}i\,\mathrm{Tr}\frac{g}{2m}i\gamma\gamma_5\partial\phi G_+{}^A. \qquad (5\text{–}9.105)$$

But we shall also find it desirable to deviate from the earlier treatment by returning to the first two terms of (5–8.7) and proceeding to make explicit the contact terms that are already incorporated in the mass operator there called $M_0$ (while removing that subscript, evocative of a null field, since it is now necessary to represent a second photon). The contact terms associated with $M$ are [Eq. (5–6.55)]

$$-m_c-\zeta_c(\gamma\Pi+m). \qquad (5\text{–}9.106)$$

Accordingly, the following contact terms are contained in the analogue of the first two parts of (5–8.7):

$$\tfrac{1}{2}i\,\mathrm{Tr}\frac{g}{2m}i\gamma\gamma_5\partial\phi G_+eq\gamma A_b(m_cG_+G_++\zeta_cG_+)$$

$$+\tfrac{1}{2}i\,\mathrm{Tr}\frac{g}{2m}i\gamma\gamma_5\partial\phi(m_cG_+G_++\zeta_cG_+)eq\gamma A_bG_+, \qquad (5\text{–}9.107)$$

or, with all photon fields united in $A$,

$$m_c\tfrac{1}{2}i\,\mathrm{Tr}\frac{g}{2m}i\gamma\gamma_5\partial\phi G_+{}^AG_+{}^A+2\zeta_c\tfrac{1}{2}i\,\mathrm{Tr}\frac{g}{2m}i\gamma\gamma_5\partial\phi G_+{}^A. \qquad (5\text{–}9.108)$$

The complete list of contact terms that is appended to

$$\tfrac{1}{2}i\,\mathrm{Tr}\frac{g}{2m}i\gamma\gamma_5\partial\phi(-G_+MG_+)^A, \qquad (5\text{–}9.109)$$

where $M$ is devoid of contact terms, is therefore

$$m_c \tfrac{1}{2} i \operatorname{Tr} \frac{g}{2m} i\gamma\gamma_5 \partial\phi G_+{}^A G_+{}^A + (\zeta_c - \zeta_{\text{pv.}}) \tfrac{1}{2} i \operatorname{Tr} \frac{g}{2m} i\gamma\gamma_5 \partial\phi G_+{}^A. \quad (5\text{–}9.110)$$

One quick approach to the first of the two contact terms in (5–9.110) is through the recognition that

$$\operatorname{Tr} i\gamma\gamma_5 \partial\phi G_+{}^A G_+{}^A = -\frac{\partial}{\partial m} \operatorname{Tr} i\gamma\gamma_5 \partial\phi G_+{}^A = 0, \quad (5\text{–}9.111)$$

since the computation of Eqs. (5–9.17–26) gives the latter Tr a value that is independent of $m$. It is more instructive, however, to repeat the calculation in the same spirit than to merely apply the known result. Accordingly, we use (5–9.19) to get

$$\operatorname{Tr} i\gamma\gamma_5 \partial\phi G_+{}^A G_+{}^A = \operatorname{Tr} \gamma^\mu \gamma_5 \phi \left[ \Pi_\mu, \frac{m^2 - 2m\gamma\Pi - \Pi^2 + eq\sigma F}{\left(\Pi^2 + m^2 - eq\sigma F\right)^2} \right]$$

$$= 4m \operatorname{Tr} \gamma_5 \gamma^\mu \gamma^\nu \phi eq\sigma F \left[ \Pi_\mu, \Pi_\nu \frac{1}{\left(\Pi^2 + m^2\right)^3} \right], \quad (5\text{–}9.112)$$

and the Euclidean momentum integral that now appears [cf. Eq. (5–9.25)] is

$$\int \frac{(dp)_E}{(2\pi)^4} \frac{\partial}{\partial p} \frac{p}{\left(p^2 + m^2\right)^3} = \frac{2\pi^2}{(2\pi)^4} \frac{p^4}{\left(p^2 + m^2\right)^3} \Bigg|_{p \to \infty} = 0, \quad (5\text{–}9.113)$$

owing to the additional factor of $p^2$ in the denominator. Having seen this, it is immediately clear that (5–9.109) also vanishes! The only information that is required concerning $M$ is the remark that, through its exponential dependence on $\Pi^2 s$, the final momentum integration over the surface of an arbitrary large sphere will enforce the limit $s \to 0$, where the convergence factor (5–9.90) vanishes. In other words, the presence of the mass operator in (5–9.109) cannot reverse the situation already encountered in (5–9.111), and both contributions are zero. What remains is the second term of (5–9.110), a multiple of the initial coupling (5–9.17), which multiple, according to (5–9.102), is $\alpha/2\pi$. Here is our version of the source-theoretic result that the initial coupling is modified by the factor

$$1 + \frac{\alpha}{2\pi}. \quad (5\text{–}9.114)$$

A reading of the operator field-theory papers will show that we are not merely quarreling about the evaluation of integrals. Error has entered these papers just at the point where renormalization is introduced, for, with the customary emphasis on the removal of divergences, it is taken for granted that two renormalization constants that have the same singular behavior are, in fact, equal. The analogue in our procedure would be to remark that

$$\lim_{\lambda\to\infty} \frac{\zeta_{\text{pv.}}}{\zeta_c} = 1 \tag{5–9.115}$$

implies the equality of $\zeta_{\text{pv.}}$ and $\zeta_c$, thereby ignoring the finite difference, of $\alpha/2\pi$. In short, the sin of the current-algebraists has been to define a significant parameter *mathematically*, rather than by examining its *physical* meaning. And that is precisely what source theory is all about.

Our discussion is completed by showing, in a rather different way, that the same conclusion follows from the consideration of pseudoscalar coupling, which is the verification of the equivalence theorem at the next dynamical level. The pseudoscalar counterpart of Eqs. (5–9.109, 110) is

$$\begin{aligned}&\tfrac{1}{2}i\,\mathrm{Tr}\,g\gamma_5\phi(-G_+MG_+)^A\\ &\qquad + m_c\tfrac{1}{2}i\,\mathrm{Tr}\,g\gamma_5\phi G_+{}^AG_+{}^A + (\zeta_c - \zeta_{\text{ps.}})\tfrac{1}{2}i\,\mathrm{Tr}\,g\gamma_5\phi G_+{}^A.\end{aligned} \tag{5–9.116}$$

Again, we begin with the $m_c$ term and first remark that

$$\begin{aligned}\tfrac{1}{2}i\,\mathrm{Tr}\,g\gamma_5\phi G_+{}^AG_+{}^A &= -\frac{\partial}{\partial m}\tfrac{1}{2}i\,\mathrm{Tr}\,g\gamma_5\phi G_+{}^A\\ &= \frac{1}{m}\tfrac{1}{2}i\,\mathrm{Tr}\,g\gamma_5\phi G_+{}^A,\end{aligned} \tag{5–9.117}$$

since the effective evaluation of this Tr given in Eqs. (5–9.10–13) shows that it is proportional to $m^{-1}$. But, alternatively, we have

$$\begin{aligned}&\tfrac{1}{2}i\,\mathrm{Tr}\,g\gamma_5\phi G_+{}^AG_+{}^A = \tfrac{1}{2}i\,\mathrm{Tr}\,G_+{}^Ag\gamma_5\phi G_+{}^A\\ &\qquad \cong \tfrac{1}{2}i\,\mathrm{Tr}\,g\gamma_5\phi\frac{1}{\Pi^2 + m^2 - eq\sigma F} = \frac{1}{m}\tfrac{1}{2}i\,\mathrm{Tr}\,g\gamma_5\phi G_+{}^A,\end{aligned} \tag{5–9.118}$$

where $\cong$ means that the $\gamma\gamma_5\partial\phi$ term that is also produced in anticommuting $\gamma\Pi$ with $\gamma_5\phi$ has been omitted, as justified by the null surface integral to which it would lead. Now note that, through the relation (5–9.96), the last two terms of (5–9.116) combine into

$$(\zeta_c - \zeta_{\text{pv.}})\,\tfrac{1}{2}i\,\mathrm{Tr}\,g\gamma_5\phi G_+{}^A = \frac{\alpha}{2\pi}\,\tfrac{1}{2}i\,\mathrm{Tr}\,g\gamma_5\phi G_+{}^A. \tag{5–9.119}$$

Hence, the equivalence between pseudoscalar and pseudovector coupling will indeed be maintained at this dynamical level if the first term of (5–9.116) vanishes:

$$\mathrm{Tr}\,\gamma_5\phi(G_+MG_+)^A = 0. \tag{5–9.120}$$

We now verify this by explicit calculation.

The procedure of Eq. (5–9.118) converts this statement into

$$\mathrm{Tr}\,\gamma_5\phi\frac{1}{\Pi^2 + m^2 - eq\sigma F}M^A = 0. \tag{5–9.121}$$

The structure of $M^A$ for an arbitrarily strong homogeneous field has been given in Section 5–6. But very little of that detail is needed here. We refer to the construction of (5–6.43), sans c.t., and remark that the odd $\gamma$-matrix term can be omitted, as can field-strength combinations of the form $F^{\mu\nu}F_{\nu\lambda}$, since only ${}^*F^{\mu\nu}F_{\mu\nu}$ is of interest to us. Accordingly, all that survives of (5–6.43) is

$$M \to -\frac{\alpha}{4\pi}m\int\frac{ds}{s}\,du\,e^{-is[u(1-u)\Pi^2+m^2u]}\gamma^\nu e^{isueq\sigma F}\gamma_\nu(1 - e^{-is(1-u)\lambda^2}), \tag{5–9.122}$$

and we have been careful to include the convergence factor. In addition, the $\gamma^\nu \cdots \gamma_\nu$ structure annihilates $\sigma F$ and we have, effectively,

$$\gamma^\nu e^{isueq\sigma F}\gamma_\nu \to -4(1 + u^2s^2\gamma_5 e^2\mathbf{E}\cdot\mathbf{H}), \tag{5–9.123}$$

or

$$M \to \frac{\alpha}{\pi}m\left[\int_0^1 du\log\frac{\lambda^2}{(1-u)\Pi^2 + m} - \gamma_5 e^2\mathbf{E}\cdot\mathbf{H}\right.$$
$$\left.\times\int_0^1 du\left\{\frac{1}{[(1-u)\Pi^2 + m^2]^2} - \frac{1}{[(1-u)\Pi^2 + m^2 + \lambda^2(1-u)]^2}\right\}\right], \tag{5–9.124}$$

where $\lambda^2$, correctly for these purposes, has been treated as a very large quantity in the first of these terms, but it has been noted that the situation of $u \to 1$ requires more careful treatment in the second term. After the $u$-integrations are performed, and with the factor of $(\alpha/\pi)m$ omitted, (5–9.124) reads

$$\log\left(\frac{\lambda^2}{m^2}\right) + 1 - \frac{x+1}{x}\log(x+1) - \gamma_5\frac{e^2}{m^4}\mathbf{E}\cdot\mathbf{H}\left[\frac{1}{x+1} - \frac{1}{x+1+(\lambda^2/m^2)}\right], \tag{5–9.125}$$

in which

$$x = \Pi^2/m^2. \tag{5–9.126}$$

The Euclidean momentum integrals that finally express the two different contributions to (5–9.121), with $\mathbf{E}\cdot\mathbf{H}$ produced by expansion of the denominator and by $M^A$, respectively, are proportional to

$$2\int_0^\infty dx\,\frac{x}{(x+1)^3}\left[\log\left(\frac{\lambda^2}{m^2}\right) + 1 - \frac{x+1}{x}\log(x+1)\right]$$
$$-\int_0^\infty dx\,\frac{x}{x+1}\left[\frac{1}{x+1} - \frac{1}{x+1+(\lambda^2/m^2)}\right]. \tag{5–9.127}$$

The first of these integrals is evaluated as

$$\log\left(\frac{\lambda^2}{m^2}\right) - 1, \tag{5–9.128}$$

and the second one precisely cancels it. All is well.

*Mathematical addendum*: The evaluation of the decisive combination of (5–9.102),

$$\zeta_c - \zeta_{\text{pv.}} = \frac{\alpha}{2\pi}, \tag{5–9.129}$$

was rather indirect. It may be instructive to see a quite direct and elementary computation of just this combination, rather than the separate consideration of the two contact terms. We first observe that the contact terms are designed to remove from $M$ [Eq. (5–9.75), with pseudovector coupling] the appropriate linear field interaction, under the physical conditions of real particle propagation [$\gamma p' + m,\ \gamma p'' + m \to 0$] and negligible momentum transfer [$(p' - p'')^2 \to 0$]. It

here to simply place $p' = p''$, and thus the ...
d from

$$\zeta_c \gamma^\mu = -ie^2 \int \frac{(dk)}{(2\pi)^4} \frac{1}{k^2} \gamma^\nu \frac{m - \gamma(p'-k)}{k^2 - 2p'k} \gamma^\mu \frac{m - \gamma(p'-k)}{k^2 - 2p'k} \gamma_\nu$$

and

$$\zeta_{\mathrm{pv.}} i\gamma^\mu \gamma_5 = -ie^2 \int \frac{(dk)}{(2\pi)^4} \frac{1}{k^2} \gamma^\nu \frac{m - \gamma(p'-k)}{k^2 - 2p'k} i\gamma^\mu \gamma_5 \frac{m - \gamma(p'-k)}{k^2 - 2p'k} \gamma_\nu, \quad (5\text{–}9.131)$$

where the left- and right-hand factor reduction, $\gamma p' + m \to 0$, is understood. The two expressions differ only through the presence of an additional factor of $i\gamma_5$ in one of them. Accordingly, whenever the central matrix, $\gamma^\mu$ or $i\gamma^\mu\gamma_5$, appears multiplied by a number, or lacks a matrix factor on one side, the resulting contributions to the respective contact terms are identical and cancel out from the difference. To exploit this property, we write

$$\gamma^\nu[m - \gamma(p'-k)] = [m + \gamma(p'-k)]\gamma^\nu + 2(p'-k)^\nu,$$

$$[m - \gamma(p'-k)]\gamma^\nu = \gamma^\nu[m + \gamma(p'-k)] + 2(p'-k)^\nu, \quad (5\text{–}9.132)$$

and then conclude that we have only to consider the respective structures

$$-ie^2 \int \frac{(dk)}{(2\pi)^4} \frac{1}{k^2} \frac{1}{(k^2 - 2p'k)^2} \gamma k \gamma^\nu (\gamma^\mu, i\gamma^\mu\gamma_5) \gamma_\nu \gamma k, \quad (5\text{–}9.133)$$

which have already received $\gamma p' + m \to 0$ simplification.

At this point, we invoke the representation [inferred from (4–14.10, 11), for example, by differentiating with respect to $m^2$, after which one sets $p'^2$, replacing $\Pi^2$, equal to $-m^2$]

$$\frac{1}{k^2} \frac{1}{(k^2 - 2p'k)^2} = -i \int_0^\infty ds\, s^2 \int_0^1 du\, u\, e^{-is[(k-up')^2 + m^2u^2]}$$

$$= 2\int_0^1 du\, u \frac{1}{[(k-up')^2 + m^2u^2]^3}. \quad (5\text{–}9.134)$$

The $k$-integration is then performed with the aid of the substitution

$$k \to k + up', \quad (5\text{–}9.135)$$

...ingly,

$$\gamma k \gamma^{\nu} \cdots \gamma_{\nu}\gamma k \to \gamma k \gamma^{\nu} \cdots \gamma_{\nu}\gamma k + u^2 \gamma p' \gamma^{\nu} \cdots \gamma_{\nu}\gamma p'. \qquad (5\text{–}9.136)$$

...st of the terms on the right becomes a multiple of $\gamma^{\lambda}\gamma^{\nu} \cdots \gamma_{\nu}\gamma_{\lambda}$ and yields ...ical contributions to the two contact terms, since $\gamma_5$ commutes with an even ...duct of $\gamma$-matrices. Note that this part of the $k$-integral requires a convergence ...actor. But the identity of the respective contributions, and their exact cancellation in $\zeta_c - \zeta_{pv}$, is obviously independent of the choice of that mathematical function. Accordingly, (5–9.133) effectively reduces to

$$e^2\, 2\int_0^1 du\, u^3(-i)m^2 \int \frac{(dk)}{(2\pi)^4} \frac{1}{(k^2 + m^2u^2)^3}(2\gamma^{\mu}, -2i\gamma^{\mu}\gamma_5)$$

$$= \frac{\alpha}{8\pi} 2\int_0^1 du\, u(2\gamma^{\mu}, -2i\gamma^{\mu}\gamma_5) = \frac{\alpha}{4\pi}(\gamma^{\mu}, -i\gamma^{\mu}\gamma_5), \qquad (5\text{–}9.137)$$

which uses the integral (5–9.14), and indeed,

$$\zeta_c - \zeta_{pv.} = \frac{\alpha}{2\pi}. \qquad (5\text{–}9.138)$$

One last remark seems to be called for. An additional piece of evidence adduced by operator field theorists, on behalf of the claim that the coupling (5–9.73) is exact, refers to the fictitious situation of massless electrodynamics, $m = 0$. And, indeed, the presence of the factor $m^2$ in the first line of (5–9.137) might seem to indicate a null result for $m^2 \to 0$. The erroneous nature of that conclusion is evident in the second line of the same equation; the momentum integral is singular in the limit $m^2 \to 0$, and the whole structure is actually independent of $m^2$.

# Index

Milton Keynes UK
Ingram Content Group UK Ltd.
UKHW020019071024
449327UK00032B/2851

9 780738 200552